ONTIERS IN NONLINEAR OPTICS

SERGEI ALEKSANDROVICH AKHMANOV 14 July 1929–1 July 1991

FRONTIERS IN NONLINEAR OPTICS

THE SERGEI AKHMANOV MEMORIAL VOLUME

Edited by
H WALTHER
Max-Planck-Institut für Quantenoptik, Garching

N KOROTEEV
Moscow State University, Moscow

M O SCULLY
University of New Mexico, Albuquerque

 **CRC Press**
Taylor & Francis Group
Boca Raton London New York

CRC Press is an imprint of the
Taylor & Francis Group, an **informa** business

First published 1993 by IOP Publishing Ltd.

Published 2021 by CRC Press
Taylor & Francis Group
6000 Broken Sound Parkway NW, Suite 300
Boca Raton, FL 33487-2742

ISBN 13: 978-0-7503-0218-0 (hbk)
ISBN 13: 978-1-00-320963-8 (ebk)

DOI: 10.1201/9781003209638

Visit the Taylor & Francis Web site at
http://www.taylorandfrancis.com

and the CRC Press Web site at
http://www.crcpress.com

British Library Cataloguing in Publication Data
A catalogue record for this book is available from the British Library

Library of Congress Cataloging-in-Publication Data are available

ACKNOWLEDGMENT OF THE EDITORS

This volume is dedicated to the memory of our distinguished colleague Sergei Aleksandrovich Akhmanov, a pioneer in the field of nonlinear optics. The editors of this book would like to express their gratitude to all the contributors of this volume, due to their effort this book represents an in depth survey of the modern trends in nonlinear optics.

The Editors

FOREWORD

The untimely death of Sergei Aleksandrovich Akhmanov came as a shock not only to his family, friends and colleagues in Eastern Europe, but also to a large number of physicists around the globe. Many of them knew him personally, or enjoyed reading his scientific papers.

It is most appropriate that this volume with writings in the broad field of quantum electronics and non-linear optics be dedicated to his memory. Sergei, as we in the West called him, was indeed one of the pioneers in this field of scientific endeavour. I consider it a privilege to have been invited to write this foreword, and in this way to pay tribute to the memory of a dear friend and colleague.

On my first visit to the Soviet Union in 1967, I recall noticing a large mural, spontaneously drawn by a number of well-wishers in a hallway at Moscow State University, depicting Rem Khokhlov and Sergei Akhmanov as intrepid young horsemen conquering the fields of non-linear optics. They had just been awarded the Lenin Prize by the Soviet State in recognition of their pioneering contributions. They had entered this new field very early, recognizing it as a natural extension of their work in radiophysics during the late fifties which dealt with parametric interactions along microwave transmission lines. In 1972 and 1963 they gave a series of lectures on problems in non-linear optics, and their lecture notes were published in 1964 in a volume sponsored by the Soviet Academy of Sciences. At the same time I had prepared my lecture notes on Non-linear Optics for the 1964 Summer School in Les Houches. These were subsequently published as an independent monograph by W A Benjamin. Rem, Sergei and I exchanged autographed copies of our respective works. From that time Sergei and I continued to work independently on many closely related topics. We saw each other regularly at international meetings such as the series of International Quantum Electronic Conferences (IQEC), at several Gordon Research Conferences, at a Vavilov Conference in Novosibirsk in 1971, to name a few.

In particular I wish to recall a few memories about Sergei Akhmanov's participation at the E Fermi Course 64, at the July 1975 Summer School in Varenna, Italy. He gave fine lectures about three different topics: 'Coherent active spectroscopy of combinatorial scattering' (better known in the West as four-wave mixing spectroscopy or CARS), 'Higher-order optical non-linearities' (including fifth-harmonic generation in calcite), and 'Statistical effects in resonant non-linear optics' (optical and temporal multimode effects with random phases). One evening we planned a party and Sergei had brought vodka and caviar from Moscow. We needed, however, a hardboiled egg and fresh lemon. The former we readily obtained from the kitchen. Then Sergei and I took an evening stroll in the garden of the Villa Monastero along the shores of Lake Como. We walked along the decorative lemon trees. In the darkness we ignored the warning signs in Italian and each of us picked a forbidden fruit, a fresh lemon. This was the beginning of a marvelous evening party. Two aspects of Sergei's life are illustrated by this story. He was a leader in science, knowledgeable in a broad range of theoretical and experimental issues. At the same time, he was kind, cheerful, modest and generous. He always gave credit and referred to work done by others.

When my wife and I visited the Soviet Union in 1971, we spent on unforgettable day at a dacha outside Moscow, where we met Sergei's wife and children. She had prepared a wonderful meal with homemade kvas. Later in the day we met Akhmanov's mother, a professor of English literature. She insisted that we have another dinner with her as well. Our deepest sympathy goes to these surviving family members.

The last times Sergei and I got together was at the IQEC 12, held in 1989 in New Hampshire, and at a binational conference on spectroscopy of condensed matter, held in Irvine, California, in January 1990. Sergei talked about diverse topics, which, as usual, were close to my own scientific interests. They included second-harmonic generation by quadrupolar interactions in metallic single crystals, femtosecond pulse interactions in gallium arsenide, and chaotic transitions in two-dimensional multimode laser patterns. This is not the place to enumerate all of his scientific accomplishments, which one may find in his bibliography, but we may mention that Professor S A Akhmanov occupied the Chair of General Physics and Wave Processes at Moscow State University from 1975. There he established and chaired the International Laser Center of the Central and East European countries. He was Vice-Chairman of the Council on Coherent and Non-linear Optics of the USSR Academy of Sciences. He also received the Lomonosov Prize. He served on organizing and program committees of various international scientific conferences.

This volume attests to the high esteem in which he was held by many colleagues in the Western scientific establishment. We all say farewell to a scientist and friend.

N Bloembergen

Harvard University
Cambridge, Massachusetts, December 1991

CONTENTS

Channelling of atoms in a standing laser light wave

V I Balykin†, V S Letokhov and Yu B Ovchinnikov

Institute of Spectroscopy, Russian Academy of Sciences, 142092 Troitsk, Moscow Region, Russia

1. Introduction

The effect of laser fields on the mechanical motion of atoms has recently been the subject of extensive studies of atomic physics (see reviews [1–5], special issues of scientific journals [6, 7], monographs [8, 9], and conference proceedings [10]).

Of special interest here is the interaction between atoms and a standing light wave— one of the simplest monochromatic light field configurations. Two interaction modes are possible in that case. The first occurs when the atoms are scattered by the light wave (see review [1]) and usually presupposes the fulfilment of the condition $kv_z < \tau^{-1}$, where $k = 2\pi/\lambda$ is the wave number, v_z is the atomic velocity component directed along the standing light wave, and τ is the time of flight of the atoms through the wave. This condition means that the atoms, while flying through the light wave, have enough time to move a distance of $\Delta z \simeq v_z\tau \leq \lambda/2\pi$ along it. In the second case where $kv_z \gg \tau^{-1}$, there may take place the channelling or one-dimensional trapping (localization) of the atoms.

The idea of atomic channelling, based on the classical atom-standing-light-wave interaction concept, was set forth in [12] the authors of which suggested that the effect should be used to eliminate the Doppler broadening of spectral lines.

This paper reviews the investigations performed at the Institute of Spectroscopy into this non-linear optic effect.

2. Gradient force and atomic potential for a standing light wave

Let us consider the interaction between a two-level atom and a quasi-resonant monochromatic standing wave of the form

$$\boldsymbol{E} = \hat{\boldsymbol{e}}\,2E_0\cos(\omega t)\cos(kz) \qquad (1)$$

where E_0 is the amplitude of the travelling light wave forming the standing wave.

So that we can consider classically the motion of an atom in such a standing light wave, the following condition must be satisfied:

$$p_z \gg \hbar k \qquad (2)$$

where $p_z = Mv_z$ is the atomic momentum component directed along the standing wave. Assuming that $\Delta p_z < p_z$, which is reasonable enough, one can clearly see that it

† Presently at the University of Konstanz, D-7750, Konstanz, Federal Republic of Germany.

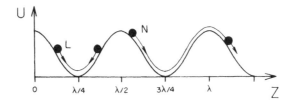

Figure 1.   Atomic localization in a plane standing light wave. L denotes a localized atom oscillating near the bottom of a potential well and N a non-localized atom.

is only in this case that, according to the uncertainty principle, the uncertainty Δz of the atomic z-coordinate is less than the optical wavelength λ. Assume that the upper bound on the z-component of the atomic velocity is set by the relations

$$kv_z < \gamma \qquad (3a)$$

$$kv_z < \Delta^2/\gamma(G_0)^{1/2} \qquad (3b)$$

where 2δ is the natural transition width of the two-level atom, $\Delta = \omega - \omega_0$ is the detuning of the laser field frequency ω with respect to the atomic transition frequency ω_0, $G_0 = d^2E_0^2/2\hbar^2\delta^2$ is the transition saturation parameter, and d is the dipole moment matrix element of the two-level atom. Inequality ($3b$) reflects the absence of Landau–Zener transitions [11]. Assume also that the atom–standing-light-wave interaction time satisfies the inequality $\tau \gg \delta^{-1}$.

Under the above assumptions, the motion of an atom in a standing light wave is governed by the gradient (or dipole) force

$$\boldsymbol{F}_{\mathbf{g}} = -(\partial U_{\mathbf{g}}/\partial z)\boldsymbol{e}_z \qquad (4)$$

of potential nature. The associated potential has the form [13]

$$U_{\mathbf{g}} = (\hbar\Delta/2)\ln\{1 + G/[1 + (\Delta/\gamma)^2]\} \qquad (5)$$

where $G = 4G_0\cos^2(kz)$ is the atomic transition saturation parameter in the standing light wave.

In the limiting case of weak saturation and large frequency detuning,

$$G \ll 1 + (\Delta/\delta)^2 \qquad (6a)$$

$$\Delta \gg \delta \qquad (6b)$$

the expression for the potential $U_{\mathbf{g}}$ reduces to the simple form

$$U_{\mathbf{g}} = (\hbar g_0^2/\Delta)\cos^2(kz) \qquad (7)$$

where $g_0 = dE_0/2\hbar$ is the Rabi frequency and the saturation parameter $G = 2g_0^2/\gamma^2$.

Expression (7) is equivalent to the following classical relation used in [12]:

$$U_{\mathbf{g}} = -\alpha E_0^2\cos^2(kz) \qquad (8)$$

where α is the polarizability of the atom at the optical frequency:

$$\alpha = e^2/m(\omega_0^2 - \omega^2) = (n_\omega - 1)/2\pi N \qquad (9)$$

where n_ω is the refractive index of the medium containing N atoms in a unit of volume.

It follows from expressions (5), (7) and (8) that the period of the potential $U_{\mathbf{g}}$ is equal to half the optical wavelength. When the field frequency detuning is positive ($\Delta > 0$), the potential wells coincide with the nodes of the standing light wave, and when it is negative ($\Delta < 0$), with the wave loops. Figure 1 illustrates this potential for the case $\Delta > 0$.

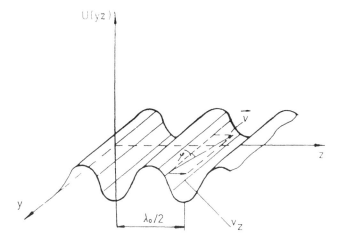

Figure 2. Potential energy of an atom in a standing light wave wherein atoms moving with a velocity of v are trapped (from [14]).

What is the character of atomic motion in such a potential? Assume that the amplitude (depth) of potential (5) is $U_g^0 \gg R$, where $R = \hbar^2 k^2 / 2M$ is the recoil energy acquired by the atom upon absorption of a single photon. The atoms flying through the standing light wave will in that case be divided into two groups according to their total energy $W = U(z) + M v^2(z)/2$. The atoms with a total energy of $W < U_g^0$ will move between the adjacent potential maxima. Such atoms are referred to in the text below as localized. In the general case, the localized atoms perform anharmonic oscillations with an amplitude of less than $\lambda/4$ about the potential wells.

The atoms with a total energy of $W > U_g^0$ move infinitely along the z-axis, the potential modulating the corresponding atomic velocity component. Such atoms will be called non-localized. The classical atomic motion in a potential similar to (5) in the weak-saturation limit has been considered in detail in [14].

3. Theory of atomic motion in a standing light wave

Let us now consider atomic motion in a one-dimensional monochromatic standing wave. In that case, atoms can be localized in one direction, but will move freely in any other direction, and so they are channelled along the light wave front (figure 2). The motion of an atom here can be determined in the quasi-classical approximation by integrating the Langevin equation [15]

$$\mathrm{d}\boldsymbol{p}/\mathrm{d}t = \boldsymbol{F} \tag{10}$$

where $\boldsymbol{p}$ is the atomic momentum and $\boldsymbol{F}$ is the total force acting on the atom. The force $\boldsymbol{F}$ may be represented in the form

$$\boldsymbol{F} = \boldsymbol{F}_g + \boldsymbol{F}_{fr} + \boldsymbol{F}_d \tag{11}$$

where $\boldsymbol{F}_g$ is the gradient force, $\boldsymbol{F}_{fr}$ is that part of the total force which depends on the atomic velocity (i.e., the friction force), and $\boldsymbol{F}_d$ is the fluctuating force due to the atomic momentum diffusion.

The gradient force for the two-level atom is given by expression (5). The friction force for such an atom is [13, 16]

$$
\begin{aligned}
\boldsymbol{F}_{\mathrm{fr}} = 2\hbar(\Delta/\gamma)G\{[1 + (\Delta/\gamma)^2 \\
- G(1 + G/2)]/[1 + (\Delta/\gamma)^2 + G]^3\} \tan^2(\boldsymbol{k} \cdot \boldsymbol{z})\boldsymbol{k}(\boldsymbol{v} \cdot \boldsymbol{k}).
\end{aligned} \tag{12}
$$

Besides, allowing for the quantum absorption and emission of photons by the atom causes the radiative force to fluctuate about its mean value. According to [13, 17], the atomic momentum diffusion coefficient, $2D = \langle \Delta p^2 \rangle / \Delta t$, which characterizes the build-up rate of the mean-square atomic momentum fluctuation about its mean value, is defined by the relations

$$
\begin{aligned}
2D_{\mathrm{i}} = \hbar^2 k^2 \gamma G(\{[1 + (\Delta/\gamma)^2]^2 + [3 - (\Delta/\gamma)^2]G + 3G^2 + G^3\} \\
\times [1 + (\Delta/\gamma)^2 + G]^3) \tan(\boldsymbol{k} \cdot \boldsymbol{z})
\end{aligned} \tag{13a}
$$

$$
2D_{\mathrm{sp}} = \hbar^2 k^2 \gamma \{G/[1 + (\Delta/\gamma)^2 + G]\} \tag{13b}
$$

where $2D_{\mathrm{i}}$ stands for the atomic momentum diffusion due to the induced emission and absorption of photons and $2D_{\mathrm{sp}}$ denotes that due to the spontaneous emission of photons.

Let us analyse briefly the above expressions. First of all, it should be noted that the motion of an atom in a standing light wave is, on the whole, determined by the three factors considered above. The motion of a single atom in a standing light wave is described by the Langevin equation given by (10) and (11), wherein $\boldsymbol{F}_{\mathrm{d}}$ is the random Langevin force, zero on average, whose fluctuations are governed by the atomic momentum fluctuations depending on the diffusion coefficient defined by (13a) and (13b) [15]. Note that the part of this force which is due to induced atomic momentum fluctuations is directed along the light-field gradient, while its other part, the one due to spontaneous transitions, has a random direction varying with a frequency equal to the rate of these transitions. Whereas the random force can, on average, only heat the atom, the friction force can both heat and cool it. As seen from (12), the friction force changes sign when $1 + (\Delta/\delta)^2 = G(1 + G/2)$. We would like to consider the situation where the friction force is directed against the atomic velocity, i.e., where it cools the atom and causes it to be localized in the vicinity of the potential well (5). Two cases can be distinguished here.

(a) 'Red shift' case

$$
\Delta < 0 \qquad 1 + (\Delta/\gamma)^2 > \overline{G(1 + G/2)}. \tag{14}
$$

The friction force in this case reaches its maximum, which does not exceed the maximum of the spontaneous light pressure force, $F_{\mathrm{sp}}^{\mathrm{max}} = \hbar k \delta$, at $\Delta \simeq \delta$ and $G \simeq 1$. The atomic momentum diffusion coefficient here is $2D \simeq \hbar^2 k^2 \delta$, and the potential depth is not very great, $U_{\mathrm{g}}^0 \simeq \hbar \delta$. The time for which the atom can be held localized within the limits of a potential well $\lambda/2$ in size (figure 1) can be taken to be approximately equal to that required for it to acquire, as a result of momentum diffusion, an energy equal to the depth of the well, i.e., $\tau_0 \simeq U_{\mathrm{g}}^0$, $M/D \simeq 1/kv_{\mathrm{r}} = 10^{-6}$ s, where $v_{\mathrm{r}} = \hbar k/M$ is the recoil velocity. This rough estimate agrees well with the more stringent estimate made in [18]. By increasing the frequency detuning and the intensity of the standing-light-wave field one can lengthen the atomic confinement time but cannot make the potential depth greater than $U_{\mathrm{g}}^0 \simeq \hbar \delta$ under the condition

$$
1 + (\Delta/\delta)^2 > \overline{G(1 + G/2)}.
$$

If this condition is not satisfied, the friction force changes sign, which leads to an additional heating of the atom.

It was precisely the 'red shift' case that was originally proposed to localize atoms in a three-dimensional standing light wave [19, 20].

(b) 'Blue shift' case:

$$\Delta > 0 \qquad 1 + (\Delta/\gamma)^2 < \overline{G(1 + G/2)} \qquad (15)$$

In contrast to the preceding case, one can achieve, with the parameters given above and $\Delta \gg \delta$, comparatively high potential depth and friction force values, $U_{\mathrm{g}}^0 \simeq \hbar\Delta$ and $F_{\mathrm{fr}} \simeq \hbar k \Delta v$. On the other hand, the induced momentum diffusion coefficient in the case of strong light fields becomes much greater than its spontaneous counterpart: $2D_{\mathrm{i}} \simeq \hbar^2 k^2 \delta G$. What is more, the analysis of this situation is more difficult to make, both the friction force and atomic momentum diffusion being strong functions of space. Figure 3 shows the spatial dependences (in the space interval 2λ) of the potential U_{g}, the friction coefficient $\beta(F_{\mathrm{fr}} = -\beta v)$ and the atomic momentum diffusion coefficient $2D = 2D_{\mathrm{i}} + 2D_{\mathrm{sp}}$ for a saturation parameter of $G_0 = 10^4$ and three different frequency detuning values: $\Delta = 20\delta, 100\delta$ and 500δ. As can be seen from figure 3(*a*), the potential reaches its maximum at an optimum frequency detuning of $\Delta_{\mathrm{opt}} \simeq 2\delta(G_0)^{1/2} = 100\delta$. The efficiency of cooling atoms by the friction force can be estimated by means of the parameter $\chi = \beta/2D$.

It has been demonstrated in the theoretical works reported in [21, 22] that the quantity $1/\chi$ represents the temperature of the atoms in the standing light wave, provided their energy $W \gg U_{\mathrm{g}}^0$, for it is only in this case that the atomic energy obeys the Boltzmann distribution law. One can see from expressions (12) and (13*a*) and from figures 3(*b*), 3(*c*) and 4 that reducing the frequency detuning will increase χ but narrow at the same time the range $2z_0$ of non-zero χ values in the vicinity of the standing-wave nodes. To illustrate, for $\Delta = 20\delta$, $\chi \simeq 0.15(\hbar\delta)^{-1}$ and $z_0 \simeq 10^{-2}\lambda$; for $\Delta = 100\delta$, $\chi \simeq 0.1\ (\hbar\delta)^{-1}$ and $z_0 \simeq 2 \times 10^{-2}\lambda$; and for $\Delta = 500\delta$, $\chi \simeq 0.06\ (\hbar\delta)^{-1}$ and $z_0 \simeq 6 \times 10^{-2}\lambda$.

The time for which an atom can be localized in a potential well of the standing light wave is $\tau_0 \simeq (\Delta/\delta)^{2/3}/kv_{\mathrm{r}}$ [22]. For $\Delta = 10^3\delta$, this time may be of the order of a few milliseconds. On the other hand, it is highly probable that once the atom has escaped from the well it will be trapped again in another one [22]. The atomic motion thus resembles the Brownian wandering across the individual potential wells produced by the standing light wave. The characteristic lifetime of the atom in each of the wells is of the order of τ_0. Such behaviour of the atom is corroborated by the analysis of the experimental results reported in [15].

As for the spatial distribution of the atoms with an energy of $W < U_{\mathrm{g}}^0 \simeq \hbar\Delta$, most of them are concentrated in regions with a characteristic size of $z_0^* \simeq (\delta/\Delta)^{1/3}/k$ near the standing-wave nodes [22]. Thus, the atomic ensemble in the standing light wave forms a lattice with period $\lambda/2$ that follows the intensity distribution period of the light field.

Let us now consider two-dimensional atomic motion in a spherical standing light wave formed by a laser beam with a Gaussian intensity distribution. Suppose that the standing wave is crossed by an atomic beam. At a certain point in the wave where the transverse kinetic energy of the atoms is equal to their potential energy in the light field, they can be localized in a potential well. The atoms will then be channelled along the nodes (or loops) of the wave, for their trajectories must follow the wave front accurate to within less than λ. Figure 5 illustrates the behaviour of the

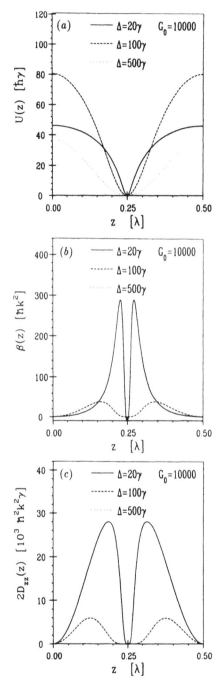

Figure 3. Spatial behaviour of (*a*) the atomic potential energy, (*b*) the friction coefficient, and (*c*) the atomic momentum diffusion in a standing light wave.

potential energy of (1) localized and (2) non-localized atoms in the spherical standing light wave, which reflects their trajectories in the laser field. The atoms, spaced a

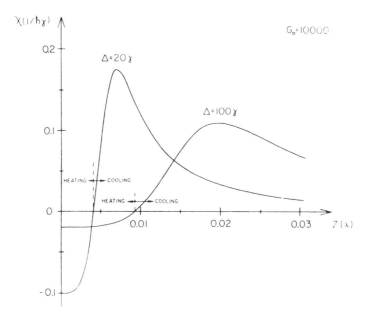

Figure 4. Spatial dependence of the parameter χ near a standing light-wave node.

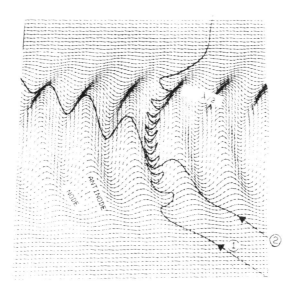

Figure 5. Atomic potential energy in a spherical standing light wave for (1) localized atoms and (2) non-localized ones. The curves reflect the trajectories of the localized and non-localized atoms.

distance of $d = \lambda/2$ apart, move parallel to one another and enter the field near the minimum of two adjacent potential wells. The laser field parameters are as follows: the laser power $P = 0.11$ W, the frequency detuning $\Delta = 400$ MHz, and the radius of curvature of the wave front, $R = 2$ m. The atoms differ in longitudinal velocity,

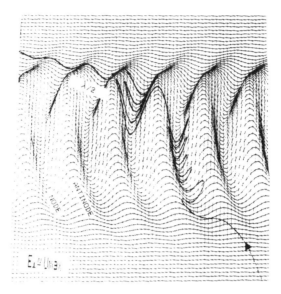

Figure 6. Effect of atomic momentum diffusion on the motion of atoms. The transverse kinetic energy of the atoms is a little lower than the potential maximum. The momentum diffusion raises this energy so that it becomes higher than the maximum potential energy at a certain point in the standing light wave, which prevents the atoms from being localized.

the localized atoms moving with a velocity of $v = 500$ m s^{-1} and their non-localized counterparts with a velocity of $v = 1200$ m s^{-1}. The atoms moving with the higher velocity will have, at a certain point in the field, their transverse kinetic energy higher than the potential maximum, and so will fly on freely.

Figure 6 shows the effect of momentum diffusion on the motion of atoms. The transverse kinetic energy of the atoms is a little lower than the potential maximum. The atomic momentum diffusion raises the kinetic energy so that it becomes, at some point in the standing light wave, higher than the potential maximum, and the atoms are thus prevented from being localized.

When the atomic transition saturation parameter is much less than unity and the atom–light-field interaction time is such that the change in the atomic momentum due to friction and diffusion is insignificant, both the friction force and atomic momentum diffusion can be disregarded. In that case, the atom sees the light wave as a spatially periodic potential field whose period is equal to that of the spatial light-field intensity distribution.

Figure 7 illustrates the change in the transverse velocity of the localized atoms in the course of their flight through the spherical standing light wave, calculated by equations (10) and (11). The atomic motion calculations have been made with and without the friction force being taken into consideration. The non-localized atoms receive a push in the region where their trajectories are tangent to the standing wave front. The friction force changes the atomic oscillation amplitude but has no effect on the character of atomic motion.

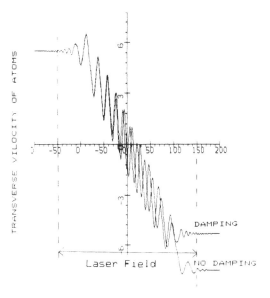

Figure 7. Transverse velocity variation of atoms localized in a spherical standing light wave in the course of their flight.

4. Experimental observation of the channelling of atoms

One of the manifestations of the atomic channelling effect in a standing light wave is the concentration of atoms near the nodes ($\Delta > 0$) or loops ($\Delta < 0$) of the wave. The first experiments in this field were aimed at detecting such spatial redistribution of atoms in a standing light wave. For example, the authors of [23] measured the fluorescence signal from an atomic beam intersecting a standing light wave at right angles. This signal was found to depend asymmetrically on the frequency detuning of the wave. These investigators attributed this asymmetry to the redistribution of atomic density in the standing wave potential produced by the gradient force. They believed that at negative frequency detunings the atoms concentrated near the loops of the standing wave. i.e., in regions where the field intensity was high, which resulted in an increased fluorescence signal, whereas at positive detunings, the atoms concentrated near the standing wave nodes, and the fluorescence signal thus decreased.

The distribution of atoms in a standing light wave was studied in more detail in [24]. The authors of this work measured the frequency dependence of the coefficient of absorption of an auxiliary weak probe wave passing through the region where an atomic beam cut across a strong standing light wave. As a result of the optical Stark shift, the frequency at which an atom absorbs the probe radiation depends on the light-field intensity at its location in the standing wave. The spatial distribution of atoms in a standing light wave can be estimated by analysing their absorption line profiles. This experiment also demonstrated that the atom concentrated in the vicinity of the standing wave nodes at $\Delta > 0$, and near the loops at $\Delta < 0$.

A number of experiments were staged to measure the changes in the microscopic atomic motion parameters following the atom–standing-light-wave interaction. The simplest experiment of this kind [25, 26] involved the scattering of atoms by the potential of a standing light wave (figure 8(a)). In these experiments, both the

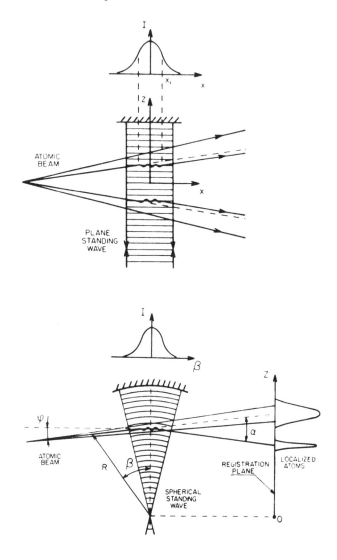

Figure 8. Localization of atoms in (*a*) a plane standing light wave (top—light-field intensity distribution along the transverse coordinate of the laser beam forming the standing wave) and (*b*) a spherical standing light wave.

divergence and shape of the atomic beam changed. These changes, however, do not allow one to speak with assurance about such details of channelling as the proportion of localized atoms and their confinement time in the potential wells.

The situation is quite different where the atomic beam interacts with a spherical standing light wave [15, 27, 28]. The experiment is illustrated schematically in figure 8(*b*). The spherical standing light wave is formed by a laser beam (with a Gaussian transverse intensity distribution) reflecting from a spherical mirror. The laser field is monochromatic, and its frequency detuning $\Delta > 0$. A beam of Na atoms meets the standing light wave tangentially to its front. The atoms localized in the

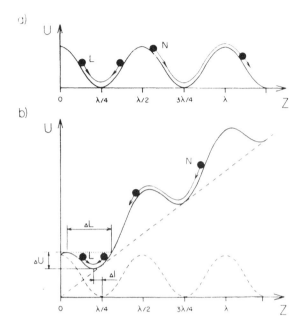

Figure 9. Effective potential for atoms in a spherical standing light wave. The centrifugal force displaces the bottom of the potential well (Δ/l) and reduces its width (ΔL) and depth (Δu).

standing wave (i.e., ones entering the wave near its potential minima) move on along the potential wells whose bend is similar to that of the nodal surfaces of the wave. As a result, the localized atoms are deflected through an angle of θ from their initial direction, θ being the angular divergence of the standing light wave. Disregarding the effect of the friction force, the non-localized atoms can be said to suffer only scattering in the standing light wave and gather a characteristic transverse velocity of $v_{\mathrm{cr}} = (2U_{\mathrm{g}}^0/M)^{1/2}$ determined by the amplitude of the standing wave potential U_{g}^0. So, for a beam of thermal atoms with a small angular divergence of $\Delta\varphi < \theta/2$, it is not very difficult to find such standing light wave parameters (θ, Δ, G) as will make the beam of localized atoms diverge from that of their non-localized counterparts.

Let us now consider the conditions necessary to localize atoms in a spherical standing light wave, and also some specific features of atomic localization in a plane standing light wave.

We will analyse the motion of an atom in a region within the standing light wave, the size of which is much smaller than the radius R of the light wave curvature in this region. If, in addition, the condition $v \gg v_{\mathrm{cr}}$ is valid for the atomic velocity, we can assume that in the polar coordinate system associated with the light wave (figure 8(b)), the atom is acted upon by the constant centrifugal force $F_{\mathrm{cf}} = Mv_{\mathrm{t}}^2/R$, where v_{t} is the tangential atomic velocity component. The atomic motion in this coordinate system is governed by the effective potential $U_{\mathrm{eff}} = U_{\mathrm{g}} + F_{\mathrm{cf}}R$ (figure 9). It can be seen from this figure that the centrifugal force reduces the size of the potential wells, displaces them along the standing light wave, and reduces the potential depth down to $\Delta U = U_{\mathrm{g}}^0 - F_{\mathrm{cf}}(\lambda/4)$, provided that $F_{\mathrm{cf}} \ll F_{\mathrm{g}}^{\mathrm{max}} = \sup(-\partial U_{\mathrm{g}}/\partial R)$. On the other hand, these potential wells can only exist if $F_{\mathrm{g}}^{\mathrm{max}} > F_{\mathrm{cf}}$.

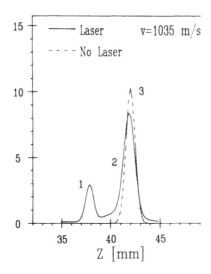

Figure 10. Experimental transverse profile of an atomic beam following its interaction with a spherical standing light wave. Peak 1 corresponds to the atoms localized in the wave and peak 2, to non-localized atoms. Dashed curve 3 corresponds to no-laser-field conditions.

An atom can be localized in a spherical standing light wave only if its radial velocity component v_R satisfies the inequality $Mv_R^2/2 < \Delta U$. This occurs in the vicinity of the point where the atomic trajectory is conjugate to the standing wave, $\beta = \varphi$ and $v_R = 0$ in the absence of the wave. The characteristic size of this region is $L \simeq (R/v)(2\Delta u/M)^{1/2}$. In the case of the $3^2S_{1/2} \rightarrow 3^2P_{3/2}$ transition in Na ($\delta = 3 \times 10^{-7}$ s^{-1}, $M = 4 \times 10^{-23}$ g) at $\Delta U = 10\hbar\delta \simeq 3 \times 10^{-19}$ erg, $v = 10^5$ cm s^{-1}, and $R = 1$ cm, the size $L \simeq 10^{-3}$ cm, i.e., the place of possible localization of the atom has been correctly determined to be the tangency point with the coordinates φ and R.

The analysis of the motion of an atom in the potential presented in figure 9 shows that the most favourable conditions for the atom to be localized occur when the atomic trajectory coincides with the curvature of standing light wave at its edge, where $F_{\mathrm{g}} \simeq F_{\mathrm{cf}}$. In the case of weak saturation, a 100% localization can be achieved if the transverse potential gradient satisfies the following conditions in the vicinity of the localization point:

$$\partial U/\partial p = F_\perp \ge (F_{\mathrm{cf}}/2)(v_{\mathrm{cr}}^*/v) \tag{16}$$

where $v_{\mathrm{cr}}^* = (F_{\mathrm{cf}}\lambda/M)^{1/2}$. This expression holds true where the contribution from the diffusion of atoms in the course of channelling is small, i.e., where $\partial U/\partial p > D/Mv$.

Consequently, two conditions must be satisfied for atoms to become localized in a spherical standing light wave. First, condition (16) must be fulfilled at the tangency point. Secondly, the potential in the vicinity of this point must obviously grow higher in the direction of atomic motion. In other words, this point must be located at the entrance half of the wave.

Figure 10 illustrates the characteristic atomic beam intensity distribution in a plane 290 mm distant from the standing light wave. The wave was formed by a single-frequency laser beam detuned by an amount of $\Delta/2\pi = 90$ MHz from the frequency of the $3^2S_{1/2}(F = 2) \rightarrow 3^2P_{3/2}(F' = 3)$ transition in Na, the laser beam intensity in the atom–laser-field interaction region being $I = 20$ W cm^{-2}. The waist of the laser beam in the interaction region was $2W = 0.6$ mm (with reference to the I/e^2 intensity level), its angular divergence θ amounting to 1.5×10^{-2} rad. The atomic beam had an angular divergence of $\Delta\varphi = 3 \times 10^{-3}$ rad and intersected the laser beam at an angle of $\varphi = 7 \times 10^{-3}$ rad. As can be seen from figure 10, the distance between the atomic beam intensity peaks is $\Delta z = 4.3$ mm, which corresponds to a deflection angle of the localized atoms of $\alpha_{\text{eff}} \simeq 1.5 \times 10^{-2}$ rad. This angle agrees well with the predicted value $\alpha = \theta = 1.5 \times 10^{-2}$ rad.

For our experiment, expression (16) is invalid. As a result, it was only about 30% of the atoms incident upon the spherical standing light wave that got localized in it (figure 10). Besides, our experimental results allowed us to conclude that all the atoms forming peak 1 in the detection zone moved in the standing light wave along certain potential wells without any jumps between them. So, the characteristic lifetime of the atoms in individual wells exceeded their time of flight through the wave, $\tau \simeq 1$ μs.

As seen from figure 9, the probability of the atoms jumping from one potential well into another in the spherical standing light wave is low. If an atom escapes from a potential well, it already possesses an energy of the order of $F_{\text{cf}}\lambda/4$ by the time it reaches the next well. With $F_{\text{cf}} \simeq F_{\text{g}}$, the centrifugal acceleration of the atom reduces its probability of being localized again. Quite a number of technical factors prevented one from attaining times of flight of the atoms through the standing light wave longer than 1 μs in the experiments reported in [15, 28]. If, however, longer times of flight are achieved, it will be possible to measure directly the confinement times of atoms in individual potential wells. It should be noted that these times differ from the confinement time in a plane standing light wave [22]. This is mainly due to the bottoms of the effective potential wells in the spherical standing light wave being displaced relative to its nodal lines (figure 9), with the result that neither diffusion coefficient (13), nor friction force (12) goes to zero at the bottom of the effective potential wells.

5. Collimation of an atomic beam by way of atomic channelling

An atom can be localized in a spherical standing light wave if its trajectory is tangent to the wave front at the entrance point, the gradient force exceeds the centrifugal force, and the field intensity grows higher in the direction of atomic motion. In the course of localization, the longitudinal atomic velocity is transformed into transverse velocity. If the law governing the light-field intensity distribution differs between the entrance and exit edges of the wave, the transverse atomic velocity at the exit from the wave may differ from its initial value, i.e., there may occur an effective transverse cooling of the atoms [29]. The relative transverse cooling of the atomic beam is defined by the relation [29]

$$T_f/T_{\text{in}} = (4e^2/\pi)(R\lambda/W^2) \tag{17}$$

which contains only the geometrical parameters of the standing light wave at the point of its intersection with the atomic beam, namely, its radius of curvature R and the laser beam radius W. Figure 11 illustrates schematically the transverse cooling of atoms by way of their channelling in a truncated spherical standing light wave. The wave was formed by shutting half the beam of a single-frequency cw laser with a safety razor blade. The spatial profile of the laser beam is shown at the top of the figure. The transverse cooling of the atoms was detected by measuring the spatial profiles of the atomic beam following its intersection with the standing light wave.

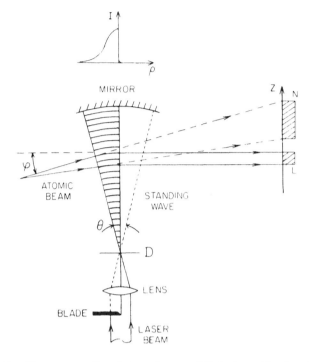

Figure 11.   Transverse cooling of atoms by way of their channelling in a truncated spherical standing light wave. The wave was formed by shutting half the beam of a single-frequency cw laser with a safety razor blade. Top—spatial laser beam profiles.

Figure 12 presents the results of such measurements. The solid curve corresponds to the whole spherical standing light wave, and the dashed one, to the truncated wave. Peak 1 is produced by the atoms localized in the whole wave, peak 2 is formed by the atoms localized in the truncated wave, and peak 3 is due to the non-localized atoms. Peak 2 due to the atoms localized in the truncated wave is narrower than the peak produced by the non-localized atoms. The transverse atomic temperature in this experiment was reduced from $k_B T_{in} = 8.5\hbar\delta$ to $k_B T_f = 2.1\hbar\delta$. The numerical modelling of the experiment yielded a value of $1.3\hbar\delta$ for the final atomic temperature $k_B T_f$, which was in good agreement with the experimental value.

6. Conclusion

A natural extension of the atomic channelling effect, i.e., one-dimensional localization, is the channelling of atoms in two (two-dimensional localization) and three (three-dimensional localization) standing light waves intersecting at right angles. In the latter case, which was already considered in [14, 20], atoms can be trapped in small regions less than $\lambda/2$ in size. The localization of an atom in a three-dimensional standing light wave in the case of red shift, $\Delta \simeq -\delta$, was observed in [30]. Based on the fact that the measured width of the elastic component of the fluorescence spectrum was $\Delta v \simeq 70$ kHz, the confinement time τ of the atom in a potential well was estimated at 2.3 μs. This value agrees well with that given in section 2 for the case of red shift.

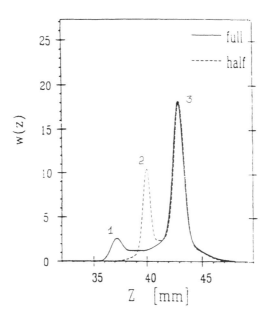

Figure 12. Experimental results of the transverse cooling of atoms by way of channelling. The solid curve corresponds to the whole spherical standing light wave and the dashed one, to the truncated wave. Peak 1 is produced by the atoms localized in the whole wave, peak 2 is formed by the atoms localized in the truncated wave, and peak 3 is due to the non-localized atoms.

This opens up a number of interesting application possibilities. First, the Doppler broadening is eliminated for all the spectral lines of localized atoms (Lamb–Dicke regime), and it was this application of the channelling effect that was proposed in [12]. Secondly, three-dimensional localization gives rise to a regular, spatially periodic arrangement of the absorbing and emitting atomic particles. In that case, it might be expected that absorption and emission anisotropy effects will become manifest similar to the Bragg diffraction.

References

[1] Ashkin A 1980 *Science* **210** 1081
[2] Letokhov V S and Minogin V G 1981 *Phys. Rep.* **73** 1
[3] Stenholm S 1986 *Rev. Mod. Phys.* **58** 699
[4] Phillips D, Gould P I and Lett P D 1991 *Science* **239** 878
[5] Cohen-Tannoudji C 1991 *Fundamental Systems in Quantum Optics* (Elsevier Science Publ. BV)
[6] Meystre P and Stenholm S (ed) 1985 *J. Opt. Soc. Am.* B **5** 11
[7] Chu S and Wieman C (ed) 1990 *J. Opt. Soc. Am.* B **6** 2109–278
[8] Minogin V G and Letokhov V S 1978 *Laser Light Pressure on Atoms* (New York: Gordon & Breach)
[9] Kazantsev A P, Surdutovich G I and Yakovlev V P 1991 *Mechanical Action of Light on Atoms* (Singapore: World Scientific)

[10]   Moi L, Gozzini S, Gabbanini C, Arimondo E and Strumia F (ed) 1991 *Light Induced Kinetic Effects* (Piza: ETS Editrice)

[11]   Kazantsev A P, Ryabenko G A, Surdutovich G I and Yakovlev V P 1985 *Phys. Rep.* **129** 75

[12]   Letokhov V S 1968 *Pis'ma Zh. Eksp. Teor. Fiz.* **7** 348 (Engl. Trans. 1968 *Sov. Phys.–JETP Lett.* **7** 272)

[13]   Gordon J P and Ashkin A 1980 *Phys. Rev.* A **21** 1606

[14]   Letokhov V S and Pavlik B D 1976 *Appl. Phys.* **9** 229

[15]   Balykin V I, Lozovik Yu E, Ovchinikov Yu B, Sidorov A I, Shul'ga S V and Letokhov V S 1989 *JOSA* B **6** 2178

[16]   Kazantsev A P, Chudesnikov D O and Yakovlev V P 1986 *Zh. Eksp. Teor. Fiz.* **63** 951 (in Russian)

[17]   Cook R J 1980 *Phys. Rev.* A **22** 1078

[18]   Baklanov E V 1987 *Pis'ma Zh. Eksp. Teor. Fiz.* **45** 247 (in Russian)

[19]   Letokhov V S 1975 *Science* **190** 344

[20]   Letokhov V S, Minogin V G and Pavlik B D 1976 *Opt. Commun.* **19** 72; 1977 *Zh. Eksp. Teor. Fiz.* **72** 1328 (in Russian)

[21]   Kazantsev A P, Smirnov V S, Surdutovich G I, Chudesnikov D O and Yakovlev V P 1985 *JOSA* B **2** 1731

[22]   Kazantsev A P, Surdutovich G I and Yakovlev V P 1988 *Opt. Commun.* **68** 103

[23]   Prentiss M G and Ezekiel S 1986 *Phys. Rev. Lett.* **56** 46

[24]   Salamon C, Dalibard J, Aspect A, Metcalf H and Cohen-Tannoudji C 1987 *Phys. Rev. Lett.* **59** 1659

[25]   Arimondo E, Lew H and Oka T 1987 *Phys. Rev. Lett.* **43** 753

[26]   Tauguy C, Reynaud S and Cohen-Tannoudji C 1984 *J. Phys. B: At. Mol. Phys.* **17** 4623

[27]   Cloppenpurg K, Henning G, Mihm A, Wallis H and Ertmer W 1987 *Laser Spectroscopy VIII* (Berlin: Springer) p 87

[28]   Balykin V I, Letokhov V S, Ovchinnikov Yu B, Sidorov A I and Shul'ga S V 1988 *Opt. Lett.* **13** 958

[29]   Balykin V I, Letokhov V S, Ovchinnikov Yu B and Shul'ga S V 1990 *Opt. Commun.* **77** 152

[30]   Westbrook C I, Watts R N, Tanner C E, Rolston S L, Phillips W D, Lett P D and Gould P I 1990 *Phys. Rev. Lett.* **65** 33

Optical nonlinearities and the Kerr-effect in phaseonium

U. Rathe,[a] M. Fleischhauer,[a,b] and Marlan O. Scully[a,b]

[a]Center for Advanced Studies and
Department of Physics and Astronomy
University of New Mexico
Albuquerque, New Mexico 87131

and

[b]Max-Planck Institut für Quantenoptik
D-8046 Garching
Federal Republic of Germany

Recent studies have shown that three-level atoms having a ground (excited) state doublet connected to an excited (ground) state via an optical frequency transition can display unusual behavior when the doublet is coherently prepared. Such a phased ensemble of atoms (phaseonium) is in a very real sense a new state of matter. For example, such a medium can display cancellation of optical absorption while at the same time retaining emission from the excited state to the ground state doublet (lasing without inversion). It has recently been shown that it is, in principle, possible to have a condition in which there is an enhancement of the linear index of refraction with vanishing absorption. In this chapter, we wish to show that phaseonium can display a large nonlinear Kerr-effect near the atomic resonance while retaining complete transparency.

1 Introduction

Systems with atomic coherence can have interesting optical properties, such as non-absorbing resonances (Alzetta et al. 1976, 1978, 1979; Gray et al. 1979; Harris 1989; Imamoğlu et al. 1989a, 1989b) or an index of refraction at a frequency of vanishing absorption many orders of magnitude larger than is otherwise possible (Scully et al. 1991, 1992; Fleischhauer et al. 1992). In the following we show that such a medium can display an ultra large Kerr-coefficient near the atomic resonance while remaining completely transparent.

Near resonance nonlinearities are usually accompanied by a large absorption of the medium. By using atomic coherence effects this absorption can be cancelled while maintaining substantial values of the nonlinear susceptibilities. In a three-level system for example (see figure 1), a strong microwave driving field coupling the two closely spaced upper levels can lead to a non-absorbing resonance with a large susceptibility $\chi^{(3)}(\omega; \omega_1, \omega_2, \nu_\mu)$, corresponding to a sum-frequency generation of two optical frequencies ω_1 and ω_2 and the

---

[*] Dedicated to Academician Akhmanov who made so many enduring contributions to our understanding of nonlinear optical physics.

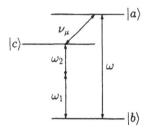

Figure 1: Level scheme for the sum-frequency generation discussed by Harris et al. (1990).

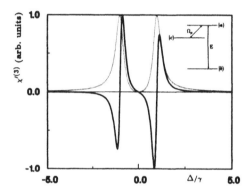

Figure 2: $\chi'^{(3)}(\omega; \omega, \omega, -\omega)$ in arbitrary units for a three-level system with a strong coherent driving field between the upper two levels as a function of the detuning (solid curve). The dotted curve shows the imaginary part of the linear susceptibility.

microwave frequency ν_μ if ω is resonant with the $|a\rangle \rightarrow |b\rangle$ transition (Harris et al. 1990; Hahn et al. 1990).

Without population in the excited state of the optical transition, however, all susceptibilities $\chi^{nl}(\omega; \pm\omega, \pm\omega, \ldots)$ containing only *one* optical frequency ω, and hence all related quantities like the Kerr-coefficient, vanish at the point of absorption cancellation as shown in figure 2. On the other hand, providing there is a small population in the upper level, the real part of the linear and nonlinear susceptibilities $\chi^{nl}(\omega; \pm\omega, \pm\omega, \ldots)$ can reach substantial values.

In the following we concentrate on one of the several proposed coherence-establishing schemes (Scully et al. 1991, 1992; Fleischhauer et al. 1992), namely the case of a three-level atom in which an initially prepared atomic coherence between a ground-state doublet, $|b\rangle$ and $|b'\rangle$, generates a point of zero absorption and a high index of refraction for the transition to an excited state, $|a\rangle$. We compare the Kerr-coefficient for this system to that of usual Kerr-type materials such as CS_2.

2 Two-level atoms

It is well known that high nonlinearities occur near atomic resonances. To illustrate this and to make a comparison to the three-level atom with an initially prepared coherence discussed in the following section, we briefly review the text-book results for a two-level system (Sargent et al. 1974).

We consider a system of two-level atoms interacting with an electric field as shown in figure 3. The atoms are initially prepared in a mixture of states described by a diagonal density matrix ρ^0. The equations of motion for the density matrix of the ith atom in a rotating frame are

$$\dot{\rho}_{aa}^i = -\gamma_a \rho_{aa}^i - \frac{i}{\hbar}\left(\wp E \rho_{ba}^i - c.c.\right), \tag{1}$$

$$\dot{\rho}_{bb}^i = -\gamma_b \rho_{bb}^i - \frac{i}{\hbar}\left(\wp E^* \rho_{ab}^i - c.c.\right), \tag{2}$$

$$\dot{\rho}_{ab}^i = -\left(\gamma_{ab} + i\Delta_{ab}\right)\rho_{ab}^i - \frac{i}{\hbar}\wp E\left(\rho_{bb}^i - \rho_{aa}^i\right), \tag{3}$$

where E is the slowly varying amplitude of the optical probe-field, $\wp$ is the dipole-matrix element of the transition $|b\rangle \to |a\rangle$, $\Delta_{ab} = \omega_{ab} - \nu$ with ν being the frequency of the optical probe, γ_a and γ_b are the longitudinal decay rates from the levels $|a\rangle$ and $|b\rangle$ to some other levels, and $\gamma_{ab} = (\gamma_a + \gamma_b)/2$. We have assumed that the decay rate from $|a\rangle$ to $|b\rangle$ is much smaller than γ_a and γ_b and can be neglected together with decay due to collisions.

If we introduce the atomic injection rate r and use the technique described in Sargent et al. (1974) to sum over all atoms, we obtain for the total steady state susceptibility defined by

$$\chi = \chi' + i\,\chi'' = -\frac{\wp^2}{\epsilon_0 \hbar V}\frac{\rho_{ab}}{\Omega}, \tag{4}$$

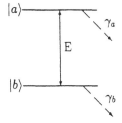

Figure 3: Level scheme for the two-level scheme.

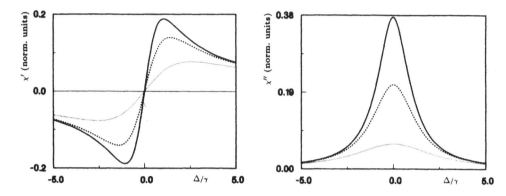

Figure 4: Real (a) and imaginary part (b) of the susceptibility as functions of the detuning in the two-level scheme. Plotted is $\chi \left(\wp^2 r/\left(\epsilon_0 \hbar \gamma^2 V\right)\right)^{-1}$. Note that in our formalism the atomic number density is given by $N = r/\gamma V$ with the cavity volume V. The parameters are $\gamma_a = 0.1\gamma$, $\gamma_b = 2\gamma$, $\rho_{aa}^0 = 0.01$, and $\rho_{bb}^0 = 0.99$. The three curves correspond to field amplitudes of $\Omega = 0.01\gamma$ for the solid curve, $\Omega = 0.2\gamma$ for the dashed curve and $\Omega = 0.5\gamma$ for the dotted curve.

where ρ is now the sum over the density matrices of all atoms in the cavity and $\Omega = \wp E/\hbar$ is the Rabi-frequency of the field,

$$\chi' = \frac{\wp^2 r}{\epsilon_0 \hbar V} \left(\frac{\rho_{bb}^0}{\gamma_b} - \frac{\rho_{aa}^0}{\gamma_a} \right) \frac{\Delta_{ab}}{\Delta_{ab}^2 + \gamma_{ab}^2 \left(1 + \frac{4|\Omega|^2}{\gamma_a \gamma_b} \right)}, \tag{5}$$

$$\chi'' = \frac{\wp^2 r}{\epsilon_0 \hbar V} \left(\frac{\rho_{bb}^0}{\gamma_b} - \frac{\rho_{aa}^0}{\gamma_a} \right) \frac{\gamma_{ab}}{\Delta_{ab}^2 + \gamma_{ab}^2 \left(1 + \frac{4|\Omega|^2}{\gamma_a \gamma_b} \right)}. \tag{6}$$

Here V is the interaction volume and ρ_{aa}^0 and ρ_{bb}^0 are the initial populations of the levels $|a\rangle$ and $|b\rangle$.

The real and imaginary parts of the susceptibility are plotted in figure 4 for different intensities. To achieve high nonlinearities, we would have to tune the probe-field near to the atomic resonance. Near resonance, however, the absorption is also very high, so that the probe-field would of course be absorbed after a short interaction length. But by employing atomic coherences, we can introduce a point of zero absorption in the spectrum near resonance and at the same time maintain the behaviour of the refractive susceptibility.

3   Phaseonium

We now consider the atomic scheme shown in figure 5. A three-level atom is initially prepared in a coherent superposition of the two lower levels $|b\rangle$ and $|b'\rangle$. This can be achieved for example by a coherent pulse excitation. It has been shown by Scully (1991), that the initially prepared coherence together with a small population in the excited state can lead

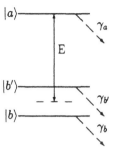

Figure 5: Level scheme for the initial coherence scheme.

to a high real part of the linear susceptibility at a point of vanishing imaginary part, i.e. zero absorption. We now investigate the nonlinear susceptibilities of this medium at such a point of vanishing absorption.

The initial density matrix for the ith atom is

$$\rho^i = \begin{pmatrix} \rho_{aa}^0 & 0 & 0 \\ 0 & \rho_{bb}^0 & \rho_{bb'}^0 \\ 0 & \rho_{b'b}^0 & \rho_{b'b'}^0 \end{pmatrix}, \tag{7}$$

and the equations of motion for the ith atom in a rotating frame read

$$\dot{\rho}_{aa}^i = -\gamma_a \rho_{aa}^i - \frac{i}{\hbar}\left[\left(\wp'\rho_{b'a}^i + \wp\rho_{ba}^i\right)E - c.c.\right], \tag{8}$$

$$\dot{\rho}_{bb}^i = -\gamma_b \rho_{bb}^i - \frac{i}{\hbar}\left(\wp\rho_{ab}^i E^* - c.c.\right), \tag{9}$$

$$\dot{\rho}_{b'b'}^i = -\gamma_{b'} \rho_{b'b'}^i - \frac{i}{\hbar}\left(\wp'\rho_{ab'}^i E^* - c.c.\right), \tag{10}$$

$$\dot{\rho}_{ab}^i = -\left(\gamma_{ab} + i\Delta_{ab}\right)\rho_{ab}^i - \frac{i}{\hbar}\wp\left(\rho_{bb}^i - \rho_{aa}^i\right)E - \frac{i}{\hbar}\wp'\rho_{b'b}^i E, \tag{11}$$

$$\dot{\rho}_{ab'}^i = -\left(\gamma_{ab'} + i\Delta_{ab'}\right)\rho_{ab'}^i - \frac{i}{\hbar}\wp'\left(\rho_{b'b'}^i - \rho_{aa}^i\right)E - \frac{i}{\hbar}\wp\rho_{bb'}^i E, \tag{12}$$

$$\dot{\rho}_{b'b}^i = -\left(\gamma_{b'b} + i\omega_{b'b}\right)\rho_{b'b}^i - \frac{i}{\hbar}\left(\wp'\rho_{ab}^i E^* - \wp\rho_{b'a}^i E\right), \tag{13}$$

where E is the slowly varying amplitude of the probe-field, $\wp$ and $\wp'$ are the dipole-matrix elements of the transitions $|b\rangle \rightarrow |a\rangle$ and $|b'\rangle \rightarrow |a\rangle$, $\Delta_{\alpha\beta} = \omega_{\alpha\beta} - \nu$ with ν being the frequency of the oprical probe, γ_α are the longitudinal atomic decay rates to other levels, and $\gamma_{\alpha\beta} = (\gamma_\alpha + \gamma_\beta)/2$. Here we have assumed as in Sec. 2 that the decay rates between the atomic levels are small compared to the decay rates to other levels.

We solve the corresponding linear system of algebraic equations analytically in the steady state. For the susceptibility defined by

$$\chi = -\frac{\left(\wp^2 \rho_{ab} + \wp'^2 \rho_{ab'}\right)}{\epsilon_0 \hbar V \Omega}, \tag{14}$$

we obtain for the simplified case where $\wp' \equiv \wp$, $\gamma_{b'} = \gamma_b$, $\Omega = \Omega^*$ and equal initial populations of the levels $|b\rangle$ and $|b'\rangle$ $\left(\rho_{bb}^0 \equiv \rho_{b'b'}^0 = (1 - \rho_{aa})/2\right)$

$$\chi = \frac{\wp^2}{\epsilon_0 \hbar V}\frac{r}{2}\frac{\left(A' + B'\,\Omega^2 + C'\,\Omega^4\right) + i\left(A'' + B''\,\Omega^2 + C''\,\Omega^4\right)}{a + b\,\Omega^2 + c\,\Omega^4 + d\,\Omega^6}, \tag{15}$$

where the real coefficients A, B, C, a, b, c, d depend only on the system parameters:

$$
\begin{aligned}
A' \;=\; & 2\delta\gamma_b \left(a_1 a_3 R_0 + 2\gamma_a\gamma_b \left[\left(a_1\gamma_b - \gamma_{ab}\omega_{b'b}^2\right) Re\left(\rho_{b'b}^0\right) + \right.\right. \\
& \qquad\qquad \left.\left. \omega_{b'b}\left(a_1 + \gamma_{ab}\gamma_b\right) Im\left(\rho_{b'b}^0\right)\right]\right)
\end{aligned} \tag{16}
$$

$$
\begin{aligned}
A'' \;=\; & 2\gamma_b \left(a_2 a_3 \gamma_{ab} R_0 + \right. \\
& \quad 2\gamma_a\gamma_b \left[\left(a_2\left(\gamma_{ab}\gamma_b - \omega_{b'b}^2/2\right) + \delta^2\omega_{b'b}^2\right) Re\left(\rho_{b'b}^0\right) + \right. \\
& \qquad\qquad \left.\left. \omega_{b'b}\left(a_2\left(\gamma_{ab} + \gamma_b/2\right) - \delta^2\gamma_b\right) Im\left(\rho_{b'b}^0\right)\right]\right)
\end{aligned} \tag{17}
$$

$$
\begin{aligned}
B' \;=\; & 4\delta \left(\left[2\gamma_b^2\gamma_{ab} + \omega_{b'b}^2\left(\gamma_{ab} + \gamma_b/2\right)\right] R_0 + \right. \\
& \quad \gamma_b \left[4\gamma_{ab}\left(\gamma_a\gamma_b - \omega_{b'b}^2\right) Re\left(\rho_{b'b}^0\right) + \right. \\
& \qquad\qquad \left.\left. \omega_{b'b}\left(4\gamma_a\gamma_b + 2\gamma_b^2 + \gamma_a^2\right) Im\left(\rho_{b'b}^0\right)\right]\right)
\end{aligned} \tag{18}
$$

$$
\begin{aligned}
B'' \;=\; & \left[8\gamma_{ab}^2\gamma_b^2 + \omega_{b'b}^2\left(4\gamma_{ab}^2 + \gamma_b^2\right)\right] R_0 + \\
& \quad 4\gamma_a\gamma_b \left[\left(4\gamma_{ab}^2\gamma_b - \omega_{b'b}^2\gamma_a/2\right) Re\left(\rho_{b'b}^0\right) + \right. \\
& \qquad\qquad \left. 2\omega_{b'b}\gamma_{ab}\left(\gamma_{ab} + \gamma_b/2\right) Im\left(\rho_{b'b}^0\right)\right]
\end{aligned} \tag{19}
$$

$$C' = 8\delta\gamma_b\left(R_0 + 2\gamma_a Re\left(\rho_{b'b}^0\right)\right) \tag{20}$$

$$C'' = 8\gamma_b\gamma_{ab}\left(R_0 + 2\gamma_a Re\left(\rho_{b'b}^0\right)\right) \tag{21}$$

$$a = a_3\gamma_a\gamma_b^2\left(\gamma_{ab}^2 + \left(\delta + \omega_{b'b}/2\right)^2\right)\left(\gamma_{ab}^2 + \left(\delta - \omega_{b'b}/2\right)^2\right) \tag{22}$$

$$b = -2\gamma_b\left(\gamma_a\gamma_b\omega_{b'b}^2\left(\gamma_{ab}^2 - \delta^2 + \omega_{b'b}^2/4\right) - 2a_2\gamma_{ab}\left(2a_3\gamma_{ab} + \gamma_a\gamma_b^2\right)\right) \tag{23}$$

$$
\begin{aligned}
c = & -2\gamma_{ab}\left[-2\gamma_b^2\gamma_{ab}\left(5\gamma_a + 4\gamma_b\right) - \omega_{b'b}^2\left(\gamma_a^2 + \gamma_a\gamma_b + 4\gamma_b^2\right)\right] + \\
& 4\delta^2\gamma_a\gamma_b^2
\end{aligned} \tag{24}
$$

$$d = 32\gamma_b\gamma_{ab}^2 \tag{25}$$

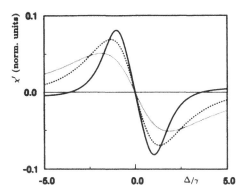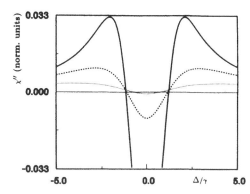

Figure 6: Real (a) and imaginary part (b) of the susceptibility as functions of the detuning in the initial coherence scheme. Plotted is the same quantity as in figure 4. The parameters are $\gamma_a = 0.1\gamma$, $\gamma_b = \gamma_{b'} = 2\gamma$, $\omega_{b'b} = 2\gamma$, $\rho_{aa}^0 = 0.01$, $\rho_{bb}^0 = \rho_{b'b'}^0 = 0.495$, and $\rho_{b'b}^0 = -i\,0.495$. The three curves correspond to field amplitudes of $\Omega = 0.1\gamma$ for the solid curve, $\Omega = 0.5\gamma$ for the dashed curve and $\Omega = 1.0\gamma$ for the dotted curve. The parameters used for this system are the same as in Scully (1991).

and

$$
\begin{aligned}
R_0 &= \gamma_a - (\gamma_a + 2\gamma_b)\,\rho_{aa}^0 \\
a_1 &= \gamma_{ab}^2 + \delta^2 - \omega_{b'b}^2/4 \\
a_2 &= \gamma_{ab}^2 + \delta^2 + \omega_{b'b}^2/4 \\
a_3 &= \gamma_b^2 + \omega_{b'b}^2 \\
\delta &= \Delta_{ab} - \omega_{b'b}/2 \equiv \Delta_{ab'} + \omega_{b'b}/2
\end{aligned}
\tag{26}
$$

The resulting susceptibility for this case is plotted in figure 6 as a function of the detuning. We first note the interesting point of vanishing absorption and very high refractive susceptibility. Furthermore, we see the decrease of the refractional part of the susceptibility with increasing field amplitudes, which implies a large Kerr-effect.

The Kerr-coefficient n_2 is defined as the coefficient of the first nonlinear correction to the index fo refraction:

$$
n(E) = n_0 + n_2 E^2 + n_4 E^4 + \cdots.
\tag{27}
$$

At a point of zero absorption ($\chi'' = 0$) the index of refraction can be expressed in terms of the real parts of the susceptibilities as

$$
\begin{aligned}
n(E) &= \sqrt{1 + \chi'^{(1)} + \chi'^{(3)} E^2 + \chi'^{(5)} E^4 + \cdots} = \\
&\quad n_0 + \frac{\chi'^{(3)}}{2n_0} E^2 + \frac{1}{2n_0}\left[\chi'^{(5)} - \left(\frac{\chi'^{(3)}}{2n_0}\right)^2\right] E^4 + \cdots,
\end{aligned}
\tag{28}
$$

where $n_0 = \sqrt{1 + \chi'^{(1)}}$. From this we obtain

$$n_2 \;=\; \frac{\chi'^{(3)}}{2n_0},\tag{29}$$

$$n_4 \;=\; \frac{\chi'^{(5)} - n_2^2}{2n_0}.\tag{30}$$

With the solution for the nonlinear susceptibility (15) we get

$$n_2 \;=\; -\frac{\wp^4 r}{4\epsilon_0 \hbar^3 V n_0}\frac{1}{a}\left[A'\frac{b}{a} - B'\right],\tag{31}$$

$$n_4 \;=\; -\frac{\wp^6 r}{4\epsilon_0 \hbar^5 V n_0}\frac{1}{a}\left[A'\left(-\frac{b^2}{a^2} + \frac{c}{a}\right) + B'\frac{b}{a} - C'\right] - \frac{n_2^2}{2n_0}\tag{32}$$

where the linear index of refraction n_0 is

$$n_0 = \left[1 + \frac{\wp^2 r}{2\epsilon_0 \hbar V a}A'\right]^{1/2}.\tag{33}$$

If $n_4 E^2$ is small compared to n_2, it is sufficient to describe the nonlinear optical properties of the medium by the Kerr-coefficient, and the preparation of an atomic beam in the described state would provide a transparent medium that shows not only a high index of refraction, but also strong Kerr-type nonlinearities.

In the following discussion we give a numerical example for the order of magnitude of this effect.

4   Discussion

Our results for the nonlinear susceptibilities clearly show the expected saturation effect. For strong fields, the real part of the susceptibility is given by

$$\chi' = -\frac{\wp^2}{\epsilon_0 \hbar V}\frac{r}{2}\frac{C'}{d\Omega^2} = \frac{\wp^2}{\epsilon_0 \hbar V}\frac{r}{2}\frac{\delta\left[(\gamma_a - (\gamma_a + 2\gamma_b)\rho_{aa}^0) + 2\gamma_a Re\left(\rho_{b'b}^0\right)\right]}{(\gamma_a + \gamma_b)^2 \,\Omega^2},\tag{34}$$

which reduces to the result for the two-level scheme (5) for vanishing real part of the atomic coherence.

For a quantitative analysis of the Kerr-effect in the discussed systems, we now turn to a numerical example and choose the atomic parameters of both systems according to figures 4 and 6, respectively. Moreover, we set the dipole-matrix elements to $\wp = a_0 e = 8.5\cdot10^{-30}\, Amp\,s\,m$, assume a decay rate of $\gamma = 10^7\,s^{-1}$, an interaction volume of $V = 10^{-6}\,m^3$, an atomic number density of $N = 10^{19}\,m^{-3}$, and an injection rate of $r = N\,V\,\gamma = 10^{13}\gamma$.

For the case of two-level atoms, we find, with $\Delta_{ab} = \gamma_{ab}$

$$n_2 = -5.5 \cdot 10^{-6}\,V^{-2}m^2.\tag{35}$$

When compared to the effect in CS_2 with the Kerr-coefficient $n_2 \simeq 10^{-20}\,V^{-2}m^2$ (Kelley 1965), we see an increase of the effect by 14 orders of magnitude. Because we are operating near the atomic resonance, however, the absorption is very high as well.

This changes dramatically if we consider Phaseonium: For the parameters listed above, we obtain

$$n_2 = -1.6 \cdot 10^{-7}\,V^{-2}m^2. \tag{36}$$

Similar to the two-level system, this is an increase of 13 orders of magnitude compared to CS_2, but now without absorption*. With Phaseonium we therefore have a transparent medium with extremely high Kerr-type nonlinearities in the electric susceptibility, which is of interest in problems such as optical phase conjugation, the generation of squeezed light via four-wave mixing, and optical computing.

Acknowledgment
This work was supported by the Office of Naval Research.

References

Alzetta G, Gozzini A, Moi L, and Orriols G 1976 *Nuovo Cimento* **36B** 5

Alzetta G 1978 *Coherence in Spectroscopy and Modern Physics* ed F T Arecchi, R Bonifacio, and M O Scully (New York, N.Y.)

Alzetta G, Moi L, and Orriols G 1979 *Nuovo Cimento* **52B** 209

Fleischhauer M, Keitel C H, Scully M O, and Su C 1992 *Opt. Commun.* **87** 109

Gray H, Whitley R, and Stroud C 1979 *Opt. Lett.* **3** 218

Hahn K H, King D A, and Harris S E 1990 *Phys. Rev. Lett.* **65** 2777

Harris S E 1989 *Phys. Rev. Lett.* **62** 1022

Harris S E, Field J E, and Imamoğlu A 1990 *Phys. Rev. Lett.* **64** 1107

Imamoğlu A and Harris S E 1989a *Opt. Lett.* **14** 1344

Imamoğlu A 1989b *Phys. Rev. A* **40** 2835

Kelley P L 1965 *Phys. Rev. Lett.* **15** 1005

Sargent M, Scully M O, and Lamb W E 1974 *Laser Physics* (Reading MA: Addison-Wesley)

Scully M O 1991 *Phys. Rev. Lett.* **67** 1855

Scully M O and Zhu S Y 1992 *Opt. Commun.* **87** 134

---

*We note that for the chosen parameters $n_4 = -3.8 \cdot 10^{-11}\,V^{-4}m^4$, i.e. the medium can accurately be described by n_0 and n_2 only if $E \ll 65\,Vm^{-1}$.

Spectroscopic studies in the far ultraviolet (80–200nm) using nonlinear tunable sources

This essay summarizes research carried out in the Department of Physics, at the University of Toronto from 1974 to 1992, in generating tunable, coherent, far ultraviolet radiation ($\lambda < 200$ nm) for use in spectroscopy. Brief reviews of the relevant theory and development of nonlinear sources are followed by discussions of a series of spectroscopic investigations of atoms and diatomic molecules. These include spectra of the rare-gas excimers Xe_2, Kr_2, Ar_2, XeKr; radiative lifetimes of selected vibronic and rovibronic levels of these excimers and of CO; second-harmonic generation in atoms of Zn and H, with reduced absorption at Lyman-α in H; and the long-standing problem of the dissociation energy of H_2.

1   Introduction

It never ceases to impress me that the once seldom used or understood nonlinear susceptibility $\chi^{(3)}$, has not only introduced us to new frontiers in the interaction of radiation with matter, but has also helped to open national borders and bring together colleagues of similar interests. It was on such an occasion in October 1967 that I had the opportunity of meeting Sergei Akhmanov, at the laser laboratories of Moscow State University. This was followed by the Third USSR Symposium on Nonlinear Optics, held in Yerevan, Armenia (October 20-27), which was the first USSR laser conference open to scientists from the West. Approximately 30 from Europe and North America attended, along with about 200 from the USSR. At that time, we were both much involved in experiments on stimulated Raman and Brillouin scattering, a field in which Akhmanov continued to make important contributions for another two decades.

As the field of nonlinear optics rapidly progressed, I became interested in the possibility of laser spectroscopy in the vacuum ultraviolet region, and emphasized its importance in a panel discussion on "The Future Course of Quantum Electronics" at the Esfahan Symposium held in 1971, another of these early international laser conferences. To quote: "One of the most difficult regions for work in spectroscopy is the vacuum ultraviolet (VUV). Materials which transmit such radiation are almost nonexistent and those with even 50% reflectance are few. This is also a region where we need new ideas for laser sources. Some advance has already been made, but this is just a beginning. New lasers in the vacuum ultraviolet will eventually lead to improved optical components, new materials, and much more spectroscopic research in this region." After these two decades, there is still a dearth of useful lasers in the wavelength region below 200 nm. However, nonlinear optics has provided radiation by second, third, and higher harmonics, and most usefully for spectroscopy, there is now available tunable radiation generated by frequency sum- and difference - mixing.

The observation of second-harmonic generation (SHG) by Franken et al (1961) was a crucial step leading to the eventual production of coherent radiation in the VUV region. This observation was quickly followed with the classic theoretical paper on second- and third-order nonlinear susceptibilities by Armstrong et al (1962). Third-harmonic radiation (THG) was demonstrated by Maker et al (1964) in crystals, glasses and liquids, and the major problem of generating even shorter wavelength radiation (due to the limited transparency of many of the nonlinear solids to the region above ~200 nm) was resolved when New and Ward (1967) succeeded in producing THG in a number of gases. Harris and Miles (1971, 1973) then demonstrated that high conversion efficiency of THG and of sum-frequency mixing could be obtained by using phase-matched metal vapors as nonlinear media, and that efficiency could be improved further by resonance enhancement. Limited tunability was achieved by the use of tunable pump lasers, but invariably with reduced efficiency, since resonance enhancement could not be maintained.

A seminal contribution providing tunability over broad regions, and using four-wave sum-mixing, $2\omega_1 + \omega_2 = \omega_3$, with the advantage of resonance enhancement, was made by Hodgson et al (1974). They used two dye lasers, one tuned to a two-photon allowed transition in a nonlinear metal vapor, and the other, tunable over a broad frequency range ω_2, such that $2\omega_1 + \omega_2$ corresponded to a transition from the ground state to a broad auto-ionizing state of the metal vapor. In this way, they succeeded in generating coherent radiation tunable over broad regions of the VUV. Such radiation is now produced from 200 to ~50 nm by this technique, using metal vapors and the rare gases as nonlinear media. The resulting radiation is coherent, monochromatic, and directional, and thus has all of the characteristics of laser radiation except high intensity. Nevertheless, the intensities achieved to date are sufficient for many applications in absorption and fluorescence spectroscopy.

It is my purpose to review the work of this laboratory in generating tunable VUV (200-100 nm) and extreme ultraviolet or XUV (<100 nm) radiation, and to discuss its application in a variety of spectroscopic problems.

2   Résumé of Theory

Laser-driven VUV sources are based on third harmonic generation (THG) or 4-wave sum mixing (4-WSM) in nonlinear media. These processes are usually described by the induced macroscopic polarization of the medium which, of course, is dependent on the polarizabilities of the atomic or molecular systems when irradiated by intense laser light. It is well known that the polarization of a medium in the presence of a monochromatic field $E(r,t) = \Sigma(\omega_i)$ can be written as

$$\overline{P}(\omega_i) = \chi^{(1)}(\omega_i)\cdot\overline{E}(\omega_i) + \sum_{jk}\chi^{(2)}(\omega_1 = \omega_j + \omega_k)\cdot\overline{E}(\omega_j)\cdot\overline{E}(\omega_k)$$

$$+ \sum_{jkl}\chi^{(3)}(\omega_i = \omega_j + \omega_k + \omega_l)\cdot\overline{E}(\omega_j)\cdot\overline{E}(\omega_k)\cdot\overline{E}(\omega_l) + ... \tag{1}$$

where $\chi^{(n)}$ are the susceptibility tensors of nth order. The lowest order term producing nonlinear effects is $\chi^{(2)}$. However, this tensor has nonzero components only in noncentro-symmetric systems; isotropic media such as cubic crystals, liquids and gases do not exhibit quadratic nonlinearities. For third-order processes such as THG and 4-WSM we need be concerned only with $\chi^{(3)}$, whose principal term may be written (Armstrong et al 1986, Orr and Ward 1971):

$$\chi^{(3)}(\omega_0 = \omega_1 + \omega_2 + \omega_3) = \frac{3e^4}{4\hbar^3} \frac{r_{ga}r_{ab}r_{bc}r_{cg}}{(\Omega_{cg} - \omega_1 - \omega_2 - \omega_3)(\Omega_{bg} - \omega_1 - \omega_2)(\Omega_{ag} - \omega_1)} \tag{2}$$

Here $r_{ga} = <g|r|a>$ is the electric dipole matrix element between the ground state $|g>$ and an excited state $|a>$, having a lifetime Γ_a, and $\Omega_{ag} = \omega_{ag} - i\Gamma_a/2$ is the energy difference (Figure 1) between states $|a>$ and $|g>$, e is the electronic charge and $\hbar = h/2\pi$, with h being Planck's constant.

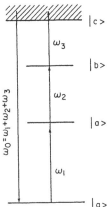

Fig. 1. Diagram of 4-WSM process $\omega_0 = \omega_1 + \omega_2 + \omega_3$ with a 2-photon resonance $\Omega_{bg} = \omega_1 + \omega_2$.

Equation (2) shows that $\chi^{(3)}$ will be resonantly enhanced whenever the applied frequencies, ω_1, ω_2, ω_3, are such that the real part of the resonance denominator vanishes, namely when $(\Omega_{ag} - \omega_1) = 0$, or $(\omega_{bg} - \omega_1 - \omega_2) = 0$, or $(\omega_{cg} - \omega_1 - \omega_2 - \omega_3) = 0$, corresponding to one, two or three photon resonance, respectively. If any of ω_1, ω_2, ω_3 is set equal to a resonance frequency (ω_{ag} etc.), $\chi^{(3)}$ will be enhanced, but the incident radiation will be strongly absorbed. If, however, $\omega_1 + \omega_2$ is equal to a 2-photon resonance (ω_{ag}), the incident radiation at $\omega_1 + \omega_2$ is expected to be only weakly absorbed by the 2-photon transition, while the resonance enhancement of $\chi^{(3)}$ could be just as strong as for the 1-photon resonances.

For third harmonic generation (THG), $\chi^{(3)}$ simplifies to

$$\chi^{(3)}(\omega_0 = 3\omega) = \frac{3e^4}{4\hbar^3} \frac{r_{ga}r_{ab}r_{bc}r_{cg}}{(\Omega_{cg} - 3\omega)(\Omega_{bg} - 2\omega)(\Omega_{ag} - \omega)} \tag{3}$$

When 2ω approaches resonance, $\chi^{(3)}$ undergoes strong ($>10^4$) enhancement. For efficient THG, collinear phase-matching is necessary, that is, the refractive index $n(3\omega) = n(\omega)$ in order to yield a maximum effective interaction length. With focused incident radiation, THG can be observed only in negatively dispersive media. Tunability is achieved by varying the incident frequency ω.

For generating tunable radiation by 4-WSM, the process $\omega_0 = 2\omega_1 + \omega_2$ is of interest. Strong enhancement is again achieved by tuning $2\omega_1$, to a parity-allowed 2-photon resonance ω_{bg}, and $\chi^{(3)}$ becomes

$$\chi^{(3)}\left(\omega_0 = 2\omega_1 + \omega_2\right) = \frac{3e^4}{4\hbar^3} \frac{1}{\Omega_{bg} - 2\omega_1} \sum_{ca} \frac{r_{ga} r_{ab} r_{bc} r_{cg}}{\left(\Omega_{ag} - \omega_1\right)\left(\Omega_{cg} - 2\omega_1 - \omega_2\right)} \tag{4}$$

Tunability and further enhancement is then obtained by selecting ω_2 so that $2\omega_1 + \omega_2$ corresponds to the ionization continuum or to broad autoionizing levels above the ionization limit. In such a situation, $\chi^{(3)}$ may be written as

$$\chi^{(3)}\left(2\omega_1 + \omega_2\right) = \frac{3e^4}{4\hbar^3} \frac{r_{ab} r_{ga}}{\left(\Omega_{bg} - 2\omega_1\right)\left(\Omega_{ag} - \omega_1\right)} C\int \frac{r_{gc} r_{bc} d\omega}{\left(\Omega_{cg} - 2\omega_1 - \omega_2\right)} \tag{5}$$

where now r_{gc} and r_{bc} are the matrix elements of dipole moment between the ground state $|g>$ to the perturbed continuum state $|c'>$, and between the states $|b>$ and $|c'>$. The integration takes into account the contribution of the auto-ionizing states as well as of the ionization continuum, and C is a normalization constant. More detailed treatments of the relevant theory including phase-matching, saturation effects, and conversion efficiencies have been given by Jamroz and Stoicheff (1983) and by Vidal (1984).

3. Sources for Generating Tunable, Coherent Radiation

A general outline of the experimental arrangement, including the primary excitation source, two dye lasers, heat pipes for producing the metal vapor, monochromator and detection systems, is shown in Figure 2.

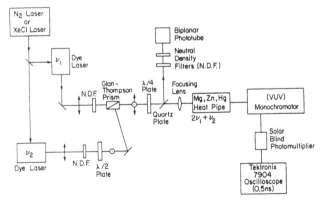

Fig. 2. Schematic diagram of the overall experimental arrangement.

3.1 Laser systems

Initially, the primary excitation source was a N_2 laser (Molectron UV-1000), which was suitable for VUV generation at λ >140 nm. For generating shorter wavelength radiation, an excimer laser with Blumlein circuitry was built (McKee et al 1977), after the design of Burnham and Djeu (1976). This was superseded by the development of a transverse-discharge excimer laser (McKee et al 1979, Andrews et al 1977), which became a commercial product for Lumonics Inc. In this application, for pumping tunable dye lasers, radiation produced by the excimers KrF at 249 nm, XeF at 350 nm and XeCl at 309 nm, was used.

Two tunable dye lasers, using gratings (of 2400 lines/mm) at grazing incidence, were operated at frequencies v_1 and v_2. Both dye lasers were pumped simultaneously, and a variety of available dye solutions provided tunable laser radiation from ~340 to 800 nm. Output powers of up to 10 kW and ~8 ns duration were obtained for the N_2-laser pumping, and up to ~40 kW and ~4 ns for the XeCl-laser pumping. Linewidths were typically <0.3 cm^{-1} in the ultraviolet region, and ~0.1 cm^{-1} for λ >400 nm (although this was reduced to 0.01 cm^{-1}, with a shorter cavity and a grating angle of ~180°, resulting in greatly diminished intensity). Both v_1 and v_2 beams were plane-polarized horizontally. The v_2 beam polarization was rotated by 90° with a half-wave plate (or with two mirrors for λ<400 nm), and then the two beams were spacially overlapped in a Glan-Thompson prism. This was the beam configuration when an S state was used for resonance enhancement. For enhancement using a D state, both beams were circularly polarized by a λ/4 plate, to eliminate THG without inhibiting the 4-WSM process.

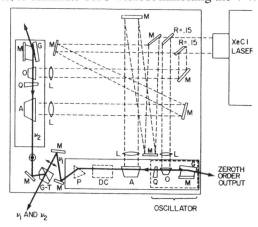

Fig. 3. Diagram of the excimer laser and tunable dye oscillator-amplifier systems used to provide radiation at v_1 and v_2 for frequency mixing in Mg, Zn and Hg vapors. All mirrors labeled M are front-surface, totally reflecting: R's are 15% reflecting mirrors. The oscillators each consist of a grating G, a dye cell O and a quartz wedge Q. The dye-cell amplifiers are labelled A, DC is a doubling crystal, P is a quartz prism to separate the fundamental and harmonic beams G-T is a Glan-Thompson prism, and L designates a cylindrical lens.

For generating VUV radiation with $\lambda < 400$ nm using XeCl pumping, amplifiers were added to each arm of the dye laser systems (Fig.3). Each oscillator-amplifier system received half of the XeCl laser power, with 15% of this allocated to the oscillator and the remainder to the amplifier. To allow for cavity buildup time of the oscillator, the pump beam to the amplifier was delayed (by traversing a longer path) for an interval of 4-7 ns. Both oscillators emitted powers of 1-10 kW, and amplified powers were typically 50-200 kW, depending on the dye efficiencies. In this way, any desired wavelength in the range 660 to 330 nm could be produced in each system with XeCl (308 nm) pumping.

Two important features, one in each dye-laser system, bear further discussion. When a fixed wavelength $\lambda_1 < 330$ nm was required for two-photon resonances, second harmonic radiation was generated in a potassium dihydrogen phosphate (KDP) crystal having a tuning range 320-350 nm, and in a barium borate (BBO) crystal which extended the tuning range to ~200 nm. A quartz prism separated the fundamental and harmonic beams. To achieve a smooth, linear scan for accurate high-resolution spectroscopy, the scanning mechanism of an infrared spectrometer (Perkin-Elmer, Model 099), was incorporated in the ν_2 oscillator to finely rotate the cavity mirror. Modifications permitted a quick change of gears, so that scan rates of 1-30 cm^{-1}/min could be obtained, with linearity accurate to 0.1%.

3.2 Metal vapors (Mg, Zn, Hg) as nonlinear media

Magnesium vapor was the first choice of this laboratory as a nonlinear medium, based on its known VUV absorption spectrum beyond the ionization limit at 162 nm, and because generation could be achieved with radiation from N_2-pumped dye lasers. Several brief reports have described this work. The first, by Wallace and Zdasiuk (1976), described third harmonic generation (THG) and 4-WSM with continuous tunability from 140 to 160 nm. The resulting VUV radiation was emitted in 4 ns pulses of ~10^{11} photons/pulse (corresponding to an efficiency of ~0.2 %) and had a linewidth of ~0.2 cm^{-1}. Tunability was extended to shorter wavelengths, 120 nm, by McKee et al (1978), and to longer wavelengths, as far as 174 nm, by Banic et al (1981). These early experiments demonstrated that 4-WSM in Mg vapor could provide an efficient source of coherent VUV radiation that is monochromatic, directional and tunable over the broad wavelength region 120-174 nm, scanning a range of ~25,000 cm^{-1}.

To generate radiation of shorter wavelengths, the prime candidates as nonlinear media were Zn and Hg, since they have higher ionization limits than Mg(1162 nm) at 132 and 119 nm, respectively, and they also have strong and broad autoionizing resonances. Zinc vapor was next selected for study. A report by Jamroz et al (1982) described the generation of continuously tunable VUV radiation from 140 to 106 nm, a range of ~23,000 cm^{-1}. Mercury vapor was found to be more efficient than Zn vapor in the region below 120 nm (Herman et al 1985) and provided XUV radiation to ~85 nm (Herman and Stoicheff 1985). Thus, Hg vapor became an important source for tunable radiation from 125 to ~85 nm, spanning a range of ~40,000 cm^{-1}.

In Figure 4 is a schematic diagram of energy levels used for the 4-WSM in Mg, Zn and Hg vapors to generate tunable radiation over the wavelength region of ~175 to 85 nm, which is represented in the chart of Figure 5.

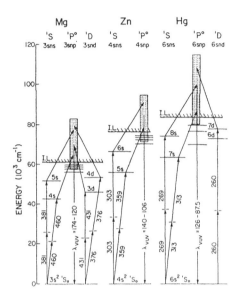

Fig. 4. Partial energy-level diagrams for atomic Mg, Zn and Hg. Levels used for two-photon resonance enhancement of 4WSM are shown (along with corresponding wavelengths in nanometers). The regions of ionization continua and broad autoionizing levels that contributre to the tunability of these sources are indicated by the hatched areas.

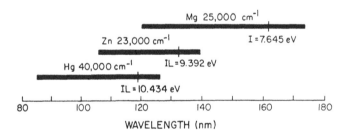

Fig. 5. Wavelength and wavenumber ranges, above and below ionization limits (IL) over which tunable coherent radiation has been obtained using Mg, Zn and Hg vapors.

The metal vapors were prepared and contained in heat pipes of the form shown in Figure 6. The concept of heat pipes for use in spectroscopy has been discussed by Vidal and Cooper (1969). For Mg and Zn vapor, heat pipes of a simple design were used at first, with heating of the central sections being provided by electrical heaters. With Mg, a temperature of ~800° C produced a vapor pressure of ~25 Torr. A helium pressure of ~100 Torr was added to confine the Mg vapor to the central 10 cm and to prevent condensation on the windows. For work with Zn vapor, a pressure of ~130 Torr (at 750° C) was used with ~80 Torr of helium.

Higher stability and more uniform vapor density over prolonged periods, was obtained by replacing the electrical heaters with a vertical heat pipe surrounding the central parts of the Mg and Zn heat pipes, as shown in Figure 6(a). Sodium metal served as the working vapor to heat the Mg and Zn, with Ar gas used to control the temperatures of the inner heat pipes. Under these conditions, stable operation could be maintained for several hours before being limited by growth of crystals near the cooled ends, and typical operating lifetimes were ~1000 h before new wicks were required.

The heat-pipe oven for Hg vapor was a simple cell of Pyrex glass, as shown in Figure 6(b). Hg vapor was confined to the central 3 cm, and used at pressures up to 95 Torr, with an equal pressure of buffer gas (Herman and Stoicheff 1985). For 4-WSM at λ ~120 to 105 nm, LiF windows ~1/2 mm thick were used; for λ <105 nm, a glass capillary array was found to be an efficient XUV window, with 50 % transmission.

The generated VUV and XUV radiation was analyzed with a grating spectrometer (McPherson 225 or 234) and detected by a solar-blind photomultiplier (EMR 510G-08-13), having a LiF window. For absolute-intensity calibration, an ionization chamber was constructed and placed between the heat-pipe oven and monochromator. The chamber was used with xylene or diethyl sulfide vapor, or Xe gas, depending on the wavelength of the radiation being generated.

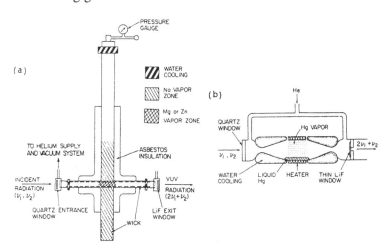

Fig. 6. Heat pipes used to produce stable densities
of (a) Mg or Zn vapors, and (b) Hg vapor.

3.3   *Molecular gases as nonlinear media for 4-WSM*

Along with the use of metal vapors for 4-WSM, preliminary experiments were carried out testing the feasibility of generating coherent radiation by resonance-enhanced 4-WSM in molecular vapors (Innes et al 1976). 4-WSM was observed near 177 nm with bromine, 163 nm with benzene, and over a broad region centered at ~143 nm in nitric oxide (NO).

Nitric oxide offered several advantages since detailed spectroscopic information was available, and a strong 2-photon allowed transition was identified which was suitable for resonance enhancement, namely, the 0-0 band of the $A^2\Sigma^+ - X^2\Pi_{3/2}$ electronic band system. Initially, a gas pressure of ~90 Torr was used, and 4-WSM resulted in emission over a region of rich rotational structure. Significant pressure broadening occurred at pressures of ~10 atm, and the discrete rotational structure was essentially eliminated. Thus, it was possible to produce tunable VUV radiation from 130 to150 nm by simple THG using a single tunable dye laser operating in the region 400 to 500 nm. The efficiency was several orders of magnitude lower than with the atomic vapors, and resulted in a beam of ~10^7 photons/pulse.

4   Spectra of Rare-Gas Excimers

The electronic spectra of the rare-gas dimers have been a subject of interest for many years, mainly because these dimers are model systems for studying van der Waals interactions, and because they are known laser media operating in the VUV. A series of investigations was begun in the early 80's in this laboratory on the spectra of Xe_2, Kr_2, and Ar_2. Two techniques were combined for this work: a pulsed supersonic jet to produce rotationally and vibrationally cold dimers, and 4-WSM to enable fluorescence-excitation spectroscopy (Lipson et al 1984). In this way, it was found possible to resolve rovibronic structures in several band systems of these dimers and their isotopes in the region of 104 to 150 nm. From these spectra, the relevant molecular constants were determined, and the potential energy curves for the ground and three lowest (stable) excited states were derived. A brief review of the main results is given here, along with examples of spectra and references to the literature.

A diagram illustrating a typical experimental arrangement is given in Figure 7, indicating all of the necessary units for 4-WSM, sample preparation, synchronization of jet and laser pulses, detection and recording of spectra, as well as wavelength calibration.

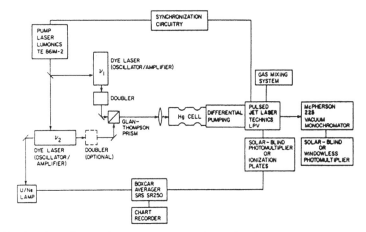

Fig. 7. Experimental arrangement for investigating spectra of Ar_2.

It is instructive to examine spectra of cold as well as warm dimers using the same experimental methods. This was possible with the gas jet by exciting the dimers at different distances from the nozzle, and the results are shown in Figure 8 for comparison. The spectrum of dimers excited as close as possible to the nozzle consists of broad bands similar to spectra obtained in cooled-cell (~150K) experiments. When excited ~15 mm from the nozzle the dimers exhibit a spectrum with considerable structure, corresponding to temperatures of ~ 10K.

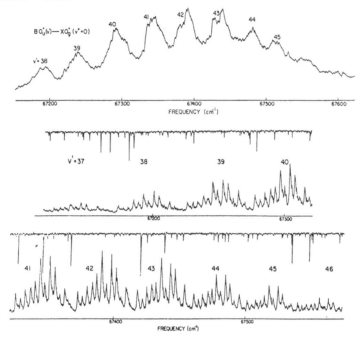

Fig. 8 Fluorescence-excitation spectrum of band system II of Xe_2, at 148.5 nm probed ~2 mm from the nozzle (upper) and ~15 mm from the nozzle (lower).

The spectrum of the cold dimers shows ten vibrational bands, each having the same structure. These resolved features are vibronic bands of isotopes of Xe_2, and their resolution and analysis has led to an unambiguous quantum numbering of the observed bands. This numbering was ,of course, essential for a determination of the molecular constants (Lipson et al 1985). Such corresponding isotopic structure was observed with Kr_2 and Ar_2 and served to identify the vibrational bands. While the available resolution was not sufficient to resolve rotational structure in the Xe_2, it was possible to resolve such structure in spectra of Kr_2 (Figure 9) for the first time and to determine the internuclear separation in the ground and excited states. Also, the very much lighter mass of Ar resulted in clear resolution of rotational structure even at low dispersion (Figure10), and the spectrum shown in Figure 11 led to the identification of the symmetry of the Ar_2 excited states. As shown in Figure 11, three rotational branches are clearly resolved, and interpreted as P- Q- R-branches. From this result it was possible to establish that Hund's case (c) is the dominant coupling for the high vibrational levels of the A state of Ar_2 and thus should be designated $A1_u$ in place of $A^3\Sigma_u^+$ (Herman et al 1987).

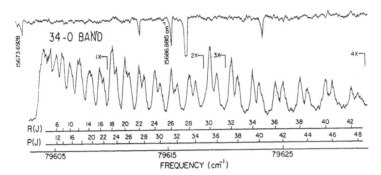

Fig. 9. Rovibronic structure in the 34-0 band of system II of Kr_2 at 125nm.

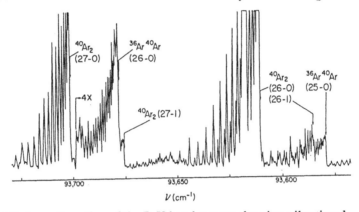

Fig, 10 A small portion of the B-X band system showing vibrational and rotational structures of two isotopic species, $^{40}Ar_2$ and $^{36'40}Ar_2$.

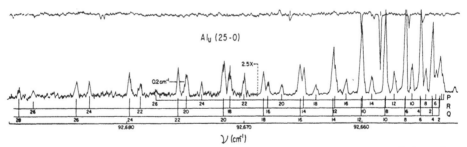

Fig.11. Spectrum of the $A1_u$–$X0_g^+$,25-0 band clearly showing P-, Q-, and R-branches.

Finally, potential energy curves for the ground state and three lowest bound states were calculated for each of the dimers(Lipson et al 1985, LaRocque et al 1986, Herman et al 1988). Those for Ar_2 are shown in Figure 12. Since data for the lowest vibrational levels were available for the ground and C states, these potential curves are deemed to be accurate. However, those for the A and B states are much less reliable since only high vibrational levels (>20) were accesssible in the experiments, and long extrapolations to v=0 were necessary.

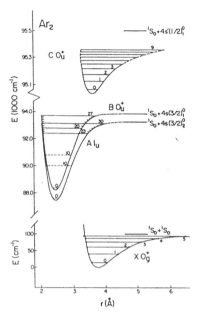

Fig. 12. Calculated potential energy curves for Ar_2 showing the observed vibronic levels and separated-atom states at the dissociation limits

5   Radiative Lifetimes of Selected Rovibronic Levels

As shown in Section 4, the extension of tunable coherent sources to VUV wavelengths has provided the possibility of high-resolution spectroscopy in this region. Moreover, since these sources also generate short pulses, of the order of a few nanoseconds, measurements of lifetimes of specific atomic or molecular levels can be made with high precision. Such measurements of radiative lifetimes of individual rotational levels of CO were demonstrated in an early work of this laboratory (Provorov et al 1977). Further experiments were carried out with CO (Maeda 1984), NO (Banic et al 1984), and the rare gases Ar_2, Kr_2, Xe_2 (Madej et al 1986, Madej and Stoicheff 1988). Some of these results will be reviewed briefly.

5.1   Lifetimes of levels in the $A^1\Pi(v=0,1)$ state of CO

State-selective VUV excitation of rovibronic levels in the $A^1\Pi(v=0,1)$ state of CO was used by Provorov et al (1977) to determine radiative lifetimes of some of the known perturbed singlet levels, and Maeda (1985) reported on more extensive measurements, including the v=1 level. Measurements of fluorescence decay were carried out for v=0 and J'=1 to 29, and for v =1 and J' = 1 to 27, of the $A^1\Pi$ state. For the v=0 levels the perturbations are caused by the $e^3\Sigma^-(v=1)$ and $d^3\Delta_i$ (v=4) states, while for the v=1 levels only the $d^3\Delta_i$ (v=5) state is involved in perturbing the lowest rovibronic levels. Thus we need only consider the theory for the interaction of a singlet state S_0 ($A^1\Pi$) with a triplet state $T_0(e^3\Sigma^-$ or $d^3\Delta_i)$. Near a perturbation there is a mixing of states S_0 and T_0, resulting in the linear combination states

$$|S> = \alpha|S_0> + \beta|T_0>; \quad |T> = -\beta|S_0> + \alpha|T_0>$$

Here, $|S>$ and $|T>$ refer to perturbed singlet and triplet states, and $|S_0>$ and $|T_0>$ to the unperturbed states. The respective decay rates Γ_S and Γ_T may be written as

$$\Gamma_S = \alpha^2\Gamma_{S_0} + \beta^2\Gamma_{T_0}; \quad \text{and} \quad \Gamma_T = \alpha^2\Gamma_{T_0} + \beta^2\Gamma_{S_0}$$

or, alternatively, in terms of the unperturbed radiative lifetimes τ_S and τ_T

$$\Gamma_S = \frac{1}{\tau_S} = \frac{\alpha^2}{\tau_{S_0}} + \frac{\beta^2}{\tau_{T_0}}, \qquad \Gamma_T = \frac{1}{\tau_T} = \frac{\alpha^2}{\tau_{T_0}} + \frac{\beta^2}{\tau_{S_0}}$$

The mixing coefficients α and β are related by $\alpha^2 + \beta^2 = 1$, and in the limit of maximum mixing $\alpha^2 \sim \beta^2 \sim 0.5$. It is known that $\tau_{S_0} \sim 10$ ns, and $\tau_{T_0} \sim 3$ μs, thus $\tau_S^{-1} \sim \alpha^2/\tau_{S_0}$ and $\tau_T^{-1} \sim \beta^2/\tau_{S_0}$, leading to maximum values of $\tau_S \sim 2\tau_{S_0}$, while $\tau_T \sim 150\tau_{T_0}$. Field (1972) and more recently, LeFloch et al (1987), have calculated values of α^2 and β^2 from analyses of observed perturbations in the high-resolution spectra of CO. With these values of α^2 and β^2, theoretical radiative lifetimes are readily obtained using the relations $\tau_S = \tau_S / \alpha^2$, and $\tau = \tau_S / \beta^2$, for the singlet and triplet levels, respectively.

For this study, VUV radiation generated in Mg vapor was focused into a stainless steel cell to excite fluorescence in flowing CO at a pressure of 50 mTorr. Radiation emitted at right angles to the incident VUV beam was detected directly with a solar-blind photomultiplier fitted with a LiF window. For each measurement, the signal was transmitted to a boxcar-averager and gated-integrator system, used in the scanning mode with a window-time of 2 ns. Initially, the fluorescence-excitation spectra of the 0-0 and 1-0 bands, $A^1\Pi(v=0,1) \rightarrow X^1\Sigma(v=0)$ were recorded by tuning the VUV radiation over the wavelength ranges 154.4 to 155.5 nm and 150.9 to 151.8 nm, respectively. The exciting radiation was then tuned to each of the unblended rotational lines and the decay of fluorescence intensity with time was measured. For each decay curve, ~2000 pulses at 8 Hz were integrated with the boxcar-integrator system. It became evident that the total instrumental response time significantly broadened the VUV excitation pulse beyond the 2 ns duration of the dye-laser pulses. Under these circumstances, it was necessary to correct for the instrumental profile, in order to determine the fluorescence decay times τ.

For the v=0 level, measurements of fluorescence decay rates were made for the unblended lines of the P-, Q- and R-branches, and radiative lifetimes for levels up to $J'=29$ were determined. Values in the range 9.8 to 10.8 ns were considered to designate lifetimes of unperturbed levels, and the average of 26 such values gave an unperturbed lifetime, $\tau_u = 10.5 \pm 0.5$ ns at a CO pressure of 50 mTorr. Values of lifetimes normalized by the unperturbed lifetime, that is τ/τ_u, are plotted as a function of rotational quantum number J' in Figure 12. Theoretical values (τ/τ_0) are also shown for comparison (Field 1972, and LeFloch et al 1987).

From Figure 13, it is clear that large perturbations occur for levels $J' = 12$ and 27 in Q-branch transitions, and for levels $J' = 9$, 16 and 27 in P- and R-branch transitions. The observed lifetimes are almost double the unperturbed lifetime of 10.3 ns. The overall good agreement with theoretical lifetimes for perturbed and unperturbed levels over the range of the $J' = 1$ to 29 is also evident. The perturbation at $J' = 27$ has been attributed to interaction of the $d^3\Delta_i$ (v =4) state with all three triplet components of the $A^1\Pi$ state. At the lower J´-values, the observed perturbations for Q-branch transitions differ from those for the P- and R-branch transitions. These variations arise from the differing selection rules for interaction with the $e^3\Sigma^-$(v =1) state: the F_2(J=N) component of the latter perturbs the Π^- component of the A state, giving rise to a perturbed Q-branch, while the F_1 and F_2 (J=N±1) components each affect the Π^+ components, resulting in perturbations of the P- and R-branches.

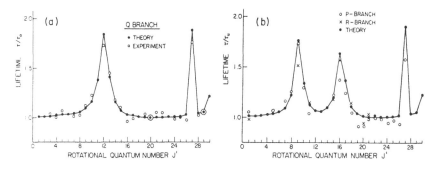

Fig. 13 Graphs of lifetimes normalized by the unperturbed lifetime τ_u plotted as a function of rotational quantum number J for the v=0 level (a) obtained from Q-branch lines, and (b) from P- and R-branch lines

The overall intensity of the 1-0 band was lower than that of the 0-0 band, and most of the R-branch lines were blended with P- or Q-branch lines. Measurements of fluorescence decay rates were made for all possible transitions from levels up to J´= 27. Those values below 12.0 ns were considered to be unperturbed, giving an average $\tau_u =$ 10.7±0.7 ns (for a CO pressure of 50 mTorr). Values of τ/τ_u are plotted versus J´ in Figure 14, along with theoretical values (τ/τ_u) provided by Field. While these

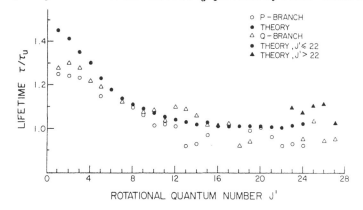

Fig. 14. Graphs of lifetimes normalized by the unperturbed lifetime τ_u plotted as a function of rotational quantum number J for the v=1 level obtained from P- and Q-branch lines.

values exhibit more uncertainty than obtained for the 0-0 band measurements, but are in reasonable agreement with the theoretical values. It is evident that perturbations occur only for the low values of J´ in the v =1 level, with τ/τ_u decreasing from ~1.3 to 1.0 in the range of J´ = 0 to 8. The interaction is caused by the v´= 5 level of the $d^3\Delta_i$ state. The selective excitation of individual rovibronic levels for v = 0 and 1 of the $A^1\Pi$ state of CO carried out with a tunable and pulsed VUV source has yielded radiative lifetimes which are in excellent agreement with theoretical values. Such good agreement attests to the validity of the theory for strongly and weakly perturbed levels of CO, as well as to the potential importance of the high-resolution state-selective excitation source used in this research.

5.2   *Dependence of Ar_2 radiative lifetimes on internuclear distance*

As a continuation of the study of the rare-gas excimers, described in Section **4**, the radiative lifetimes of vibronic levels of the $A1_u$ states of Ar_2, Kr_2 and Xe_2 were measured (Madej and Stoicheff 1988). Because of the relative positions of the potential energy curves for the strongly bound excimer states and the shallow ground states, only high vibronic levels of the $A1_u$ states are accesible for fluorescence excitation. Thus only the levels v´ =24 to 30 for Ar $_2$, v´ = 32 to 38 for Kr_2 , and v´ =36 to 43 for Xe_2 could be investigated. For all of these levels, the fluorescence decay curves exhibited single-exponential decays with time. The measured lifetimes were found to be essentially constant for these high vibronic levels, with average lifetimes of τ=160±10 ns for Ar_2, 55±5 ns for Kr_2, and 47±5 ns for Xe_2. These results differ significantly from the values of 2.9µs, 264 ns, and 99 ns, respectively, quoted by earlier investigators who used charged particles and synchrotron radiation for fluorescence excitation of these excimers formed at relatively high pressures in cell experiments. At these pressures, rapid vibrational relaxation occurs, resulting in fluorescence emission from low vibrational levels of the excited states. Thus the differences in lifetimes imply reductions by factors of 20 for Ar_2, 5 for Kr_2, and 2 for Xe_2, in going from v'=0 to v'~20-40 in the $A1_u$ states.

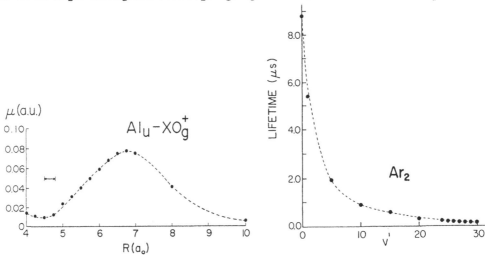

Fig. 15. Calculated transition moment µ(R) for the A-X transition in Ar_2.

Fig 16. Calculated radiative lifetimes of vibrational levels in the AI_u state of Ar_2.

While differences of a factor of two in radiative lifetimes for vibrational levels of the same electronic state are not uncommon in molecular spectroscopy, a factor of 20 is unique. Madej et al (1986) explained this large difference for Ar_2 by calculating the dependence of the electronic transition moment $\mu(R)$ on internuclear distance, given in Figure 15. In the united- and separated- atom limits, transitions are forbidden; but at intermediate distances, spin-orbit coupling causes transitions to be weakly allowed. It was found that the derived values of $\mu(R)$ increase rapidly, from $\sim 1 \times 10^{-2}$ a.u. at 4.5 a_0 (corresponding to $v' \sim 0$) to $\sim 8 \times 10^{-2}$ a.u. at 6.8 a_0 (for $v' \sim 30$). These values were used to calculate spontaneous emission probabilities, which give the reciprocals of the radiative lifetimes. Figure 16 shows a comparison of the calculated and measured lifetimes. There is excellent agreement for high vibronic levels, and a discrepancy of a factor of three for $v'=0$, which can be corrected by an increase of $\sim 5\%$ in the equilibrium internuclear distance in the $A1_u$ state (presently only known to an accuracy of $\sim 10\%$). Similar calculations for Kr_2 and Xe_2 yielded good agreement with measured values at high vibronic quantum numbers.

6   Second-Harmonic Generation (SHG) in Atoms

It is well known that the second-order susceptibility, $\chi^{(2)}$, vanishes in the dipole approximation for isotropic media. Nevertheless, the experimental generation of second-harmonic radiation in the absence of external fields has been reported for a number of atomic gases, namely; barium, lithium, mercury, potassium, sodium, zinc, and most recently, hydrogen. These observations have been explained by the contributions of quadrupole or magnetic dipole moments, by intensity gradients of the incident laser beam, by electric-field-induced harmonic generation (with the field caused by ions in 3-photon ionization), and by collision effects. For some atoms there is good agreement between theoretical calculations and experimental observations; for others, the observed results remain unexplained or even appear to contradict the proposed models. The generation of SH radiation with the application of electric or magnetic fields to remove the symmetry of free atoms or nonpolar molecules has also been observed, and appears to be well understood. Two series of experiments have been carried out in our laboratory and these have been explained by the creation a charge-separation field arising from 3-photon ionization.

6.1   *Experiments with zinc vapor*

While investigating 4-WSM with zinc vapor in our laboratory, and using the 4^1D_2 and 5^1S_0 states for 2-photon resonance, intense second-harmonic (SH) radiation was observed at 160.1 and 179.25 nm, respectively (Jamroz et al 1982). The experimental arrangement and heat pipe were the same as described in Section *3.2*. Tunable radiation (of ~ 0.12 mJ, 0.4 cm^{-1} linewidth, and 12 ns duration) was obtained from a dye laser pumped by a KrF excimer laser. This radiation was focused to a spot size of $\sim 5 \times 10^{-3}$ cm^2 in the center of the Zn vapor region of the heat pipe, which was operated at ~ 30 Torr of Zn vapor with He buffer gas at ~ 30 Torr. When the dye laser was tuned in the region of 320.2 nm, corresponding to the 2-photon transition $4d^1D_2 \leftarrow 4s^1S_0$, SHG at 160.1 nm was observed with a conversion efficiency of $\sim 10^{-5}$ to 10^{-6}. At the same time, THG was monitored at 106.7 nm, and the ratio of SHG to THG was found to be 2×10^3. Similarly, SHG at 179.25 nm (with efficiency $\sim 10^{-6}$ to 10^{-7}) was observed when radiation at 358.5 nm was incident on the Zn vapor (corresponding to the 2-photon transition $5s^1S_0 \leftarrow 4s^1S_0$). THG at 119.5 nm was also observed, with the ratio SHG /THG being ~ 2.

With the ionization limit of Zn corresponding to 132 nm, it was clear that the third-harmonic radiation at 106.7 and 119.5 nm would result in ionization and creation of a charge-separation field. Thus the above observations with Zn vapor were explained in terms of the creation of static electric fields by free electrons and ions produced by multiphoton ionization, which induce SHG through the third-order susceptibility $\chi^{(3)}$, as discussed by Bethune (1981), and by Okada et al (1981). The third-order susceptibility for resonant SHG arising from a dc electric field ($\omega=0$), and that corresponding to resonant THG, can be written as

$$\chi^{(3)}(2\omega) \sim \frac{3}{2\Gamma_b} \sum_{ac} \frac{r_{ga}r_{ab}r_{bc}r_{cg}}{(\omega_{cg}-2\omega)(\omega_{ag}-\omega)}$$

$$\chi^{(3)}(3\omega) \sim \frac{1}{4\Gamma_b} \sum_{ac} \frac{r_{ga}r_{ab}r_{bc}r_{cg}}{(\omega_{cg}-3\omega)(\omega_{ag}-\omega)}$$

Here the symbols have the meanings given in Section **2**. The ratio of second-harmonic photon number $n(2\omega)$ to third-harmonic number $n(3\omega)$ is given by

$$\frac{n(2\omega)}{n(2\omega)} \sim \frac{2}{3} \frac{\left|\chi^{(3)}(2\omega)\right|^2}{\left|\chi^{(3)}(3\omega)\right|^2} \frac{E^2(0)}{E^2(\omega)}$$

where $E(0)$ is the dc electric field, and $E(\omega)$ is the incident laser field. With only the 5^1P level (at 62 910 cm^{-1}) used in the sum over the final states, since it is closest to the 2-photon resonant levels 5^1S (at 55 789 cm^{-1}) and 4^1D (at 62 458 cm^{-1}) and therefore makes the dominant contribution, the ratio simplifies to:

$$\frac{n(2\omega)}{n(3\omega)} \sim \frac{2}{3} \left|6\frac{\omega(5'P)-3\omega}{\omega(5'P)-2\omega}\right|^2 \frac{E^2(0)}{E^2(\omega)}$$

For doubling from the ^1D state, our results of $n(2\omega)/n(3\omega) = 2\text{x}10^3$ for an incident power of 10 kW, or $E(\omega) = 39$ kV/cm, lead to a dc field of $E(0) = 5.2$ kV/cm. For doubling from the 5^1S state, the observed $n(2\omega)/n(3\omega) = 2$ gives a dc field of 4.0 kV/cm, in agreement with that obtained for doubling from the 4^1D state. Such dc fields can be shown to arise from plasma densities of $\sim 5\text{x}10^{15}$ ions/cm^3, corresponding to ~ 1-2% of the atoms in the focal volume being ionized. This small percentage of ionization is plausible to achieve by 3-photon ionization of Zn with the use of a 10 kW laser beam of ~ 12 ns duration.

6.2 *Experiments with hydrogen*

Almost a decade later, we returned to the problem of SHG in atoms, with the simplest system, atomic hydrogen (Marmet et al 1991). With incident radiation at 243 nm, corresponding to the 2-photon transition $2^1S \leftarrow 1^1S$, Lyman-α radiation at 121.6 nm was generated. The choice of atomic hydrogen was motivated by the simplicity of the theory and calculations, by the forbidden quadrupole and magnetic dipole transitions, and by the availability of an efficient nonlinear crystal, β-barium borate (having transmission just below 200 nm) which was used to generate the incident radiation at 243 nm. The transition matrix elements for the n = 2--1 transition of H are large, thus permitting the use of low gas density with the reduction or elimination of pressure effects. In this way, it was possible to demonstrate that the dominant process for inducing SH radiation in H is a charge-separation field, arising from 3-photon ionization.

Moreover, it was possible to measure this field, and to compare its value with that calculated from theory.

The experimental arrangement is shown in Figure 17. A dye laser, pumped by a XeCl excimer laser emitted tunable radiation near 486 nm, which was doubled to 243 nm. Hydrogen atoms were generated in a dc glow discharge of H_2 gas, and directed into the experimental chamber as a beam, through a 0.9 mm nozzle. The gas density at the nozzle was $\sim 10^{14}$ cm^{-3}, with the background pressure of the chamber maintained at 2×10^{-6} Torr. The tunable UV beam was focused midway between the (nozzle) electrode and a fine grid, spaced 1.3 mm apart, and used to apply a variable dc electric field.

Measurements of SH radiation intensity were made with applied electric fields, E_a, that ranged from 0 to $\pm$ 14 kV/cm. The results are shown in Figure 18, for incident laser power of 43 kW. The SH intensity increased quadratically with E_a up to ~ 5 kV/cm, approached saturation at ~ 7 kV/cm (with output of ~ 40 mW), and then decreased at higher fields. It is clearly shown that the SH intensity is not zero when $E_a = 0$, but ~ 850 μW. As expected for SHG, the intensity with as well as without an applied electric field, increased quadratically with increasing incident laser power. All measurements were carried out with linearly polarized laser radiation; the SH intensity dropped by a factor >20 when circularly polarized radiation was incident on the H beam.

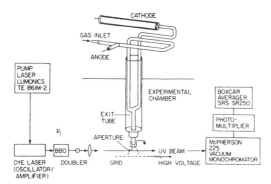

Fig. 17. Experimental arrangement for generating SH radiation at 121.6 nm in a hydrogen atom beam with and without an applied electric field.

An important clue to the mechanism leading to SHG was the observation of the SH intensity profile at $E_a = 0$. The profile was a doubly peaked structure with a dip in the center. Such nonuniformity is a sign of a radial electric field produced by the laser radiation, which in turn points to the creation of a charge-separation field (CSF). This CSF was attributed to the motion of free electrons and protons that result from 3-photon ionization of the hydrogen atoms. Indeed the presence of ions was confirmed experimentally by the detection of an ion current only when 243 nm radiation was incident on the H beam. Measurement of ion current exhibited a cubic dependence on laser power, and provided an estimate of the instantaneous ion density ($\sim 4.5 \times 10^{13}$ cm^{-3}), for use in calculating the CSF under the experimental conditions.

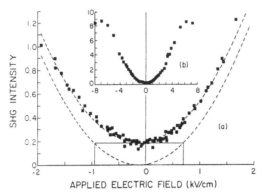

Fig. 18. SHG at 121.6 nm plotted as a function of applied electric field
(a) for fields between -2 and +2 kV/cm and (b) for fields up to ±8 kV/cm.

An estimate of the CSF was made from the experimental data of Figure 18. While the observed low-field data were fitted to a parabola, as shown in Figure 18(a), it was also possible to fit the high-field data to a parabola with apex at zero SH intensity for $E_a = 0$. This could be thought of as a representation of SH intensity dependence on the applied dc field, assuming that no other fields were present. A line drawn through the experimental measurements near $E_a = 0$ and parallel to the field axis, meets the latter parabola at 820±70 V/cm. This value is the average CSF in the interaction region, created by the laser beam through 3-photon ionization. A value for this CSF was calculated based on the mechanism of charge separation arising from the initial kinetic energy of ejected photoelectrons, which carries them away from the laser beam axis (Bethune 1981). In this calculation it was also necessary to include the motion of the protons. A calculated CSF of ~1 kV/cm was found, in good agreement with the value derived experimentally.

This experiment with atomic hydrogen has turned out to be of some consequence on two accounts. Firstly, it is the only known example of SHG for which the CSF has been measured experimentally, as well as confirmed by calculation. Secondly, as discussed in Section 6.3, it led to a direct experiment on electric-field-induced SHG of Lyman-α radiation with reduced absorption, which is relevant to a topic of current interest in quantum optics, namely, gain and lasing without population inversion.

6.3   SHG of Lyman-α radiation with reduced absorption
Recently there has been considerable discussion concerning the possibility of obtaining stimulated emission without population inversion. In 1989 Harris developed a theory to show that when two upper levels of a four-level system are purely lifetime broadened and decay to the same continuum, there will be a destructive interference in the absorption profile of lower-level atoms which is not present in the stimulated emission profile of upper-level atoms, thus resulting in laser gain without inversion. The upper levels are coupled by spontaneous decay to the same final continuum or discrete level, and the broadening can be caused by autoionization, tunneling, or radiative decay. It was also predicted (Imamoglu and Harris 1989) that such transparency could be induced by active coupling of the two upper levels with a strong electromagnetic field, and that in nonlinear media, this process could resonantly enhance the nonlinear susceptibility

and at the same time induce transparency and a zero in the contribution of the resonance transition to the refractive index (Harris et al 1990). Work in this laboratory demonstrated that coupling of metastable and upper levels by an applied dc electric field yields the same characteristics. In addition, experimental evidence was found using the behaviour of SHG, that coupling of the 2s and 2p levels in atomic hydrogen leads to reduced absorption at the center of the Stark-split components (Hakuta et al 1991).

In an external field, SHG in atomic systems can be treated by the third-order nonlinear susceptibility, or as a second-order process where the electric field is explicitly included in the expression for $\chi^{(2)}$. The latter was selected, and calculations of the linear and nonlinear susceptibilities describing single-photon absorption and SHG were carried out using bare 1s, 2s and 2p states as a basis set. These are shown in the energy-level diagram of Figure 19. A dc electric field was applied to couple the metastable 2s with the 2p state which radiatively decays to the 1s ground state. A laser field of frequency ω_a, incident on the system coherently drives the dipole at $\omega_b = 2\omega_a$ and generates SH radiation. The susceptibilities are given by

$$\chi^{(1)}(-\omega_b;\omega_b) = \frac{N}{2\varepsilon_0\hbar} \frac{\Delta\overline{\omega}_{21}}{\Delta\overline{\omega}_{21}\Delta\overline{\omega}_{31} - |\Omega_{32}|^2} \mu_{13}\mu_{31}$$

$$\chi^{(2)}(-\omega_b;\omega_a,\omega_a) = \frac{N}{\varepsilon_0\hbar^2} \frac{\Omega_{32}}{\Delta\overline{\omega}_{21}\Delta\overline{\omega}_{31} - |\Omega_{32}|^2} \mu_{13} \sum_j \frac{\mu_{2j}\mu_{jl}}{\omega_j - \omega_a}$$

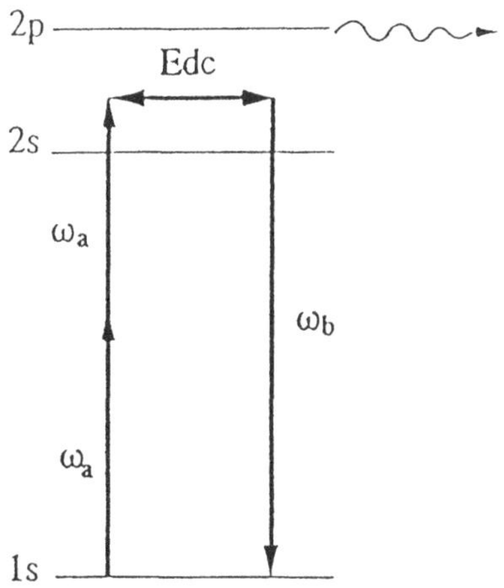

Fig. 19. Energy levels of atomic hydrogen involved in the SHG process.

Here, $\Delta\overline{\omega}_{21} = \omega_2 - \omega_1 - 2\omega_a$, and $\Delta\overline{\omega}_{31} = \omega_3 - \omega_1 - \omega_b - i\Gamma/2$, and Ω_{32} denotes the Stark shift. From these formulae it can be shown that if the decay rate of the 2s state is zero, then $\chi^{(1)}$ becomes zero at the bare 2s position ($\Delta\overline{\omega}_{21}=0$), which is due to the destructive interference between two Stark-mixed states via spontaneous emission processes. On the other hand, $\chi^{(2)}$ shows a completely different behaviour due to the constructive interference in the SHG process. This situation is illustrated in Figure 20 where the behaviour of $|\chi^{(2)}|^2$ and Im $\chi^{(1)}$ are given for two values of the dc field. At the high field, Im $\chi^{(1)}$ or the absorption of ω_b becomes nearly zero at the center of two lines while $|\chi^{(2)}|^2$ or second-harmonic radiation maintains a reasonably large value.

The experimental arrangement was the same as that used for SHG in hydrogen (Figure 17), but with applied dc electric fields in the range of 0-25 kV/cm. Second-harmonic radiation at 121.6 nm was detected by a solar-blind photomultiplier after being dispersed by a monochromator, and photoion current was observed simultaneously, as a monitor of the photon absorption. The measurements confirmed the basic characteristics predicted theoretically, as shown in Figure 21. In conclusion, this investigation has provided evidence for enhanced SHG with reduced absorption at Lyman-α. A conversion efficiency of $\sim 10^{-6}$ was obtained in a 2 mm interaction length, at a hydrogen density of 10^{14} atom / cm^3. It should be possible to improve the efficiency by several orders of magnitude by increasing the interaction length and the atomic density, and so produce a powerful source of coherent Lyman-α radiation.

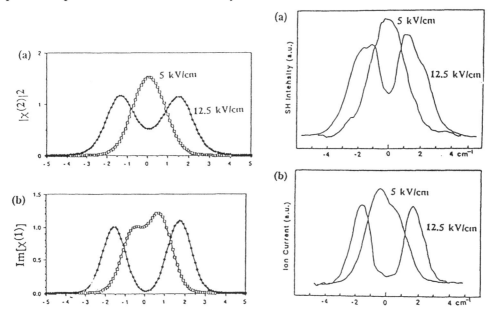

Fig. 20. Calculated characteristics of $|\chi^{(2)}|^2$ and $\chi^{(1)}$.

Fig. 21. Observed characteristics of SH intensitiy and ion current.

7   Dissociation Energy of Molecular Hydrogen

The experimental determination of a precise value for the dissociation energy of molecular hydrogen in its ground electronic state has long remained a challenge for spectroscopy. This parameter is important as a test of basic molecular theory, which today includes calculations of relativistic, radiative, and nonadiabatic corrections. While experimental values were available as early as 1926, all attempts to observe the onset of the dissociation continuum, whether in emission or absorption, were plagued by overlapping molecular bands in this region. Prior to the application of laser techniques to this problem, the best values were those reported by Herzberg (1970), $D_0(H_2) = 36$ 118.6 cm^{-1}, and by Stwalley (1970), $D_0(H_2) = 36$ 118.3 $\pm$ 0.5 cm^{-1}. More recently, McCormack and Eyler (1991) have used high-resolution, double resonance, laser spectroscopy and reported a value of 36 118.26 $\pm$ 0.20 cm^{-1}.

Experiments on this problem have recently been carried out in this laboratory (Balakrishnan et al 1992) based on fluorescence-excitation spectroscopy near 84.5 nm, with detection of Lyman-α radiation at 121.6 nm (Figure 22). Spectra were obtained as the XUV radiation was tuned through the higher levels of the B and B' states and into the so-called second dissociation continuum to yield: $H(B'^1\Sigma_u) = H(1s) + H(2s)$.

The experimental arrangement for generating coherent, tunable radiation in the region of 84.5 nm, was essentially the same as that shown in Figure 2, except that here the nonlinear medium was a pulsed jet of Kr gas. Radiation at ~424 nm with a bandwidth of 0.2 cm^{-1} was amplified and frequency-doubled in a β-barium borate crystal to provide radiation at $2\nu_1$ which is resonant with the level $4p^55p[0,1/2]_0$ of Kr. For ν_2, a single-longitudinal-mode dye laser was used to generate radiation with a bandwidth of 0.02 cm^{-1}, and tunable in the region of 414 nm. The generated radiation was focused with a lens of 25 cm focal length just below the nozzle of a pulsed jet of Kr gas, to generate tunable XUV radiation at $2\nu_1 + \nu_2$, in pulses of ~4 ns duration. This radiation was incident (at right angles) on a nearby vertical beam of H_2 gas, in the region of supersonic expansion, ~50 mm below the nozzle of a second jet. A solar-blind photomultiplier was positioned perpendicular to the XUV and H_2 beams to detect the 121.6 nm fluorescence radiation.

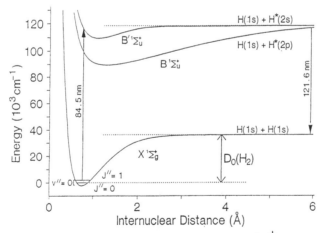

Fig. 22. Energy level diagram showing the ground $X(^1\Sigma_g^+)$ and excited states $B(^1\Sigma_u^+)$ and $B'(^1\Sigma_u^+)$ involved in the photodissociation of H_2.

In preliminary experiments, the spectra revealed intense molecular lines which completely obscured the threshold of the continuum, as shown in Figure 23(a). In order to obtain an unobstructed view of the threshold region, the fluorescence-excitation spectrum was investigated with delayed detection of fluorescence radiation. For this purpose, the metastable H(2s) atoms formed in the photodissociation process were quenched with an electric field which coupled the 2s and 2p levels, resulting in the emission of Lyman-α radiation. The field was switched on ~200 ns after the excitation pulse, a delay sufficient to allow molecules in the B and B' states to decay by radiation, before the long-lived 2s atoms were quenched. In this way, it was possible to observe the threshold of the second dissociation limit clearly, as illustrated in Figure 23(b).

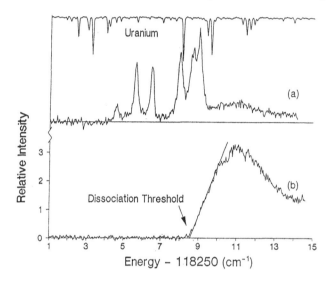

Fig. 23. Fluorescence-excitation spectra in the vicinity of the dissociation limit of ortho-H_2: in (a) is shown the fluorescence spectrum obtained immediately after laser excitation, and in (b) is the spectrum of delayed fluorescence. At the top is the opto-galvanic uranium spectrum used for calibration of v_2.

A value of $118\ 258.57 \pm 0.04$ cm^{-1} was obtained for the dissociation threshold measured from J"=1, v"=0 of the ground state, by drawing a straight line through the linear portion of the onset using least squares, as shown in Figure 23(b). To this value was added the rotational spacing of 118.49 cm^{-1} between J"=0 and J"=1, to give the second dissociation limit of $118\ 377.06$ cm^{-1}. On subtracting the 2s-1s separation of $82\ 258.95$ cm^{-1}, we obtain for the dissociation energy of H_2 in its ground electronic state, $X(^1\Sigma_g)$, the value $D_0(H_2) = 36118.11 \pm 0.08$ cm^{-1}. This experimental value is in agreement with the latest theoretical calculation of $36\ 118.088 \pm 0.10$ cm^{-1}, which includes nonadiabatic, relativistic, and radiative corrections (Kolos et al 1986). When the present value of $D_0(H_2)$ is combined with experimental values for the ionization potentials of H ($109\ 678.764 \pm 0.01$ cm^{-1}) and of H_2 ($124\ 417.512 \pm 0.016$ cm^{-1}), a value of $D_0(H_2) = 21\ 379.36 \pm 0.08$ cm^{-1} is obtained for the dissociation energy of the ion H_2. Again, this is in good agreement with the theoretical value of $21\ 379.348$ cm^{-1} (Wolniewicz and Orlikowski 1991).

8 Conclusions

The summary of results presented here is given as an indication that a little progress has been made in the past two decades in opening up the VUV and XUV regions for laser spectroscopy. These examples have depended on the tunability and monochromaticity of the radiation, and in some cases, on the short duration of the pulses. It would have been difficult to obtain some of these spectra and this information by other means at this time. Other groups and laboratories have developed complementary techniques using the rare-gases as nonlinear media, including pulsed supersonic jets instead of gas cells. It is clear that any progress has come about because of the ubiquitous $\chi^{(3)}$ to which our friend and colleague Sergei Akhmanov contributed so much. Higher-order nonlinear susceptibilities have also been used for generating XUV radiation and have helped the push towards 50 nm. Obviously, this remains a difficult region of the spectrum for research. and still, much of a challenge.

9 Acknowledgements

The research reported here was carried out in the period 1974 to 1992, by my many colleagues who join me in lauding the scholarly activities and contributions of Sergei Akhmanov: A. Balakrishnan, J. R. Banic, P. Dubé, T. Efthimiopoulos, K. Hakuta, P. R. Herman, K. K. Innes, W. Jamroz, A. Jares, W. J. Jones, M. J. Kiik, P. E. LaRocque, R. H. Lipson, A. A. Madej, M. Maeda, C. G. Mahajan, L. Marmet, T. J. McKee, A. C. Provorov, V. Smith, R. I. Thompson, S. C. Wallace, and G. A. Zdasiuk. This research was supported by the Natural Sciences and Engineering Research Council of Canada, the Ontario Premier's Council Technology Fund, and the University of Toronto.

10   References

Andrews A J, Kearsley and Webb C E 1977 Opt. Commun. **20** 265
Armstrong J A, Bloembergen N, Ducuing J and Pershan P S 1962 Phys. Rev. **127** 1918
Balakrishnan A, Smith V and Stoicheff B P 1992 Phys. Rev. Lett. **68** (in press)
Banic J, Lipson R H, Efthimiopoulos E and Stoicheff B P 1981 Opt. Lett. **6** 461
Banic J R, Lipson R H and Stoicheff B P 1984 Can. J. Phys. **62** 1629
Bethune D S 1981 Phys. Rev. **A23** 3139
Burnham R and Djeu N 1976 Appl. Phys. Lett. **29** 707
Field R W 1972 PhD Thesis Harvard University
Franken P A, Hill A E, Peters C W and Weinreich G 1961 Phys. Rev. Lett. **7** 118
Hakuta K, Marmet L and Stoicheff B P 1991 Phys. Rev. Lett. **66** 596
Harris S E 1989 Phys. Rev. Lett. **62** 1033
Harris S E, Field J E and Imamoglu A 1990 Phys. Rev. Lett. **64** 1107
Harris S E and Miles R B 1971 Appl. Phys. Lett. **19** 385
Harris S E and Miles R B 1973 IEEE J. Quantum Electron. **QE-9** 470
Herman P R, LaRocque P E, Lipson R H, Jamroz W and Stoicheff B P 1985 Can. J. Phys
  63 1581
Herman P R, LaRocque P E and Stoicheff B P 1988 J Chem. Phys. **89** 4535
Herman P R, Madej A A and Stoicheff B P 1987 Chem Phys. Lett. **134** 209
Herman P E and Stoicheff 1985 Opt. Lett. **10** 502
Herzberg G 1970 J. Mol. Spectry. **33** 147
Hodgson R T, Sorokin P P and Wynne J J 1974 Phys. Rev. Lett. **32** 343
Imamoglu A and Harris S E 1989 Opt. Lett. **14** 1344
Innes K K, Stoicheff B P and Wallace S C 1976 Appl. Phys. Lett. **29** 715
Jamroz W, LaRocque P E and Stoicheff B P 1982a Opt. Lett. **7** 148
Jamroz W, LaRocque P E and Stoicheff B P 1982b Opt. Lett. **7** 617
Jamroz W and Stoicheff B P 1983 *Progress in Physics XX* ed E Wolf (Amsterdam:
  North-Holland) pp 325-380
LaRocque P E, Lipson R H, Herman P R and Stoicheff B P 1986 J. Chem. Phys. **84** 6627
LeFloch A C, Launay F, Rostas J, Field R W, Brown C M and Yoshino K 1987 J. Mol.
  Spectry. **121** 337
Lipson R H, LeRocque P E and Stoicheff B P 1985a Opt. Lett. **9** 402
Lipson R H, LaRocque P E and Stoicheff B P 1985 Jb. Chem. Phys. **82** 4470
Madej A A, Herman P R and Stoicheff B P 1986 Phys. Rev. Lett. **57** 1574
Madej A A and Stoicheff B P 1988 Phys. Rev. A38 3456
Maeda M 1985 Jpn. J. Appl. Phys. **24** 717
Maker P D, Terhune R W and Savage C M 1964 *Quantum Electronics III* eds
  P Grivet and N Bloembergen pp 1559
Marmet L, Hakuta K and Stoicheff B P 1991 Opt. Lett. **16** 261
McKee T J, Stoicheff B P and Wallace S C 1978 Opt. Lett. **3** 207
McKee T J, Banic J, Jares A and Stoicheff B P 1979 IEEE J. Quantum Electron. **QE-15**
  332
McKee T J, Stoicheff B P and Wallace S C 1977 Appl. Phys. Lett. **30** 278
New G H C and Ward J F 1967 Phys. Rev. Lett. **19** 556
Okada J, Fukuda Y and Matsuoka M 1981 J. Phys. Soc. Jpn. **50** 1301
Orr B J and Ward J F 1971 Mol. Phys. **20** 513
Provorov A C, Stoicheff B P and Wallace S C 1977 J. Chem. Phys. **67** 5393
Stwalley W C 1971 Chem. Phys. Lett. **6** 241

Vidal C R 1984 *Tunable Lasers* eds I F Mollenauer and J C White (Heidelberg: Springer-Verlag)
Wallace S C and Zdasiuk G 1976 Appl. Phys. Lett. **28** 449
Wolniewicz L and Orlikowski T 1991 Mol. Phys. **74** 103

CARS in plasma physics

M. Lefebvre, M. Péalat, JP. Taran
Office National d'Etudes et de Recherches Aérospatiales
BP 72 - 92322 CHATILLON CEDEX - FRANCE

The usual low temperature plasmas are generally strongly displaced from equilibrium and are the siege of many chemical reactions. Their microscopic properties are best studied optically; optical probing does not perturb the flow or the chemistry. For a long time, the only optical measurements possible were emission spectroscopy and absorption spectroscopy, the first one giving information about excited molecular and atomic states, the second one about the ground states. However, absorption and emission suffer from a lack of spatial resolution. Many usual molecules like N_2, O_2, H_2 require, in addition, operation of sources and detectors in the VUV.

The development of laser-based light scattering methods in the early seventies has revolutionized the field. Certainly one of the most useful methods is the coherent Anti-Stokes Raman Scattering (CARS) which was originally developed for combustion studies (Régnier and Taran, 1973; Moya et al., 1975). It requires two mechanically stable, synchronous, pulsed lasers, one fixed in frequency and the other tunable over part of the visible range. It offers good spatial resolution and sensitivity. It can operate against strong stray light emissions. It is quantitative and precise. Virtually all molecules can be detected and their quantum state populations measured. The detection sensitivity reaches 10^{11} - 10^{13} cm^{-3} per quantum state depending on the chemical species and the static temperature. Therefore, the method can render invaluable services to plasma modellers. It is now believed that CARS has a considerable role to play in plasma physics; of particular interest is the application to Plasma-Enhanced Chemical Vapor Deposition (PECVD), which presents an enormous industrial importance.

Following the original demonstration of Nibler and coworkers in a D_2 glow discharge (Nibler et al., 1976; Tolles et al., 1977), numerous applications have followed, principally on N_2 (Osin et al., 1981; Dreier et al., 1982., Valyanskii et al., 1984; Massabieaux et al., 1987), and H_2 (Péalat et al., 1981, 1985; Otchkin et al., 1988; Bornemann et al., 1990), some work on atomic ions in laser breakdown plasmas (Gladkov et al., 1989), on copper atoms in a metal vapor laser (Gladkov et al., 1991) and on O_2 glow and microwave discharges (Lefebvre et al., 1991) has also been published.

This work presents some CARS plasma physics applications recently performed at ONERA. We shall not dwell on the physics of the CARS nonlinear optical interaction, which has been thoroughly reviewed (Druet and Taran, 1981; Levenson, 1990; Nibler and Pubanz, 1988; Eckbreth, 1988) and we shall only briefly summarize the technical characteristics of the CARS instrument in the first section. Many precautions had to be taken to ensure good data quality ; those will be pointed out.

The second and third section cover the applications. The first of these application is a study of a low-pressure magnetic multipolar plasma of H_2 developed for H^- generation, the second is the analysis of a nitriding glow discharge.

1 CARS setup

The setup is built compactly for easy transport and installation. A passively Q-switched single frequency Nd:YAG laser chain delivers the 200 mJ energy needed at 532 mm in a 12 ns pulse at 5 Hz. The second laser is a scanning dye laser delivering 1-5 mJ in the range 560-800 nm, with a 0.08 cm^{-1} bandwidth. Both laser beams are made as close as possible to the diffraction limit in order to optimize the signal generation efficiency (Druet and Taran, 1981). Depending on the sensitivity and the spatial resolution required, either collinear CARS or BOXCARS (Eckbreth, 1978) are used. Only a fraction of the 70 mJ really available for the spectroscopy at 532 nm is applied to the samples in order to limit saturation and Stark shifts of the Raman lines (Lucht et al., 1988; Péalat et al., 1988). This places stringent constraints on all of the optical components to preserve the wavefront quality.

Signal normalization by means of a reference cell is used to reduce the shot to shot fluctuations and to improve measurement accuracy. Referencing is performed in series, i.e the laser beams are first focused in the reference cell, then refocused in the discharge. This enables one to use a rare gas like argon in the reference cell, at a pressure low enough, viz 1 bar, that the phase matching conditions are identical at both focal volumes. Should they be different, shot to shot fluctuations would not be cancelled as completely.

The signals from the reference cell and from the sample are isolated by means of dichroic filters and spectrographs, and detected by phototubes. The electronics and computer take the ratio of the signals at each pulse, calculate the square root of this ratio and, if necessary, an average over a set number of pulses fired at a fixed spectral position.

More information about this setup will be found also elsewhere (Druet and Taran, 1981).

In combustion, the spectra which are recorded are generally processed by global comparison with theoretical spectra computed assuming realistic values for the rotational and vibrational temperature. In rarefied gases, this procedure is not frequent. The lines are well separated and can be processed individually, giving detailed access to quantum state populations. The lineshapes are assumed to be Doppler-broadened, which is valid given the low pressures at which the spectroscopy is done, and the static temperature is taken equal to the rotational temperature derived from the lowest rotational states. A notable example is H_2, for which the rotational distribution does not obey the Bolzman law at $J \geqslant 4$.

2   Magnetic multipolar H_2 discharge

Magnetic-multipolar hydrogen plasmas have received considerable attention for positive and negative ion production (Kunkel, 1979; Bacal et al., 1984; Leung et al., 1985; Bonnie et al., 1988; Stutzin et al., 1988; Eenshuistra et al., 1989; Bacal, 1989; Hiskes and Karo, 1989). The kinetic behaviour of these plasmas remains difficult to study. CARS has already given some insight into the problem (Péalat et al., 1981; 1985) but does not have the sensitivity required to detect the highest vibrational states. Multiphoton ionization has also been used (Bonnie et al., 1987; 1988) The latter is more sensitive, but does not permit *in situ* measurements because the discharge interferes with the collection of the ionization current. Also promising is straight VUV absorption, which shows good sensitivity on atoms, but lacks spatial resolution (Stutzin et al., 1988). This chapter presents a study of the dynamics of the plasma.

We recall, as a preamble, that the preceding studies (Péalat et al., 1981; 1985) had shown the medium to be highly displaced from Boltzmann equilibrium:

1 - rotational equilibrium in the first four to five rotational states is indeed established, at a temperature of 550 K typically;

2 - however, higher rotational states rapidly deviate from the Boltzmann law as J increases;

3 - this rotational distribution is the same for the vibrational states v = 0 - 2 within measurement error;

4 - the vibrational states follow the Boltzmann law up to v = 3, highest state detected, and the vibrational temperature is about 2500 K;

5 - the H_2 density corresponds to a partial pressure of only 70% of the total pressure as a consequence of dissociation and ionization;

6 - the pressure balance is contributed for the most part by atoms at low density, but high temperature; the latter, originally assessed to be ≈ 0.4 eV, was subsequently corrected and placed at ≈ 0.25 eV (Bruneteau et al., 1987; 1990).

Some of the key figures are grouped in Table 1.

Table 1. Main results concerning H_2 molecules and electrons from Péalat et al. (1985) and H atoms from Péalat et al. (1985) and Bruneteau et al. (1990), at 5.3 Pa and 10 A discharge current.

H_2				H		electrons	
density	T_{rot}	T_{vib}	p_{H_2}	density	temperature	density	temperature
(10^{14} cm^{-3})	(K)	(K)	(Pa)	(10^{13} cm^{-3})	(K)	(10^{12} cm^{-3})	(K)
5.4	530	2390	3.9	4.6	2700	1	9400

2.1 *Experimental setup and data analysis*

The discharge is produced in a 16 cm-diam, 20 cm-long cylindrical copper chamber, serving as the anode, cooled to 220 K and mounted vertically under a bell jar. The magnetic multicusp arrangement is produced by means of 12 ceramic magnets attached to the outside chamber wall. The discharge is created by thermionic electrons emitted by two 0.5 mm-diam, 10 cm-long, thoriated tungsten filaments which are Joule-heated; these filaments are connected in parallel and mounted side by side. They are placed near the axis of the chamber. One of their ends is grounded. The positive voltage (90V) is applied to the copper wall of the chamber. The main innovation with respect to the previous H_2 plasma experiment (Péalat et al., 1985) is that the discharge can be pulsed if necessary for kinetics studies; the positive voltage applied to the wall then can be switched on and off using a pulse generator which gives a square 90 V pulse. The pulses are delivered by a capacitor bank switched by a thyratron. They have a duration which is deliberately limited to 1 ms so that only 5 % of the stored charge is removed, which guarantees good stability of the discharge voltage. The rise and fall times are 10 µs. The pulses are repeated at a rate compatible with the CARS laser firings. Their timing is controlled by a clock running at 5 Hz, which also triggers the laser after an adjustable delay. The discharge current is set at 10 A in all cases. This is achieved by adjusting the filament current; depending on the H_2 gas pressure and on the filament characteristics, both for continuous and pulsed discharge operations, the current per filament falls in the range 18-20 A.

The H_2 gas is introduced at the bottom of the generator through a small injector. The tests were conducted for two values of the pressure, namely 5.3 and 0.53 Pa. The pressure is monitored in the bell jar, but outside of the generator, by means of an MKS Baratron gauge. The flow rate is 0.23 cm^{3}s^{-1} at standard temperature and pressure, i.e 4.4 dm^{3}s^{-1} at the pressure of 5.3 Pa. These figures are to be compared to the generator's volume of 5 dm^3 and to the bell jar's (40 dm^3). The operation of the discharge is done at pressures which are comparable to

or one order of magnitude larger than those used for H⁻ production. In
practice, this is acceptable for our kinetics studies.

The BOXCARS (Eckbreth, 1978) optical configuration, which offers better
spatial resolution at a cost of a factor of 5 - 10 in sensitivity, was
used most of the time. The probe volume is $\approx$ 3 cm long and 150 μm in
diameter, with an appreciable fraction of the signal coming from a
shorter 1 - 2 cm-long segment. The axis of this probe volume is kept
fixed in space and positioned so as to intercept the generator axis near
the center, and 4 cm below the filaments. The probe volume can be moved
along this line of sight in order to explore the radial distributions.
This is done by translating the focussing and recollimating lenses. The
total travel of the lenses is about 8 cm, which allows one to explore a
little more than one full radius of the generator.

2.2   Results

Two classes of experiments were conducted:
(i) measurement of radial profiles of temperature and H_2 quantum-state
populations with a continuous discharge;
(ii) monitoring of rise and decay of temperatures and densities
following switch-on and then switch-off of a 1 ms-long square discharge
pulse.
Both types of experiments give insight into the dynamics of the plasma.
Most of the data were collected at a pressure of 5.3 Pa, which is higher
than the normal operating conditions for this sort of plasma (Bacal et
al., 1984) but remains amenable to CARS density measurements.
Occasionally, measurements were also taken at the pressure of 0.53 Pa to
complement the results in a regime where the molecular mean free path
($\approx$ 6 cm for this case) becomes comparable with the reactor size; this
reduces the impact of diffusion and leaves the molecular kinetics
dominated by wall collisions.

2.2.1 Continuous discharge

• *rotation*

Our previous measurements (Péalat et al., 1985) had shown the first five
rotational states to be populated according to the Boltzmann law in the
discharge. We have verified this point using BOXCARS for good spatial
resolution. The new results, while roughly confirming the old ones,
actually reveal some unexpected effects (Fig.1). An apparent breakdown
of the 1 to 3 ratio of populations between the para and ortho forms of
H_2 is observed systematically at 5.3 Pa. The data are presented for the
position r = 5 cm from the generator axis. A population defect of about
20 % is seen in the para densities for states J = 0 and 2 (or,
equivalently, an excess in ortho states J = 1 and 3) of v = 0. The

effect is seen in this figure to be even more pronounced in v = 1 with state J = 2 (unfortunately, J = 0 is below detection limit). This phenomenon is not an artefact, although the difference is barely in excess of instrumental detection error, viz. ≈ ± 5 %. An example of the typical data precision is given by the Boltzmann plot obtained without discharge, where the data point alignment is near perfect. In that case, one finds from the slope a temperature of 231 ± 2 K, which is close to the temperature of the circulating wall coolant (223 K). Note in passing that the measurement accuracy quoted (± 2 K) is the accuracy given by the least-mean-squares routine used to derive the rotational temperature from the Boltzmann plots. This accuracy was also demonstrated to be equal to the standard deviation of half a dozen temperature measurements taken under the same conditions. Finally, one notes that, if the temperatures found for the ortho and para forms in the discharge are here equal within instrumental error, viz. 512 and 518 K respectively, one frequently finds the value to be slightly larger for the para.

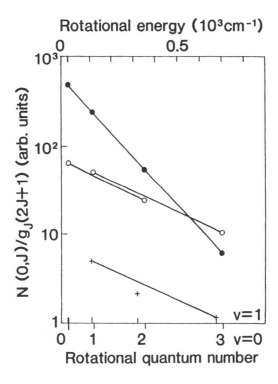

Fig. 1 Boltzmann diagram of rotational populations of v = 0 (○) and v = 1 (+) at 5.3 Pa and position r = 5 cm in the generator, with discharge, and for v = 0 without discharge and filament heating (•). g_j is the nuclear spin factor. Horizontal scales gives energies and markers for rotational state positions in v = 0 and 1.

The most probable explanation of the ortho/para anomaly is that low-energy electronic collisions cause the population to climb the ladder of rotational states, and that the pumping rate is faster in the para form, in part because its rotational states are more closely spaced. It has been shown theoretically and also experimentally that, for the range of energies 1 - 10 eV, the rotational-excitation cross-section is 50-100 % larger for the para form than for the ortho (Lane, 1980). Confirmation of these experimental results and of the interpretation based on the difference in ortho/para electron collision cross-sections will be given in the transient experiments.

In view of the results of Fig. 1, it has been decided to take for T_{rot} and T_t the T_{rot} value found from the populations of the odd states $J = 1$ and 3 which are always easily detected, rather than the result of a least-squares fit to the states $J = 0-3$. In vibrational state $v = 1$, T_{rot} is similarly derived from $J = 1$ and 3 only.

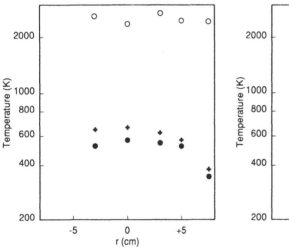

Fig. 2 Radial distribution of rotational temperatures in vibrational states $v = 0$ (•) and $v = 1$ (+), and of vibrational temperature (○) as determined from the ratio of populations in these two states; pressure: 5.3 Pa.

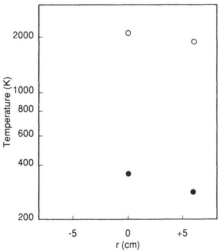

Fig. 3 Radial distribution of temperatures at 0.53 Pa:
• - rotational temperature in $v = 0$;
○ - vibrational temperature.

Temperature profiles obtained for $v = 0$ and 1 in the 5.3 Pa case are plotted *vs* distance r from the generator axis in Fig. 2, and in Fig. 3 for the 0.53 Pa case. Since the travel of the focussing optics is limited, the data could not be recorded over the full diameter of the reactor. Several conclusions are deduced from these profiles:

1 - The rotational temperature is not uniform, as had been suspected in previous investigations (Péalat et al., 1985). This is attributed to the combination of cooling on the plasma boundaries by the walls and heating throughout the volume by collisions with electrons, ions and atoms, the electrons being probably the main agent of excitation. The phenomenon is more pronounced at 5.3 than at 0.53 Pa. This is not surprising because the mean free path ℓ (Table 2) is, in the latter case, comparable to the reactor radius (6 cm *vs* 8 cm, respectively). In other words, the "heat conductivity" to the walls is then higher.

Table 2. Diffusion constant D, mean velocity $\vartheta = 0.92\sqrt{\dfrac{3RT_t}{m}}$ and mean free path ℓ for H_2 under conditions of Table 1 (R: ideal gas constant and m: molar mass).

D $(m^2 s^{-1})$	ϑ (ms^{-1})	ℓ (cm)
5.6	2400	1

2 - The rotational temperature in state v = 1, T_{rot} (v=1), is significantly higher than that in v = 0, T_{rot} (v=0) . This is quite pronounced near the center of the generator, as shown in Fig. 2 (650 *vs* 550 K). This result requires some discussion.

a - It can be explained in part by a difference in pure rotational pumping (*i.e.* of the type $\Delta v = 0$, $\Delta J \geq 2$) of the two states by the electrons, because the rotational levels in v = 1 have closer spacing ($B_1 = 56.3$ cm^{-1}) than those of v = 0 ($B_0 = 59.3$ cm^{-1}). It is likely that the rotational-excitation cross-section is larger in v = 1 than v = 0 because the electronic charge distribution is more elongated.

b - Rotational-vibrational heating by electrons causing transitions like:

$$H_2 (v=0,J) + e \longrightarrow H_2 (v=1,J+\Delta J) + e, \text{ with } \Delta J \geq 2 \qquad (1)$$

must play a role. The cross-section for the process with $\Delta J = 2$ is quite significant in the collision energy range 1-10 eV (Henry and Chang, 1972). One may therefore suspect that this is one of the prime causes for the rotational temperature to be larger in v = 1 than in v = 0.

c - The phenomenon can also be explained by a quasi-resonant rotation-rotation (R-R) pumping mechanism, causing rotational population transfer from the vibrational state v = 0 to the state v = 1, also because of the different rotational level spacings which make the reaction:

$$H_2 (v=1,J) + H_2 (v=0,J+2) \longrightarrow H_2 (v=1,J+2) + H_2 (v=0,J) \qquad (2)$$

exothermic. Similar phenomena are well-known in vibrationally excited gases and lead to anharmonic V-V pumping (Treanor et al., 1968) and to selective energy transfers in quasiresonant vibrational modes, *e.g.* in the N_2-CO_2 or the N_2-CO discharge lasers (Basov et al., 1969). These processes are observed in molecular systems when the translational temperature is different from the vibrational temperature. For the harmonic oscillator case, a closed-form expression between the relevant temperatures is easily derived (Basov et al., 1969). Unfortunately, there exists, to our knowledge, no closed-form equation linking T_t and the T_{rot} values of two near-resonant rotors. Thus, one should resort to computer calculations based on state-to-state rotational transfer rates. Some of the rates are already known (Valley and Amme, 1968; Wolfrum, 1987; Farrow and Chandler, 1988) for H_2 (Table 3). We have not attempted to perform these calculations, but did examine if R-R pumping is effective in our plasma (Lefebvre et al., 1991). By comparing the distance travelled by molecules between rotational inelastic collisions during time $\tau \approx 1/kn$ (with k rate constant as given in Table 3 and n, density as given in Table 1) to the reactor size, we conclude it to be negligible here. At pressures higher than those used in this study, for instance 10 Pa or more, one would certainly have to pay due consideration to R-R pumping. If this process was able to develop, then one would have $T_t < T_{rot}$ (v=0) by analogy with the V-V pumping results, and our assumption of Section II that these two temperatures are equal should be revised.

Table 3. Comparison of rotational energy transfer rates in H_2, in units of 10^6 s^{-1} amagat^{-1} (= 3.72 10^{-14} cm^{3}s^{-1}). Reverse rates ($k_{2\rightarrow0}$ and $k_{3\rightarrow1}$) are obtained from detailed balance.

v = 0, at 300 K[a]		v = 1, at 295 K[b]	
$k_{0\rightarrow2}$	$k_{1\rightarrow3}$	$k_{0\rightarrow2}$	$k_{1\rightarrow3}$
95	6.5	180 ± 27	50 ± 8

[a] Rates within state v = 0, from Farrow and Chandler (1988), based on the results of Valle and Amme (1968).

[b] Rates for v = 1 in a bath of v = 0 molecules, from Farrow and Chandler (1988).

3 - Near the walls of the generator, the temperatures in v = 0 and v = 1 are lower and become equal within experimental error (353 and 360 K, respectively). This is a consequence of translational-rotational cooling by the wall. We observe that, because the gas is approximately 100 K hotter than the wall at a distance of the order of one mean free path or less, the accommodation must be incomplete. In fact, the wall is mainly covered with tungsten evaporated from the filament, and, for tungsten, the accommodation coefficient of H_2 is in the range 0.06 - 0.2 according to Tamm and Schmidt (1969), Thomas (1981) or Rettner et al., (1985). Two

of us also have measured this coefficient and found 0.1 for a polycrystalline surface at 1170 K (Herlin et al., 1991).

• *vibration*

The figure 2 also presents the radial profile of the vibrational temperature, T_{vib}. The latter is determined from the two states v = 0 and 1, which are the only states detectable using BOXCARS under these conditions. This, however, is sufficient for determining a temperature applicable to the states v = 0-3, since Péalat et al. (1985) have shown that the vibrational population distribution obeys the Boltzmann law up to v = 3; Young and coworkers (1990) have confirmed this to hold up to v = 8. Other recent results, obtained however with hot filaments but without the main discharge, seem to indicate that the higher states could depart from it (Hall et al., 1988). The radial distribution is seen to be uniform, which is at variance to the behaviour of the rotation. This difference is significant. It indicates that the deactivation of vibration at the walls is not as rapid as the other volume-distributed mechanisms of excitation and loss. These mechanisms have been compiled in the communications of Gorse and coworkers (1985, 1987). The volume loss mechanisms in v = 1 are, among others:

1 - Dissociation and dissociative attachment, both of which are weak processes.

2 - The quasi-resonant reaction:

$$H_2(v=1) + H_2(v=1) \Leftrightarrow H_2(v=0) + H_2(v=2), \qquad (3)$$

which normally restores Boltzmann equilibrium.

3 - The very effective inelastic collisions with H-atoms (V-T exchange):

$$H_2(v=1) + H \longrightarrow H_2(v \neq 1) + H. \qquad (4)$$

With their kinetic energies of ≈ 0.3 eV, (Péalat et al., 1985; Bruneteau et al., 1987, 1990) the latter have a large cross-section in collisions with H_2 for atom exchange (Gorse et al., 1987; Buchenau et al., 1988), so-called reactive collisions, and for causing straight vibrational population change without atom exchange (non-reactive collisions). The rates for V-T exchange in H_2-H collisions have been calculated by Gorse et al., (1987). These rates are given in Table 4 assuming that the temperature is 1500 K for both species, which yields a fair representation of the relative motion of the 2700 K atoms and 550 K molecules (see Table 1). Some interpolations had to be made using the tables at 500, 1000 and 4000 K given by these authors. With a net rate constant of $k_{v-T} \approx 0.5 \times 10^{-10}$ cm^3 atom^{-1} s^{-1} out of v = 1, the lifetime of state v = 1 calculated using the H density reported in Table 1 is only 4.4×10^{-4} s at 5.3 Pa.

4 - The very effective electronic collisions (e-V collisions):

$$H_2(v=1) + e \longrightarrow H_2(v \neq 1) + e. \qquad (5)$$

The corresponding rates have been computed by Wadehra (1986). Those pertinent to v = 1 are given in Table 5 for 1 eV electrons. The sum of those out of v = 1 yields a rate constant K_{e-v} (v=1) = 1×10^{-9} cm^3 electron^{-1} s^{-1}. Given the electron density of n_e = 10^{12} cm^{-3}, the lifetime of state v = 1 imposed by electrons alone is about 1 ms. This lifetime is comparable to that imposed by atomic collisions (Eq. 4). Finally these results are in reasonable agreement with those recently computed by Bacal et al (1989).

Table 4. Reactive (with atom exchange) and non-reactive vibration-vibration rate constants in units of 10^{-12} cm^3 atom^{-1} s^{-1}, for H + H$_2$ (v=1) $\longrightarrow$ H + H$_2$ (v') at translational temperature 1500 K.[a]

	v' = 0	v' = 1	v' = 2
non-reactive	10	150	-
reactive	14	22	3

[a] Gorse et al. (1987); although H is at translational temperature of $\approx$ 2700K in the discharge, one must bear in mind that H$_2$ is at 550 K only; to model relative motion in center of mass of H + H$_2$ system, we calculate rate constants from their tables 2-9 using an interpolation.

Table 5. Rate constants in units of 10^{-9} cm^3 electron^{-1} s^{-1} for reaction H$_2$ (v) + e $\longrightarrow$ H$_2$ (v') + e.

v \ v'	0	1	2
0	49.2	0.16	-
1	0.45	56.1	0.48

It is quite interesting to compare the net resulting "volume" lifetime imposed by the combined action of atomic and electronic collisions, viz about 0.3 ms, to the sole effect of the walls. We shall see below that the "wall vibrational lifetime" at 0.53 Pa is of the order of 1.5 ms after discharge switch-off, which is significantly longer than the "volume lifetime". This is consistent with the fact that the radial distribution of T_v is found to be flat at both 5.3 Pa and 0.53 Pa, as shown in Fig. 3. It is also interesting to point out that the lifetime of state v = 1 ($\approx$ 0.3 ms) is of the same order of magnitude as the lifetimes of any of the first four rotational levels of v = 0 (see Section 2.2.1, table 3).

• *number density and pressure*

The H$_2$ number density is obtained by summing the number densities of all the rotational states in the v = 0 vibrational manifold. Appropriate corrections for the undetected states are made. The densities of molecules in vibrational states v > 0 are also calculated and added to that of v = 0. This calculation is based on the assumption that the vibrational populations follow the Boltzmann law. All these corrections remain small, *i.e.* a total of less than 10 %.

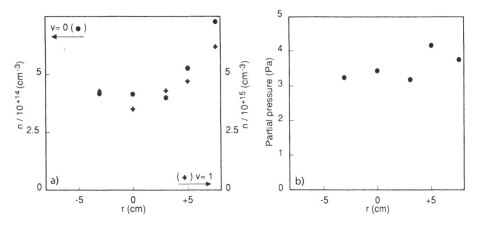

Fig. 4 Radial distribution of number densities in v = 0 and v = 1 (a), and H_2 partial pressure (b).

The resulting density profile N(r) is shown in Fig. 4a for the 5.3 Pa case. As expected, this profile shows the density to increase where the static temperature is lower, *i.e.* close to the walls. Also shown is the partial pressure of H_2 molecules, $p_{H_2}(r)$, which is proportional to the product $N(r) \times T_t(r)$, where we take $T_t(r) = T_{rot}(v=0,r)$ (Fig. 4b). The pressure profile is found to be flat within experimental error. Uncertainties are larger than for N(r) or $T_t(r)$ alone, since $p_{H_2}(r)$ is proportional to their product. We also note that $p_{H_2}(r)$ is about 70% of the pressure of 5.3 Pa which is maintained in the bell jar (the latter pressure is read by the Baratron gauge at a point outside the generator). This phenomenon had been reported previously (Péalat et al., 1985). It can be explained by dissociation. Dissociation of 15 % of the molecules would cause the observed drop in H_2 partial pressure, assuming the H and H_2 species had the same translational temperature. However, several reports reveal that the dissociation is less than 15 % in similar reactors (Péalat et al., 1985; Stutzin et al., 1988; Eenshuistra et al., 1989; Bruneteau et al., 1990). One therefore concludes that the translations of H and H_2 do not obey the Boltzmann equilibrium or that their velocity distribution functions do not follow the Maxwell-Boltzmann law. This is also suggested by the non-Boltzmann character of the rotation and of the vibration in H_2. If, for instance, the velocity distribution of H_2 had a broad, high energy tail, the CARS lines would present weak, broad wings; such wings are hard to detect and some of the molecular population would remain undetected. Non-Maxwellian distributions have already been reported in such plasmas (Ortobaev et al., 1981; Stutzin et al., 1988). In addition to this phenomenon, coexistence of translationally cold molecules and hot atoms is a possibility. In the previous study, we brought evidence based on Balmer β emission spectroscopy of this coexistence (Péalat et al., 1985); Stutzin *et al.* (1988) confirmed this result using VUV absorption spectroscopy.

2.2.2 Pulsed discharge

Measurements performed with the pulsed discharge provide additional insight. The spectra have been collected by repetitively firing the CARS set up at fixed delays with respect to the leading edge of the discharge pulse. The parameters of interest have been monitored through the entire duration of the pulse, as well as before and after it. The time evolutions of the rotational and vibrational temperatures and of the density have been followed at the center of the generator and near the wall. For clarity, the results are presented at switch-on (thus revealing mostly the kinetics resulting from electron excitation), and then switch-off, showing the effects of collisional relaxation and transport of rotational and vibrational energies to the walls with considerably reduced density and activity of the charged species.

The transients at switch-on and switch-off also provide excellent opportunities for testing computer models of the plasma because only a limited number of physical processes are at play to bring the system to its new equilibrium, especially at switch-off; rate constants are more easily calculated. Conversely, the models can also be employed to assist in the data analysis, particularly by showing which phenomena are important. Comparison has been performed with a model developed at Laboratoire de Physique des Milieux Ionisés of Ecole Polytechnique. This model assumes the plasma to be homogeneous, the rotational and vibrational temperatures to be constant (as well as the different static temperatures for the atoms, molecules and ions). The code calculates self-consistently the electron energy distribution function (EEDF) through the time-dependent Boltzmann equation (Bretagne et al., 1985, 1986) and the densities of H, H_2, H^+, H_2^+, H_3^+, from kinetic equations including diffusion and recombination on the walls (Jacquin et al., 1989). The model also calculates the kinetics of vibrationally excited molecules. This model was first developed by Gorse and coworkers (1985). The vibrational excitation of molecules $H_2(v)$ by electronic collisions and their relaxation by atomic and wall collisions are included. A newer version of the code, faster and more accurate over long time scales, has been used (Skinner et al., 1991). Part of the cross-sections involved have been updated, including vibrational excitation by energetic electrons and V-T relaxation by atomic collisions (Capitelli et al., 1991). There are two adjustable parameters, namely the wall quenching of vibration which was adjusted to match the decay of $N_2(v=1)$ at switch-off (at 0.53 Pa) and the atomic recombination coefficient at the walls. The code computes the electron density and temperature; under stationary conditions, comparison with experiment is possible and good agreement is obtained for these two parameters.

switch-on

At switch-on, the disturbances caused to the gas by the previous discharge pulse applied 200 ms earlier have decayed and some "dischargeless" equilibrium prevails.

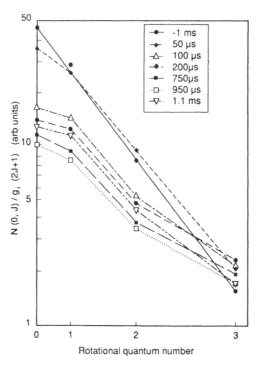

Fig. 5 Boltzmann plots of rotational state densities in v = 0 for different times and at 2.76 Pa.

• **rotation.** At all pressures, the rotational temperature of v = 0 rises in a few hundred µs (Figs. 5, 6). It is quite interesting to monitor the change in density of each rotational state (Fig. 5). Before switch-on, a slight deficit in para H_2 density is seen. At 50 µs after switch-on, J = 0 suffers a serious drop while J = 1 drops slightly and the other states grow slightly. The substantial population displaced from J = 0 is evidently already gone to higher para states, since the population gain in J = 2 is quite small. At later times, the process accelerates and steady state is practically achieved at 200 µs. Complementary results at

5.3 and 0.53 Pa are shown in Figs. 6a and b for the temperature as measured from J = 1 and 3. The fast rise and decay are evident. The rotational temperature overshoots the steady-state value of 550 K by 100 K for a short period of time shortly after the center of the current pulse; it has returned to the 550 K value at approximately 1 ms. The value of T_{rot} also rises near the wall, also immediately at switch-on, but is less pronounced and does not present an overshoot. The main agents of excitation are the electrons. The atoms, which cause vibrational cooling, appear slowly according to the predictions of the model and have only modest influence at the beginning. The present results are also consistent with measurements in a similar source which showed establishment of steady-state in under 200 µs for the kinetics of electron density and temperature (Hopkins and Graham, 1991).

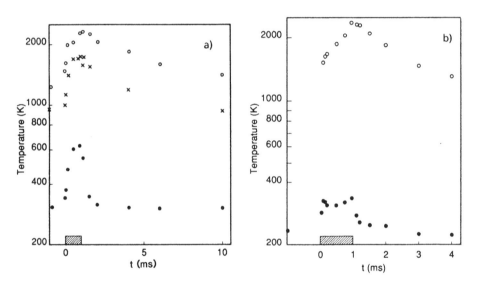

Fig. 6 Rise and decay of temperatures of rotation T_{rot} (v=0) (•) and of vibration T_{vib} (○) on the generator axis (r = 0) and for 5.3 Pa (a) and 0.53 Pa (b), with 1-ms-long discharge pulse (shown by hatched rectangle). T_{vib} is also shown at r = 7.5 cm and 5.3 Pa (x).

• **density.** The rapid heating is accompanied by a compression effect, associated with the fact that the gas suddenly heated by the discharge does not have the time to expand through the openings provided in the reactor walls for optical access and passage of the filaments; in effect, the relaxation of a pressure imbalance between reactor and bell jar is estimated to take place in 10-20 ms at 300 K, given the total area of these openings ($\approx$ 10 cm^2). In the 5.3 Pa case (Fig. 7a,b), the H_2 partial pressure is seen to reach about 6 Pa near 0.5 ms, then to begin to drop toward its steady-state value of $\approx$ 4 Pa as the dissociation and the atomic temperature and partial pressure are increasing. The discharge pulse is not long enough for steady state to be attained, but, as expected, the pressure history is identical at r = 0 and 7.5 cm within experimental uncertainty. The H_2 partial pressure increases only by 20 % and does not reach its potential maximum of $\approx$ 10 Pa, which one would expect from the doubling of the static temperature observed near the filaments. This is because the molecules dissociate, also because a sizeable fraction of the volume of the generator is not as strongly excited and heated as the vicinity of the filaments and because a small fraction of the gas escapes through the openings in the reactor walls.

Also interesting is the fact that, before switch-on, the H_2 pressure is slightly less than the nominal value by 5-10 % at both r = 0 and r = 7.5 cm. This pressure deficit is comparable to or slightly larger than the experimental error, *i.e.* 5 %, and the result is systematic. Unless the H_2 pressure measurement is biased as, *e.g.*, if $T_t \neq T_{rot}$ (v=0), the pressure defect indicates the presence of other species such as H-atoms. The latter can be produced by dissociation of the molecules when they collide with the filament, as shown by the spatially-resolved measurements of Meier et al. (1990).

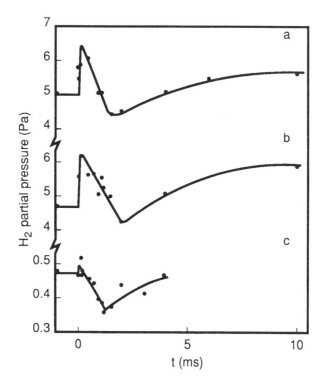

Fig. 7 Plots of H_2 pressure variations for the 5.3 Pa case at r = 0 (a), r = 7 cm (b), and for 0.53 Pa, r = 0 (c).

• **vibration.** An important phenomenon, revealed by the Fig. 7, is that, at 5.3 Pa before switch-on, the vibrational temperature has a value of 1275 K which is well above T_{rot} (v=0). Furthermore, this value is <u>lower</u> near the walls (911 K) than on the axis close to the filaments. (The latter property is more striking when one considers the density of N_2 (v=1), which is typically six times lower at the wall than the center.) This rules out wall recombination of atomic hydrogen as the dominant source of vibrationally excited H_2 in our generator during afterglow, since one would then expect T_{vib} to be uniform. One explanation of the high vibrational temperature near the filaments is, again, that the filaments create between their two ends a confined, "mini-discharge" which collisionally populates vibrational states. We do not conclude, however, that formation of vibrationally excited H_2 by recombinative desorption is absent here, or that it is negligible in other sources of similar nature (Bonnie et al., 1987; Hall et al., 1988), yet the "mini-discharge" constitutes a satisfactory interpretation of the "dischargeless" results obtained by these groups, including the presence of vibrational states as high as v = 7. As a matter of fact, as long as the filaments are heated under a voltage of $\gtrsim$ 1 V, they can sufficiently accelerate the thermionic electrons which they emit to form vibrationally excited H_2.

The rise of T_{vib} at r = 7.5 cm and for 5.3 Pa is less pronounced than at r = 0 (Fig. 8), although the steady-state discharge values are the same at the two positions (Fig. 2). It is quite likely that this is because the majority of the v = 1 molecules are produced away from the walls, closer to the filament and take some time to diffuse through the rest of the reactor. At r = 0 for 0.53 Pa (Figs. 6b,7c), little change is seen in either the time response or the numerical values compared with the 5.3 Pa case. Note that our detection sensitivity is insufficient to monitor N_2 (v=1) just before switch-on. We assume $T_v \lesssim$ 1000 K at that position.

switch-off

At switch-off, the electron temperature and density drop very rapidly, on time-scales of 50 - 200 μs (Hopkins and Graham, 1991). Important phenomena are visible from the figures.

• **rotation.** The decay of rotational excitation is very fast , and takes place in 100 - 200 μs regardless of pressure (Fig. 6); this is comparable to those of ion density and electron temperature and density (Hopkins and Graham, 1991). This result is readily interpreted at 0.53 Pa (Fig. 6c) because this time scale is also comparable to the mean time between wall collisions. The result at 5.3 Pa is surprising, since the heat diffusion to the wall requires times of the order of 0.5 - 1 ms. The fast drop in the population of J = 3 which is actually seen could result from the superelastic collisions with low-energy electrons. The latter are far more mobile than the molecules and could evacuate the heat to the colder molecules close to the walls.

• **density**. The temperature drops very quickly to its pre-discharge level, which permits direct comparison of p_{H_2} values before and after the discharge and clearly reveals the deficit in H_2 molecules resulting from dissociation and from leaks through the openings in the walls driven by the discharge overpressure (Fig. 7). The drop in H_2 pressure with respect to the dischargeless steady state reaches 20 % at 0.5 - 1 ms after switch-off at both 5.3 and 0.53 Pa, indicating that some dissociation has been accomplished in 1 ms. The ensuing pressure recoveries are very similar for the three cases presented. Also notable is the overshoot visible at 5 - 10 ms which stays at the same level until 20 ms (data points not shown in the figures), which is reproducible and slightly exceeds measurement uncertainties. This overshoot may be caused by wall recombination (recombinative desorption) of the excess free H atoms produced by the dischage and by the outgassing of H atoms implanted below the metal surface by ion bombardment.

• **vibration**. The vibrational population is characterized by slow decays. The relaxation times are typically in the range of 1.4 ms at 0.53 Pa (Fig.8c) to 4 ms at 5.3 Pa (Fig.8a) and r = 0. Because diffusion complicates the study of molecular interaction with the wall, the results at 5.3 Pa are difficult to analyze. However, it is possible to extract some information from those at 0.53 Pa. Because the molecules rarely collide with other molecules between two wall bounces, the decay in vibrational population is mainly caused by wall deactivation. A simulation of the result of Fig. 8c was performed, adjusting the probability b(v) of wall deactivation of state v. The rate of decay of the population N_v in state v used in the model is given by (Gorse et al., 1987):

$$\frac{dN_v}{dt} = \frac{1}{4} \frac{S}{V} \bar{v} \left[\sum_{v' > v} \frac{N_{v'}}{v'b(v')} - \frac{N_v}{b(v)} \right], \qquad (6)$$

where $\bar{v}$ is the mean molecular velocity, V the reactor volume and S the area of its walls. Following Karo and Hiskes (1985), a linear variation is assumed for $b^{-1}(v)$, with b(14) = 1 and b(1) adjusted to fit the present experimental result. To simulate the production of vibrationally excited molecules at 1000 K near the filament and in the absence of discharge, a dc discharge of ~ 0.3 A is assumed to be maintained before and after the square pulse. A satisfactory agreement is obtained (Fig. 9) for b(1) = 1/16. Allowing for the various experimental and geometrical uncertainties, we conclude that the probability of survival

of state v = 1 is 16 ± 5 collisions. The figure also shows the
experimental and simulated behaviours of the vibrational population to
be in fair agreement with each other during the discharge.

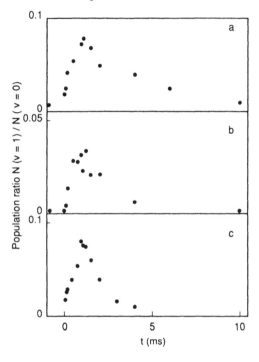

Fig. 8 Plots of v = 1 density
normalized to v = 0 at r = 0 (a),
r = 7 cm (b), and for 0.53 Pa,
r = 0 (c).

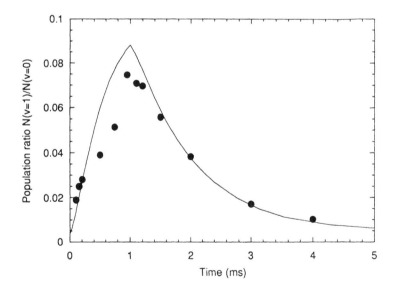

Fig. 9 Numerical simulation for conditions of Fig. 9c,
and experimental measurement of time dependence of N_2(v=1)
taken from Fig. 9c.

3   Nitrogen discharge

The understanding of the nitrogen vibrational kinetics in gas discharges is of fundamental interest for plasma chemistry studies. $N_2(X, v)$ molecules are thought to play an important role during processes such as excitation, relaxation, ionisation and dissociation. The first use of CARS for the investigation of an N_2 discharge at 4mA, 130 Pa was reported by Shaub and coworker (1977). In a similar discharge Smirnov (1978) has observed vibrational states up to v = 6. Also, pulsed discharges have been studied by Valyanski et al. (1984) and Devyatov et al. (1985) at pressures of the order of 10^4 Pa.

Vibrational distribution and rotational temperatures of N_2 (X) have also been measured by CARS in a low pressure DC glow discharge and in its post discharge region (Massabieaux et al., 1987). Rotational temperatures of N_2 (X, v) are deduced from Boltzmann plots of the rotational populations of each vibrational state while the vibrational excitation is characterized using the θ_1 parameter with:

$$\theta_1 = \frac{E(v = 1) - E(v = 0)}{k \ \ell n \left[\dfrac{N_2 \ (v = 0)}{N_2 \ (v = 1)} \right]}$$

where E(v) is the vibrational energy, $N_2(v)$ the population of the v state and k is the Boltzmann constant.

In the positive column of a 260 Pa, 80mA DC glow discharge, it has been shown that: i) the rotational temperature is T_R = 530 K for the first 10 vibrational levels which could be probed, ii) the vibrational distribution never has a Boltzmann-like behavior, iii) the θ_1 parameter is equal to 5300 K. Further, we could measure the decay of these parameters in the post discharge region. For instance, we observed that T_R has decreased to 340 K and θ_1 to 3350 K at a point located 30 cm downstream from the cathode i.e. about 15 ms after leaving the discharge.

We report here some new results obtained in the vicinity of a steel sample undergoing nitriding and located in a similar N_2 discharge. Using CARS, spatial profiles of rotational temperature and vibrational excitation are recorded under two distinct experimental conditions. First, the sample is located in the post discharge region, 30 cm downstream from the cathode; second the sample itself is the cathode of the discharge. These two cases are called later on the PD (post discharge) case and the D (discharge) case, respectively.

We first describe the discharge reactor and then compare the rotational and vibrational distributions obtained in the D and PD cases.

3.1   Reactor

The plasma reactor is shown in figure 10. A discharge tube of inner diameter 2 cm is connected to a vessel of diameter 15 cm. The N_2 gas (purity > 99,99 %) passes through a mass flowmeter and then enters the discharge tube. The N_2 pressure is 260 Pa and the flow velocity is 20 ms^{-1} .The gas is pumped out of the vessel by a primary pump. The susceptor is a 3 cm-diameter, 0.8 cm-thick steel disk placed in the vessel perpendicular to the discharge tube axis. The susceptor can be translated over 10 cm along the discharge axis.

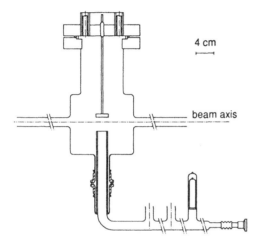

4 cm

beam axis   Fig. 10 Diagram of the discharge reactor.

Three nickel-chrome electrodes mounted in side-arms are placed along the discharge tube, 20 cm from each other. In the PD case, the N_2 glow discharge is struck between two of those electrodes: the length of the discharge is 28 cm and the discharge current is 50 mA. The steel sample is heated by the Joule effect up to 820 K in order to produce γ' Fe_4N. It is located between 20 cm and 30 cm downstream from the cathode. In the D case, the discharge is struck between one electrode and the sample acting as the cathode. The discharge length varies from 40 cm to 50 cm depending on the position of the sample. In the D case, the discharge current is reduced to 30 mA. The sample is heated by ion bombardment and neutral gas collisions. Its temperature depends on the flow rate; it reaches 720 K for a flow rate of 0.9 ℓ min^{-1} at standard temperature and pressure.

To reduce saturation of the vibrational transitions by the CARS process (Péalat et al., 1988), 500 mm focal length achromats are used to focus the laser beams. As a result, the optical ports have to be installed away from the focus, at the end of two long arms to avoid burning the windows. Using such a configuration, the CARS probe volume is 10 mm long and 50 μm in diameter. This probe volume is located in front of the susceptor with its axis parallel to the susceptor surface.

3.2   *Rotational temperature*

Rotational temperatures are deduced from Boltzmann plots of the rotational populations of the v = 0 vibrational state. Rotational temperatures of higher vibrational states would not differ significantly as already discussed. Figure 11a presents the temperature profile in the PD case. The gas exits the discharge at 350 K, in equilibrium with the wall temperature, and then reaches the sample temperature close to the steel surface, indicating a complete rotational accommodation of N_2 on the nitrided surface. Neglecting the presence of excited N_2 molecules and N atoms, the N_2 flow is modelled using a FORTRAN code called MOCROI (Dupoirieux et al., 1985). The differential equations representative of mass, momentum, and energy conservations are resolved locally. The physical quantities which are calculated at every step are: momentum, density and enthalpy. They are directly related to the velocity and the temperature. The flow in the discharge tube downstream from the cathode and in the vessel has been modelled assuming the following boundary conditions: at the input section, the N_2 temperature is in equilibrium with the wall temperature of 350 K; in the boundary layer surrounding the steel sample, the surface and the gas temperatures are equal. The over-all agreement shows that energy deposition by atoms and excited N_2 can be neglected in the PD case when calculating rotational temperature. As shown in figure 11b, the same conclusion cannot be extended to the D case. Although the rotational temperature near the surface is still equal to the sample temperature, it increases higher than predicted in the cathode fall region demonstrating that ground-state N_2 molecules are there efficiently heated by collisions with electrons and excited particules. The most probable process explaining the temperature increase in the cathode fall region is the efficient exchange between fast ions and ground state molecules:

$$N_{2\,fast}^+ \;+\; N_{2\,slow} \;\longrightarrow\; N_{2\,faster} \;+\; N_{2\,slower}^+$$

then followed by fast R-T transfer. The latter is possible because the mean free path of N_2, viz 90 μm, is smaller than the thickness of the cathode fall region.

3.3   *Vibrational temperature*

In the D case (Figure 12a), θ_1 reaches 3200 K and does not change significantly for positions comprised between 100 mm and 5 mm from the cathode. Closer to the surface, θ_1 decreases to 2200 K. Note that the experiment being done by moving the cathode, the length of the discharge changes but the residence time between the anode and the probe volume is constant. In the PD case, the behaviour of θ_1 differs totally. θ_1 is constant, around 3000 K, whatever the distance from the surface. θ_1 is close to the value observed in the positive column which the molecules are coming from (Figure 12b).

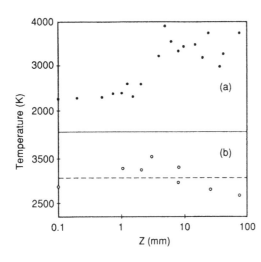

Fig. 11 Rotational temperature profiles obtained in the PD case (a) and in the D case (b); • CARS measurements; full line: MOCROI calculation for flow rate $0.14\ mn^{-1}$ (STP); Z: distance between the CARS probe volume and the surface of the sample.

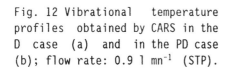

Fig. 12 Vibrational temperature profiles obtained by CARS in the D case (a) and in the PD case (b); flow rate: $0.9\ l\ mn^{-1}$ (STP).

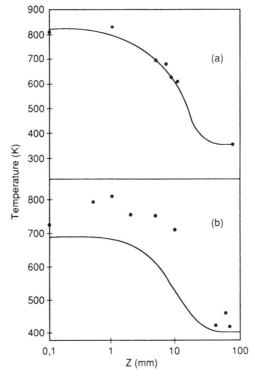

The PD case results demonstrate that N_2 vibrational deexcitation over the nitrided surface is negligible. In the D case, we observe a sharp decrease of θ_1 in the cathode fall and negative glow regions. The specific properties of these regions remain to be studied. The presence there of the fast ions may be responsible for the observed results. The cross-section for the vibrational deexcitation of $N_2(X, v)$ colliding with N_2^+ possibly increases when the velocity of the ions increases. However, no data is available, to our knowledge. More experiments and finer modelling are required.

4   Conclusion

Thanks to its unique properties of sensitivity, temporal and spatial resolutions, CARS permits a very fine analysis of the physics of low temperature plasmas. The experimental results it provides are accurate and can be used for the modelling of the microscopic properties of these plasmas. In the form we have presented here, CARS does not quite have the absolute sensitivity of two other optical techniques used for plasma diagnostics, viz Laser Induced Fluorescence and optogalvanic spectroscopy, but it affords a considerable advantage over them, namely it is impervious to stray light interference and is not disturbed by the electric current which maintains the discharge. In the near future one may expect exciting results to come from (i) introduction of resonance enhanced CARS, which affords gains in detection sensitivity of the order $10\text{-}10^3$ (Druet and Taran, 1981), and (ii) development of a new field of application, namely chemical vapor deposition.

REFERENCES

Bacal M, Bruneteau A M and Nachman M 1984 J. Appl. Phys. **55** 15

Bacal M 1989 Nucl. Instrum. and Methods in Phys. Res. **B 37/38** 28

Bacal M, Berlemont P and Skinner D A 1989 "volume production of Hydrogen Negative Ions", OE/LASE'89 SPIE Meeting, Los Angeles, USA

Basov N G, Mikhailov V G, Oraevskii A N and Scheglov V A 1969 Sov. Phys. Techn. Phys. **13**, 1630

Bonnie J H M, Eenshuistra P J and Hopman H J 1988 Phys. Rev. **A 37** 1121

Bonnie J H M, Granneman E H A and Hopman H J 1987 Rev. Sci. Instrum. **58** 1353

Bornemann T, Kornos V, Schultz-von der Gathen V and Döbele H F 1990 Appl. Phys. **B51** 307

Bretagne J, Delouya G, Gorse C, Capitelli M and Bacal M 1985 J. Phys. **D18** 811

Bretagne J, Delouya G, Gorse C, Capitelli M and Bacal M 1986 J. Phys. **D19** 1197

Bruneteau A M, Hollos G, Bacal M and Bretagne J 1987 IAEA Technical Committee Meeting on Negative Ion Beam Heating Culham Laboratory G. B

Bruneteau A M, Hollos G, Bacal M and Bretagne J 1990 J. Appl. Phys. **67** 7254

Buchenau H, Herrero V J, Toennies J P and Vodegel M 1988 VII European Molecular Conference on Dynamics of Molecular collisions Assisi (Perugia), Italy

Capitelli M, Gorse C, Berlemont P, Skinner D A and Bacal M 1991 Chem. Phys. Letters **179** 48

Devyatov A A, Dolenko S A, Rakhimov A T, Rakhimova T V and Suetin N V 1985 XVIIth Int. Conf. on Phenomena in Ionized Gases Budapest p. 1078.

Dreier T, Wellhausen V, Wolfrum J and Marowsky G 1982 Appl. Phys. **B29** 31

Druet S A J and Taran J P 1981 Progr. Quant. Electr. **7** 1

Dupoirieux F and Scherrer D 1985 La Recherche Aérospatiale **5** 301

Eckbreth A C 1988 "Laser Diagnostics for Combustion Temperature and Species" *Energy and Engineering Science series* Ed A Gupta and D Lilley Cambridge: Abacus Press

Eckbreth A C 1978 Appl. Phys. Letters **32** 421

Eenshuistra P J, Kleijn A W and Hopman J H 1989 Europhys. Letters **8** 423

Eenshuistra P J, Heeren R M A, Kleyen A W and Hopman H J 1989 Phys. Rev. **A40** 3613

Farrow R L and Chandler D W 1988 J. Chem. Phys. **89** 1994

Gladkov S M, Zheltikov A M, Koroteev N I, Rychev M V and Fedotov A B 1989 Sov. J. Quant Electron. **19** 923

Gladkov S M, Zheltikov A M, Il'Yasov O S, Isaev A A and Koroteev N I 1991 Sov. J. Quant. Electron **21** 659

Gorse C, Capitelli M, Bretagne J and Bacal M 1985 Chem. Phys. **93** 1

Gorse C, Capitelli M, Bacal M, Bretagne J and Lagana 1 1987 Chem. Phys. **117** 177

Hall R I, Cadez I, Landau M, Pichou F and Scherman C 1988 Phys. Rev. Letters **60** 337

Henry R J W and Chang E S 1972 Phys. Rev. **A5** 276

Herlin N, Péalat M, Lefebvre M and Alnot P 1991 Surface Sc. **258** 381

Hiskes J R and Karo A M 1989 Appl. Phys. Letters **54** 508

Hopkins M B and Graham W G 1991 J. Appl. Phys. **69** 3461

Jacquin D, Bretagne J and Ferdinand R 1989 Plasma Chem. Plasma Proc. **9** 165

Karo A M, Hiskes J R and Hardy R J 1985 J. Vac. Sci. Technol. **A3** 1222

Kunkel W B 1979 IEEE Trans. Nucl. Sci. **26** 4166

Lane N F 1980 Rev. Mod. Phys. **52** 29

Lefebvre M, Péalat M and Taran J-P 1991 "Diagnostics of Plasmas by CARS", presented at the 10th International Symposium on Plasma Chemistry, Bochum

Lefebvre M, Péalat M, Taran J-P and Bacal M 1991 "Study of a hydrogen plasma by CARS" ONERA Technical Report n° 62/7131 PN

Leung K N, Ehlers K W and Pyle R V 1985 Rev. Sci. Instrum. **56** 364

Levenson M D 1982 "Introduction to Nonlinear Laser Spectroscopy" *Quantum Electronics - Principles and Applications Series* Ed P Liao and P Kelley (New York: Academic Press)

Lucht R P and Farrow R L 1988 J. Opt. Soc. Am **B5** 1243

Massabieaux B, Gousset G, Lefebvre M and Péalat M 1987 J. Phys. (Paris) **48** 1939

Meier U, Kohse-Höinghaus K, Schäfer L and Klages C P 1990 Appl. Optics **29** 4993

Moya F, Druet S A J and Taran J-P-E 1975 Optics Comm. **13** 169

Nibler J W, Mc Donald J R and Harvey A B 1976 Opt. Comm. **18** 371

Nibler J N and Pubanz G A 1988 Coherent Raman Spectroscopy of Gases, *Advances in Non-linear Spectroscopy,* Ed R J H Clark and R E Hester 1988 (Chichester: Wiley)

Ortobaev D K, Ochkin V N, Preobrazhenskii N G, Savinov S Yu, Sedelnikov A I and Sobolev N N 1981 Sov. Phys. JETP **54** 865

Osin M N, Pashinin P P, Smirnov V V, Fabelinskii V I and Tsai N S 1981 Sov. Phys. Techn. Phys. **26** 58

Otchkin V H, Savinov S Yu, Sobolev N N and Tsai S N 1988 J. Technical Physics (USSR) **58** 1283

Péalat M, Taran J-P, Taillet J, Bacal M and Bruneteau A M 1981 J. Appl. Phys. **52** 2687

Péalat M, Taran J-P, Bacal M and Hillion F 1985 J. Chem. Phys. **82** 4943

Péalat M, Lefebvre M, Taran J-P and Kelley P L 1988 Phys. Rev. **A38** 1948

Régnier P R and Taran J-P 1973 Appl. Phys. Letters **23** 240

Rettner C T, Delouise L A, Cowin J P and Auerbach D J Farady Discuss. Chem. Soc. **80** 127

Shaub W M, Nibler J W and Harvey A B 1977 J. Chem. Phys. **67** 1883

Skinner D A, Berlemont P and Bacal M 1991 Proc. 4th European Workshop on the Production and Application of Light negative Ions. Graham W G Ed Queen's University of Belfast p. 1

Smirnov V V and Fabelinskii V T 1978 JETP Letters **28** 427

Stutzin G C, Young A T, Schlachter A S, Stearns J W, Leung K N, Kunkel W B, Worth G T and Stevens R R 1988 Rev. Sci. Instrum. **59** 1363

Tamm P W and Schmidt L D 1969 J. Chem. Phys. **51** 5352

Thomas L B 1981 "Rarefied Gas Dynamics", Progress in Astronautics and Aeronautics Series Ed M Summerfield (New York: AIAA) **74** 83

Tolles W M, Nibler J W, Mc Donald J R and Harvey A B 1977 Appl.
  Spectroscopy **31** 253
Treanor C F, Rich J W and Rehm R G 1968 J. Chem. Phys. **48** 1798
Valley L M and Amme R C 1968 J. Acoust. Soc. Am. **44** 1144
Valyanskii S I, Verchagin K A, Vernke V, Volkov A Yu,
  Pashinin P P, Smirnov V V, Fabelinskii V I and Chapovskii P L
  1984 Sov. J. Quant. Electron. **14** 1226
Valyanskii S I, Vereshchagin A, Vernke V, Volkov A Yu,
  Pashinin P P, Smirnov V V, Fabelinskii V I and Chapovskii P L
  1984 Sov. J. Quant. Electron. **14** 1226
Wadehra J 1986 "Nonequilibrium Vibrational Kinetics" Topics
  in Current Physics **39** Ed M Capitelli (Berlin: Springer-Verlag)
Wolfrum J 1987 Faraday Discuss. Chem. Soc. **84** 191
Young A T, Stutzin G C, Kunkel W B and Leung K N 1990 AIP
  Conf. Proc. **210** 450

Velocity distribution of electrons for tunnel ionization by a light field with polar asymmetry

N B Baranova and B Ya Zel'dovich

Technical University, Lenin prospekt 76, Chelyabinsk, 454080, Russia

Abstract. Free electrons appearing from the ionization process oscillate in an intense laser polychromatic field with large velocity. Smooth decrease of the field amplitude results in a residual velocity, which depends on the particular amount of the electron liberation from the atom. Polar asymmetry of the resultant angular distribution is predicted and calculated for the tunnel ionization by the light field with $\langle E^3 \rangle \neq 0$.

1. Introduction

Consider the movement of a free electron in a spatially homogeneous oscillating electrical field $E(t)$. The equation of Newton's second law

$$m \mathrm{d}v/\mathrm{d}t = e \cdot \boldsymbol{E}(t) \equiv -(e/c)\partial \boldsymbol{A}/\partial t \tag{1}$$

has evident solution

$$\boldsymbol{v}(t) = -(e/mc)[\boldsymbol{A}(t) - \boldsymbol{A}(t_0)] + \boldsymbol{v}(t_0). \tag{2}$$

Here

$$\boldsymbol{A}(t) = -c \int_{-\infty} \boldsymbol{E}(t') \, \mathrm{d}t' \tag{3}$$

is the vector potential of the field. For an optical pulse with no DC component to the field one has $\boldsymbol{A}(t = -\infty) = 0$, $\boldsymbol{A}(t = +\infty) = 0$, and therefore the velocity of an electron which remains after the action of the light pulse is

$$\boldsymbol{u} \equiv \boldsymbol{v}(t \to +\infty) = \boldsymbol{v}(t_0) + (e/mc)\boldsymbol{A}(t_0). \tag{4}$$

Thus the resultant velocity $\boldsymbol{u}$ is determined by $\boldsymbol{A}(t_0)$ and $\boldsymbol{v}(t_0)$ at the moment t_0 of liberation of electron, if starting from t_0 the electron may be considered as free.

These simple considerations have important consequences for the tunnel ionization of an atom (or a molecule) by an intense quasimonochromatic laser field.

The classification of multiphoton ionization processes has been known since the paper by Keldysh [1], (see also [2] and §77 of [3]). If the ionization potential I, frequency of light ω and the field strength $|\boldsymbol{E}|$ satisfy the 'adiabaticity' condition

$$\gamma = \omega\sqrt{2mI}/|eE| \ll 1 \tag{5}$$

then ionization may be considered quasistatically with the instantaneous probability

$$W(t)(\mathrm{s}^{-1}) \approx (I/h)\exp\{-2E_{\mathrm{at}}/3|\boldsymbol{E}(t)|\}. \tag{6}$$

Here $E_{\mathrm{at}} = (2I)^{3/2}m^{1/2}|e\hbar|^{-1}$ is the characteristic value of intra-atomic field. Typically $|\boldsymbol{E}(t)| \ll E_{\mathrm{at}}$ and therefore the main contribution to the ionization process is due to those moments t_n, when $|\boldsymbol{E}(t)|$ reaches its maximum. Moreover, the velocity $\boldsymbol{v}(t)$ of the released electron must be almost zero to get the lowest potential barrier: $\boldsymbol{v}(t_n) \approx 0$. Consider the simplest example of a quasimonochromatic linearly polarized field

$$\boldsymbol{E}(t) = \boldsymbol{c}_x F(t)\cos\,\omega t \quad \boldsymbol{A}(t) = \boldsymbol{c}_x c\omega^{-1}F(t)\sin\,\omega t \tag{7}$$

where $F(t)$ is a slowly varying envelope. The moments t_n corresponding to the maxima of the field strength are

$$t_n = (\pi/\omega)n \tag{8}$$

where n is integer. It is seen from (4), (7) and (8), that in this case $u \equiv v(t \to +\infty) = 0$.

However, for a circularly polarized quasimonochromatic wave

$$\boldsymbol{E}(t) = (\boldsymbol{e}_x\cos\,\omega t + \boldsymbol{e}_x\sin\,\omega t)F(t) \quad \boldsymbol{A}(t) = c\omega^{-1}F(t)(\boldsymbol{e}_x\sin\,\omega t - \boldsymbol{e}_y\cos\,\omega t) \tag{9}$$

there is no preferred moment, because $|\boldsymbol{E}(t)| = \mathrm{constant}$ during the period. It means, that in the case

$$\boldsymbol{u} \equiv \boldsymbol{v}(t \to +\infty) = \frac{eF}{m\omega}(\boldsymbol{e}_x\sin\,\omega t_n - \boldsymbol{e}_y\cos\,\omega t_n) \tag{10}$$

with an arbitrary value of t_n. Such a dramatic difference in the residual velocity distribution under switching the polarization type was predicted earlier and observed experimentally, see e.g. [4].

The aim of the present paper is to discuss the specific effects which arise for tunnel ionization by an optical field with polar asymmetry, or $\langle E^3 \rangle \neq 0$.

2. Bichromatic field

The simplest and the most popular example of a field with polar asymmetry is the radiation of a laser pulse and its second harmonic

$$\boldsymbol{E}(t) = \left(\boldsymbol{E}_1 e^{-i\omega t} + \boldsymbol{E}_1^* e^{i\omega t} + \boldsymbol{E}_2 e^{-2i\omega t} + \boldsymbol{E}_2^* e^{2i\omega t} \right)/2 \tag{11}$$

The same type of expression is valid for $\boldsymbol{A}(t)$ with $\boldsymbol{A}_1 = -i(e/c\omega)\boldsymbol{E}_1$, $\boldsymbol{A}_2 = -i(e/2c\omega)\boldsymbol{E}_2$. It is important that, even for a laser pulse with non-ideal monochromaticity, the phase of complex amplitude E_2 is rigidly connected with the phase of E_1^2 due to the nature of the process of second-harmonic generation. Consequently, the value $\langle E_1^2 E_2^* \rangle \neq 0$ even after the time or ensemble averaging, and $\langle E^3 \rangle$ characterizes the polar asymmetry of the field, see [5].

An example of such a field with linear polarizations, $\boldsymbol{E}_1 = E_1\boldsymbol{e}_x$, $\boldsymbol{E}_2 = E_2\boldsymbol{e}_x\exp(i\pi/4)$ for the case $E_1 = E_2 = 0.01E_{\mathrm{at}}$ is shown in the figure 1(*a*). The second graph, figure 1(*b*) shows the time behaviour of $(\omega/cE_{\mathrm{at}})A_{\mathrm{real}}(t) = -\frac{\omega}{E_{\mathrm{at}}}\int_{-\infty} E_{\mathrm{real}}(t')\,\mathrm{d}t'$ which is proportional to the residual velocity of an electron, which was released at the moment t with zero kinetic energy. The third graph, figure 1(*c*), shows the time

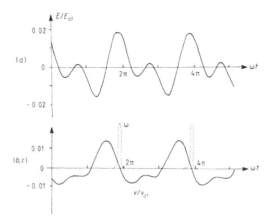

Figure 1. (a) Time dependence of the real field strength for superposition of linearly polarized fundamental field ω and its second harmonic with $\pi/3$ phase shift, $E_x = E[\cos \omega t + \cos(2\omega t + \pi/3)]$. Such a field possesses polar asymmetry, which is characterized by $\langle E^3 \rangle \neq 0$. ($b$) Time dependence of the value $-\frac{\omega}{E_{\text{at}}} \sin E_x(t')\,dt'$, proportional to the residual velocity of the electrons, produced at the moment t by the field, shown in (a), $E/E_{\text{at}} = 0.01$. (c) Time dependence of the probability of tunnel ionization $W = \exp(-2E_{\text{at}}/3|E|)$ for the field in (a), $E/E_{\text{at}} = 0.01$.

dependence of the function $\exp(-2E_{\text{at}}/2|E(t)|)$, which characterizes the instantaneous ionization probability $W(t)$. It is clearly seen, that the polar asymmetry of the distribution of residual velocity for the electrons appears. In particular, $\langle V_x \rangle$ is positive in that particular example.

It is very easy to calculate by computer the velocity distribution, $\langle v_i \rangle$, $\langle v_i v_k \rangle$ etc. using the instantaneous probability (6). Moreover, it is not difficult to take into account the change of the pulse envelopes in time and in space near the focal waist. That is the reason why we will not make any attempt in this short paper to present any analytical expressions for the asymmetry effects. It is evident, that polar asymmetry vanishes for a monochromatic field, i.e. in the cases when either $\boldsymbol{E}_1 = 0$ or $\boldsymbol{E}_2 = 0$. Contrary to intuition, the field

$$\boldsymbol{E}(t) = E\boldsymbol{e}_x(\cos \omega t + \cos 2\omega t) \qquad (12)$$

shown in figure 2 for $E = 0.01E_{\text{at}}$, also gives no polar asymmetry in the approximation of $W(t)$ from (6). Moreover, for $|\boldsymbol{E}| \ll E_{\text{at}}$ the residual velocity will be almost zero for the field (12).

It is curious, that the field with orthogonal linear polarizations,

$$\boldsymbol{E}(t) = \boldsymbol{e}_x \cos \omega t + \boldsymbol{e}_y \cos(2\omega t + \phi) \qquad (13)$$

produces the distribution of residual velocities with $\langle v_x \rangle = 0$, but $\langle v_y \rangle \neq 0$.

The most characteristic feature of the polar asymmetry under the discussion is its periodic dependence on the phase shift ϕ between the fundamental field and its second harmonic. Figure 3 shows the dependence of $\sqrt{\langle v_x^2 \rangle}$ and $\langle v_x \rangle$ on the phase ϕ for the field $\boldsymbol{E}(t) = E\boldsymbol{e}_x[\cos \omega t + \cos(2\omega t + \phi)]$ and $E/E_{\text{at}} = 0.01$.

It is worth mentioning that contrary to intuition the polar asymmetry appears even in the case when both waves, $\boldsymbol{E}_1$ and $\boldsymbol{E}_2$, are circularly polarized.

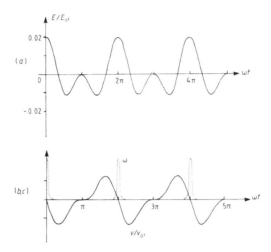

Figure 2. Time dependence of the values analogous to figure 1, but for the phase shift $\phi = 0$. It is clear, that the field possesses polar asymmetry, $\langle E_x^3 \rangle \neq 0$, but this field does not give the polar asymmetry of the electrons by tunnel ionization.

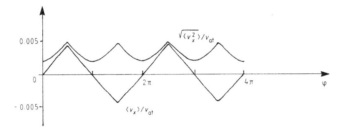

Figure 3. The dependence of the average residual velocity of the electrons $\langle v_x \rangle$ and $\sqrt{\langle v_x^2 \rangle}$ on the phase shift ϕ for the field $\boldsymbol{E} = E\boldsymbol{e}_x[\cos \omega t + \cos(2\omega t + \phi)]$ and $E/E_{\mathrm{at}} = 0.01$.

The last comment is connected with the possibility of using intersecting optical beams. For monochromatic radiation this results in an interference (or maybe speckle) pattern. For the ideally coherent monochromatic field the polarization at any given point is of an elliptical type with some certain plane of that ellipse. However, for bichromatic field the planes of the corresponding ellipses generally need not be coincident. This is the case where the three-dimensional nature of the electric-field vector in light waves may be revealed.

In conclusion, we have shown that the residual velocity for the tunnel ionization by a field with $\langle E^3 \rangle \neq 0$ possesses polar asymmetry. Contrary to intuition, the maximum of the velocity asymmetry does not coincide with the maximum of $\langle E^3 \rangle$, but with the maximum of something like $\langle E^2(t) \int E(t') \, \mathrm{d}t' \rangle$. The physical reason for this is the time asymmetry of the detachment process in contrast to the attachment process.

Professor S A Akhmanov has been a teacher to both authors of the present paper. In the last years he was very interested in the problem of ultrastrong optical fields, which would be able to detach an electron not only from an atom or an ion, but from a virtual electron–positron pair in vacuum. It is a great honour for us to contribute this paper to the book dedicated to his memory.

References

[1] Keldysh L V 1965 *Sov. Phys.–JETP* **20** 1307
[2] Reiss H R 1980 *Phys. Rev.* A **22** 1786
[3] Landau L D and Lifshitz E M 1987 *Quantum Mechanics (Nonrelativistic theory)* (Oxford: Pergamon)
[4] Corcum P 1991 Multiphonon Ionization from a Plasma Perspective *XIV Int. Conf. on Coherent and Nonlinear Optics (Leningrad, 1991)*
[5] Baranova N B and Zel'dovich B Ya 1991 *JOSA* B **8** 27

Hot carriers in semiconductors studied via picosecond optical nonlinearities in the infrared

T. Elsaesser and W. Kaiser

Physik Department E 11, Technische Universität München, D-8046 Garching, Germany

1.   Introduction

During the past four decades, hot carriers in semiconductors have been the subject of numerous experimental as well as theoretical studies. Early experiments have concentrated on the electric field dependence of charge transport, measuring macroscopic parameters such as the electrical conductivity. More recently, detailed insight into the microscopic scattering dynamics of hot carriers has been gained by optical spectroscopy (see, e.g., Shah 1989, Göbel 1990, Kash 1991). Photoexcitation of electrons and holes leads to strong changes of the linear optical properties of the material and - in many cases - to substantial optical nonlinearities. Nonlinear changes of the near-bandgap absorption due to excitonic effects and/or band-filling by non-equilibrium free electrons and holes were found in bulk III-V (e.g. GaAs, InSb) and II-VI (CdS/Se, ZnS) compounds and in layered quasi-two-dimensional heterostructures. In the spectral range below the bandgap, hot carriers created by intraband or inter-valence-band excitation give rise to large third-order nonlinearities.

Hot carriers frequently exhibit fast relaxation that is governed by pico- or subpicosecond time constants of different scattering processes. Ultrashort laser pulses are required for time resolved investigations of nonlinear optical effects which give direct access to the underlying scattering dynamics of the carriers. In most experiments with femto- or picosecond time resolution, relaxation phenomena of electron-hole plasma were monitored via the time evolution of interband absorption or luminescence. Here the relaxation processes of both electrons and holes determine the transient behavior. Experiments with a single type of carriers, e.g. in doped semiconductors, provide specific information on their relaxation mechanisms.

In this article, we discuss picosecond infrared studies where hot carriers are investigated via nonlinear changes of absorption or of the refractive index. Population changes in the valence and conduction bands of the different materials are induced by the incident pump pulses and the subsequent relaxation kinetics of the carriers is monitored by weak probe pulses in spectrally and temporally measurements. The article is organized as follows. We first give a brief account on the generation of pico- and subpicosecond pulses which are tunable over a wide range in the mid-infrared (section 2). Nonlinear changes of the inter-valence band absorption and optical gain due to hot holes in p-type germanium are discussed in section 3. Section 4 is devoted to hot electrons and holes in quasi two dimensional $Ga_{0.47}In_{0.53}As/Al_{0.48}In_{0.52}As$ multiple quantum well (MQW) structures. We report on inter-subband scattering and absorption of hot electrons, on band-filling by hot electron-hole plasma, and on excitonic absorption changes in higher subband systems.

2.   Generation of tunable pico- and subpicosecond pulses in the infrared

Intense pico- or subpicosecond pulses tunable over a wide wavelength range are required for infrared studies of ultrafast processes in semiconductors. Lasers emitting in the infrared (Mollenauer 1984, Ippen 1989, Zhu 1990, Islam 1989, Polland 1983, Elsaesser 1984a, Beaud 1986, Roskos 1986) and techniques of nonlinear frequency conversion (Akhmanov 1968a,b, Laubereau 1974, Seilmeier 1978, Elsaesser 1984b, 1985, Bakker 1991, Bareika 1981, Edelstein 1989, Laenen 1990, Moore 1987, Jedju 1988) provide pulses in the wavelength range between 1 μm and 15 μm. In the following, we briefly describe the infrared sources used in our present experiments.

Picosecond infrared pulses are generated with the setup depicted schematically in Figure 1 (Elsaesser 1985). Single picosecond pulses of a peak intensity in the range of several

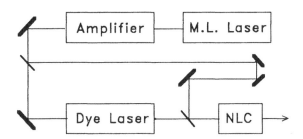

Fig. 1   Generation scheme for tunable pico- and subpicosecond pulses in the mid-infrared. Ultrashort pulses of high intensity are generated in a mode-locked laser oscillator in conjunction with an amplifier. Part of the amplifier output pumps a dye laser emitting at longer wavelengths. The output of the dye laser and the remainder of the amplified pulses are mixed in a nonlinear crystal (NLC) to generate the difference frequency.

GW/cm^2 are produced with active-passively mode-locked Nd:YAG or Nd:glass oscillators in conjunction with a single pulse selector and an amplifier. The pulses at a wavelength around 1 μm have a duration of 4 ps and 20 ps for the Nd:glass and the Nd:YAG laser, respectively. Two infrared dye lasers are pumped in a traveling wave geometry with the Nd:laser output, emitting pulses tunable between 1.1 and 1.4 μm with a narrow bandwidth of 7 cm^{-1} ($\simeq$ 1 meV) (Elsaesser 1984a). The pulses from the dye lasers have an energy of up to 50 μJ (peak intensity 50 MW/cm^2) and a duration close to that of the pumping pulse. The pulses from the two dye lasers are synchronized with a temporal jitter of less than one tenth of the pulse duration. Thus experiments with pump and probe pulses independently tunable in the near infrared can be performed with high time resolution.

These near infrared pulses are down-converted to longer wavelengths by parametric frequency conversion (Elsaesser 1985). The dye laser output and part of the pulse from the

Nd:laser are mixed in a nonlinear $AgGaS_2$ crystal to generate the difference frequency. The mid-infrared pulses are tunable over a wide range between 3.5 and 10 μm by changing the wavelength of the dye laser output and adjusting the phase-matching angle of the nonlinear crystal. In Figure 2, the photon conversion efficiency of the parametric process is plotted as a function of the mid-infrared wavelength for a system pumped by a Nd:YAG laser. The photon conversion efficiency represents the fraction of incoming photons ($\simeq 10^{16}$ per pulse at 1.064 μm) which is converted to photons at the respective wavelength. The nonlinear frequency mixing shows high conversion efficiencies of up to several percent over a wide wavelength range. The drop of conversion at short wavelengths is due to the decrease in the dye laser output at the corresponding wavelengths in the near infrared. At long wavelengths, the effective nonlinearity for the parametric process and thus the conversion efficiency decreases. The duration of the mid-infrared pulses was measured by recording the cross-correlation with the second harmonic of the Nd:laser. Pulse durations of 2 ps and 8 ps at 5 μm were found with Nd:glass and Nd:YAG pump lasers, respectively. Two independently tunable mid-infrared pulses were generated by parallel operation of two traveling-wave dye lasers and of two down-converters. The jitter between the two infrared pulses is less than 0.5 ps for pumping by a Nd:glass laser and less than 2 ps for the Nd:YAG laser.

Recently, a similar system was designed to generate mid-infrared pulses of subpicosecond duration (Elsaesser 1991a). First, 100 fs pulses at 620 nm were generated in a colliding-pulse-modelocked dye laser and a six-pass dye amplifier pumped by a copper vapour laser with a repetition rate of 8 kHz. Red-shifted femtosecond pulses around

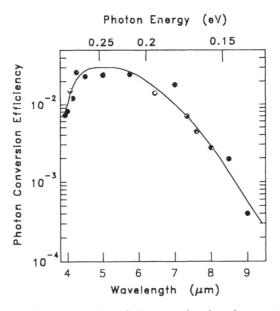

Fig. 2 Parametric down-conversion of picosecond pulses from a traveling-wave dye laser in a nonlinear $AgGaS_2$ crystal. The photon conversion efficiency is plotted as a function of the wavelength of the generated infrared pulses which are tunable between 3.5 and 10 μm.

700 nm were produced in a single-cell traveling wave laser pumped by the pulses at 620 nm (Hebling 1989). The output of the traveling wave cell and a second part of the pulse at 620 nm generate the difference frequency in a thin LiIO₃ crystal (thickness 0.3 cm). The spectrum of the mid-infrared pulses is presented in Figure 3 (a). For a center wavelength of 5.2 μm, the spectrum extends from 4.5 to 5.5 μm with a bandwidth of 180 cm^{-1}. Figure 3 (b) gives the cross-correlation of the mid-infrared and the 620 nm pulses. A duration of approximately 400 fs is derived for the infrared pulses. The energy per pulse has a value of 10 nJ corresponding to 2.5×10^{11} infrared photons.

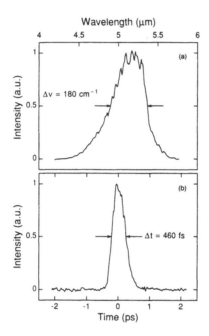

Fig. 3 (a) Spectrum of subpicosecond infrared pulses generated by difference frequency mixing in a thin LiIO₃ crystal. The decrease of intensity at short wavelengths is due to a decrease of the output of the traveling wave dye laser. The long wavelength part is determined by the absorption of the nonlinear mixing crystal. (b) Cross-correlation of the infrared pulses and femtosecond pulses at 620 nm. The sum frequency signal is plotted as a function of delay time. Deconvolution gives a duration of the infrared pulses of approximately 400 fs.

The experiments reported in the following sections are based on the pump and probe technique, where a first intense pump pulse excites the sample and the resulting change of transmission is monitored by tunable probe pulses of variable time delay.

3.   Hot holes in p-type germanium

The valence band structure of various semiconductors, e.g. silicon, germanium, and III-V compounds, consists of three bands : the heavy hole, the light hole, and the split-off band. For germanium, the band parameters necessary to describe effective masses around **k**=0, nonparabolicity and warping, are well known from bandstructure

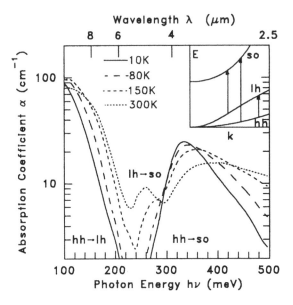

Fig. 4 Steady-state absorption spectra due to inter-valence band transitions in p-type germanium (doping density 7×10^{16} cm^{-3}). The valence band structure (hole energy versus k-vector) is depicted in the insert. The three inter-valence band transitions are indicated by arrows.

calculations and experimental studies (Kane 1956, Fawcett 1965). The strong, dipole-allowed transitions between the different valence bands occur at wavelengths longer than 2 μm, below the indirect band gap of germanium. In Figure 4, the steady-state infrared absorption of p-type Ge is depicted for four different lattice temperatures T_L (Kaiser 1953, Kahn 1955, Vasileva 1967). Three distinct bands are observed at T_L=300 K which

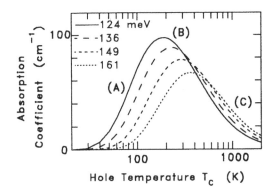

Fig. 5 Calculated inter-valence band absorption of the heavy-hole→light-hole transition for four different photon energies used in the picosecond experiments. The absorption coefficient α for a hole density of 7×10^{16} cm^{-3} is plotted as a function of the carrier temperature T_C.

are due to transitions from the heavy hole to the light hole (hh→lh) and to the split-off band (hh→so), and from light hole to split-off states (lh→so). In k-space, the valence band states giving rise to absorption in this spectral range are approximately 10 meV above the extrema of the heavy and light hole band at k=0. The shape and the strength of the three absorption bands is strongly influenced by the distribution function of holes in the different valence bands, resulting in a well-pronounced change of the spectra with temperature. For instance, the absorption of the lh→so transition disappears at low temperatures because of the depletion of the light hole band. As a result, the sample becomes transparent in this wavelength range.

In Figure 5, the hh→lh absorption is plotted as a function of the hole temperature for several fixed photon energies. The absorption coefficient exhibits a distinct maximum around room temperature. Heating of holes which are initially at 300 K (B) leads to a reduced absorption (C) due to the smaller population of the initial states with rising temperature. The situation is completely different when starting at low temperature (A). The steep Fermi tail at low temperature, i.e. the carrier freeze out, gives rise to a smaller initial absorption but leads to larger absorption values as a result of carrier heating. For very high temperatures, the thermal broadening of the hole distribution leads to a strongly reduced absorption that is even lower than the initial value.

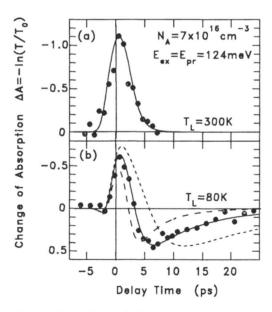

Fig. 6 Transient change of the heavy hole→light hole absorption after picosecond excitation at $E_{ex}=E_{pr}$=124 meV. The absorption change $\Delta A=-\ln(T/T_0)$ is plotted vs the time delay between pump and probe pulses (T_0,T : transmission before and after excitation). The data points were taken at lattice temperatures of (a) T_L=300 K and (b) 80 K (hole density 7×10^{16} cm^{-3}). The solid lines represent the result of a calculation with an optical deformation potential of $D_0 = 6.3 \times 10^8$ eV/cm. For comparison, the short and long-dashed lines are calculated with $D_0 = 5 \times 10^8$ and 9×10^8 eV/cm, respectively.

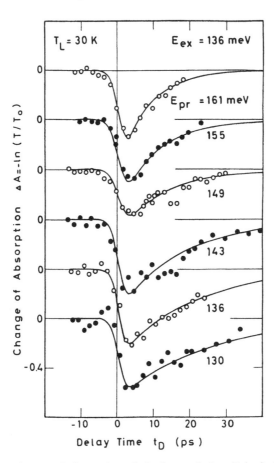

Fig. 7 Transient change of absorption of the heavy-hole→light-hole transition at six
different probe frequencies 130 meV$\leq E_{pr} \leq$161 meV after excitation at E_{ex}=136 meV
(lattice temperature T_L=30 K). The measured absorption change is plotted versus delay
time (points). The solid lines are calculated from a theoretical model discussed below.

We now present picosecond pump-probe studies of the transient inter-valence band absorption (Woerner 1990, 1991, 1992). Spectrally and temporally resolved experiments
in the mid-infrared reveal strong changes of the inter-valence band absorption due to
hot holes. The data give direct insight into the time evolution of hot hole distributions,
allowing a quantitative comparison with theoretical calculations of carrier scattering by
optical phonons. We first discuss picosecond absorption changes of the hh→lh transition. In a second type of experiment, carriers are excited from the heavy hole to the
split-off band (hh→so) and the transition from the light hole to the split-off band
(lh→so) is monitored. Holes are excited from the heavy hole to the light hole band
with intense pulses around 10 μm (E_{ex}=124 meV). In Figure 6, the absorption change
probed at the same spectral position is plotted as a function of delay time. At room
temperature (T_L=300 K), one observes a bleaching of the sample that rises within the
time resolution of the experiment and decays within the first 10 ps (Figure 6 a). At
lower lattice temperature (T_L=80 K), this fast bleaching is followed by an increase of

absorption which shows a slower relaxation on a time scale of several tens of picoseconds (Figure 6 b). The relative strength of bleaching and induced absorption depends on the intensity of the excitation pulses I_{ex}. The data of Figure 6 were measured with $I_{ex} \simeq 100$ MW/cm^2. For a reduction of I_{ex} by a factor of 10, the fast decrease of absorption disappears completely, whereas the induced absorption changes only slightly (Woerner 1990). Temporally and spectrally resolved data are presented in Figure 7 for $I_{ex} \simeq 20$ MW/cm^2. The different transients rise with the excitation pulse and recover slowly with decay times decreasing for higher photon energies of the probe pulses.

The following relaxation processes of hot holes are important for the transient absorption changes in the mid-infrared. At carrier concentrations exceeding 10^{16} cm^{-3}, carrier-carrier and inter-valence band scattering occur on a subpicosecond time scale establishing a quasi equilibrium distribution of hot heavy holes within our time resolution of 1 ps. The energy supplied by the exciting photons is redistributed among the carriers and very high carrier temperatures of up to 1000 K occur at very early times after excitation. The hot heavy holes subsequently cool down to the respective lattice temperature by emission of longitudinal optical phonons via the deformation potential, a process occurring on a time scale of several tens of picoseconds. Thus the picosecond changes of the hh→lh absorption which are directly monitored in our experiments, are governed by the transient temperature of hot heavy holes. For an initial carrier temperature of $T_C = T_L = 300$ K, the hh→lh absorption becomes smaller with increasing T_C, as was explained with the help of Figure 5. Correspondingly, a bleaching of the sample is found in the time resolved experiments (Figure 6 a). For an initial $T_C = 80$ K, the initial absorption coefficient is lower (c.f. Figure 5). At early delay times, when the hole gas is very hot, the absorption is even less than this small initial value. However, the bleaching at early delay times is followed by an absorption increase at later times, when the holes have temperatures of several hundred Kelvin and the transient absorption is higher than the starting value (Figure 6 b). At very low initial temperatures of the carriers, the transient absorption exceeds the original value for the whole temperature range covered during the cooling process, leading to an enhanced absorption at all delay times (Figure 7). For a fixed initial carrier temperature, excitation with lower intensities reduces the maximum carrier temperature at early delay times and therefore directly affects the relative amplitudes of bleaching and enhanced absorption.

Different periods of the picosecond time evolution of the carrier temperature T_C are monitored by the absorption changes at the different photon energies plotted in Figure 7. The absorption at high photon energies mainly gives information on the regime of high T_C, as is evident from the relationships plotted in Figure 5. Here fast cooling results in a rapid recovery of the transient absorption. At smaller photon energies, cooling of carriers at lower temperatures determines the signal, leading to a slower decay of the absorption change.

The solid lines drawn through the data points in Figures 6 and 7 are calculated for carrier cooling via optical phonon emission. The agreement between simulation and experiment is excellent for an optical deformation potential of $D_0 = 6.3 \times 10^8$ eV/cm which is in fair agreement with values derived from transport measurements (Reggiani 1976). The dashed lines in Figure 6 b which were calculated with values of $D_0 = 5 \times 10^8$ and 9×10^8 eV/cm deviate markedly from the measured points. Thus our data determine the absolute value of the deformation potential with an accuracy of ± 30 percent.

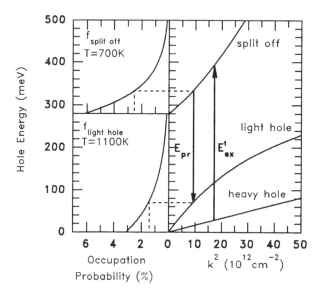

Fig. 8 Left hand side : Calculated distribution functions of light and split-off holes during picosecond excitation at E^1_{ex}=363 meV. At delay zero, a transient population inversion, i.e. optical gain, is found for the optically coupled states around 260 meV. Right hand side : Valence band structure of germanium. The hole energy is plotted as a function of k^2 (k : wavevector). The arrows indicate the relevant optical transitions.

We now discuss transitions involving the split-off band (Woerner 1991). On the right hand side of Figure 8 part of the valence bands of germanium is depicted and the transitions for excitation and probing are indicated. In Figure 9 a, experimental data for pump photons of E^1_{ex}= 363 meV ($\simeq 3\mu$m) and probing photons of E_{pr}= 263 meV ($\lambda \simeq 4.7\mu$m) are presented. During the excitation of the sample the transmission increases but decreases quite rapidly at the end of the pumping pulse. We recall that the sample is transparent at the probing wavelength prior to excitation. The enhanced transmission at early times corresponds to amplification of the probing pulse. For $\Delta T/T_0$=0.1 and a sample thickness of 0.02 cm one derives a substantial gain coefficient of 5 cm^{-1} from the data. It should be noted that gain in this wavelength range has never been anticipated or seen before. At later times, inter-valence band scattering and thermalization establish a hot hole distribution in the heavy and light hole bands. The hot carriers give rise to an enhanced lh→so absorption as shown in Figure 9 a for delay times longer than 3 ps. This absorption increase decays by carrier cooling within 10 ps.

For comparison, we performed a measurement with picosecond excitation of the hh→lh transition at E^2_{ex}=211 meV creating a hot carrier distribution in the heavy and light hole band. Probing the lh→so transition at 263 meV, we solely find an increase of absorption, as shown in Figure 9 b. The kinetics of this signal is close to the enhanced absorption observed at late delay times after excitation to the split-off band (c.f. Figure 9 a).

The optical gain at 263 meV (4.7μm) is favored by two facts : (i) The dipole matrix element of the so→lh transition is large (six times larger than for transitions between the heavy and light hole bands). (ii) The split-off band and the light hole band have identical slopes at those states which are optically coupled by the probe pulses at $E_{pr}=263$ meV. The joint density of states has a singularity in this region of the Brillouin zone. The singularity is lifted on account of the fast scattering processes, i.e. carrier-carrier and LO-deformation potential scattering. The first scattering process is expected to establish very rapidly (< 200 fs) a quasi-equilibrium distribution in the split-off band with an approximate temperature of 700 K (see Figure 8 left). Via the second scattering mechanism the holes in the split-off band return very rapidly to the two lower valence bands with a time constant of roughly 200 fs. The high excess energy of the light holes gives rise to a higher temperature of approximately 1100 K which results in a smaller occupation probability in the terminal states of the so→lh transition. In other words, during the pumping pulse a small quasi-stationary inversion (of approximately 0.01) between split-off and light hole band is realized explaining the observed gain at 4.7 μm.

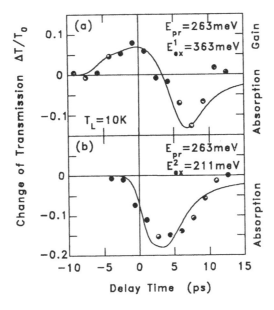

Fig. 9 Picosecond transmission changes of the light-hole→split-off transition at $E_{pr}=263$ meV after inter-valence band excitation. The transmission change is shown as a function of delay time. (a) Heavy holes excited to the split-off band at $E_{ex}^1=363$ meV give rise to optical gain. (b) Excitation of the heavy-hole→light-hole transition at $E_{ex}^2=211$ meV exclusively leads to an absorption increase.

Detailed measurements showed that the gain exists only over a narrow spectral range of 20 meV around 263 meV. When the split-off band is depleted of excited holes, the transient population of the light hole band leads to an absorption at 263 meV. The latter is seen in Figure 9, several picoseconds after delay zero. Cooling of the hot hole distribution reestablishes the original transparency of the sample in this wavelength range.

In conclusion, the relaxation processes of hot holes in p-type germanium were studied via nonlinear changes of the inter-valence band absorption. Our results demonstrate that the transient heavy-hole→light hole absorption is governed by the time dependent temperature of heavy holes. Strong picosecond excitation of heavy holes to the split-off band leads to optical gain on the split-off→light hole transition. The large amplitude and the fast recovery of the absorption changes on a time scale below 100 ps suggest application of p-type germanium for fast optical modulation in the infrared spectral range.

4.   Hot carriers in quasi-two-dimensional $Ga_{0.47}In_{0.53}As/Al_{0.48}In_{0.52}As$ multiple-quantum-well structures

In recent years, the optical properties of semiconductor quantum-well (QW) and superlattice (SL) structures have attracted much interest (for a review, see Schmitt-Rink 1989). The quasi two-dimensional character of electrons and holes in those systems results in valence and conduction subbands and in optical spectra substantially different from the corresponding bulk materials. Transitions from continuum states in the different valence subbands to the respective conduction subbands give rise to a step-like absorption spectrum reflecting the quasi-two-dimensional density of electronic states. Excitonic excitations result in strong absorption lines superimposed on the onset of the band-to band absorption. In the infrared spectral range, intense inter-subband absorption bands occur that are due to dipole-allowed transitions between two different subbands of electrons **or** holes. Novel nonlinear optical phenomena like the screening of the excitonic absorption by free carriers, the quantum-confined Stark effect, or large second order susceptibilities associated with inter-subband transitions in asymmetric QW's (Fejer 1989) have been found. Those effects are of relevance for optical switching devices.

The relaxation of hot carriers in QW and MQW structures made of III-V semiconductors has been the subject of numerous investigations (for a review, see Shah 1986). Ultrashort laser pulses have been used to study intra- and inter-subband scattering of electrons and holes, mainly in GaAs/AlGaAs quantum well structures. In the following, we present experimental studies of hot carrier relaxation in $Ga_{0.47}In_{0.53}As/Al_{0.48}In_{0.52}As$ MQW structures. This material system with a fundamental band gap in the wavelength range from 1.2 to 1.6 μm is of substantial interest for laser diodes, modulators and receivers in conjunction with optical (fiber) transmission lines. In contrast to the GaAs/AlGaAs and the $Ga_{0.47}In_{0.53}As/InP$ material systems, $Ga_{0.47}In_{0.53}As/Al_{0.48}In_{0.52}As$ shows a large conduction band offset of approximately 0.5 eV resulting in relatively deep potential wells for the strongly confined electrons (People 1983). This fact facilitates model calculations of scattering processes of hot carriers and - consequently - the interpretation of the experimental data.

4.1 Inter-subband processes of hot electrons

4.1.1 Stationary inter-subband absorption of electrons

Optical transitions from the n=1 to the n=2 subband of electrons or holes are dipole-allowed and give rise to intense absorption bands in the infrared (West 1985, Asai 1991). We investigated the inter-subband absorption of electrons in an n-type modulation-doped $Ga_{0.47}In_{0.53}As/Al_{0.48}In_{0.52}As$ MQW structure. The sample was grown by molecular beam epitaxy on a (100) InP substrate and consists of 50 $Ga_{0.47}In_{0.53}As$ QW of 8.2

nm thickness separated by 23 nm thick $Al_{0.48}In_{0.52}As$ barriers (Stolz 1987a,b). The electron density per QW has a value of 4.2×10^{11} cm^{-2}. The dipole moment of the subband transition is oriented along the stack axis of the MQW sample. The incoming infrared light must have a component of the electric field parallel to this axis to interact with the inter-subband transition dipole. For this reason, the polarization of the infrared light is adjusted parallel to the plane of incidence and the sample is oriented under Brewster angle. In Figure 10, the steady-state inter-subband absorption of our sample is plotted for a lattice temperature of T_L=10 K.

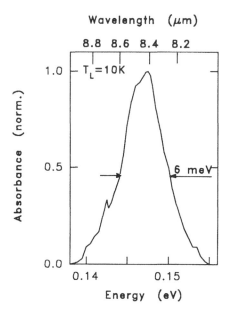

Fig. 10 Steady state inter-subband absorption spectrum of an n-type modulation-doped $Ga_{0.47}In_{0.53}As/Al_{0.48}In_{0.52}As$ multiple-quantum-well structure (lattice temperature T_L=10 K). The absorbance A=$-$ln(transmission) is plotted vs photon energy.

The spectrum peaks at a wavelength of 8.2 μm (0.148 eV) with a spectral width (FWHM) of 6 meV. It should be noted that the energy separation of the two subbands is substantially larger than the optical phonon energy in $Ga_{0.47}In_{0.53}As$ of 31 meV. The quantitative analysis of the absorption band gives an overall oscillator strength of f$\simeq$22, in agreement with theoretical calculations for this material system (Lobentanzer 1988).

The spectral position and the spectrally integrated strength of the inter-subband absorption change only slightly with temperature between 10 K and 300 K. In contrast, the linewidth increases substantially from a value of 6 meV at 10 K to 12 meV at 300 K. A similar behavior has been found in n-type modulation-doped $GaAs/Al_{0.3}Ga_{0.7}As$ MQW's (West 1985, von Allmen 1989). There are several mechanisms that contribute to the linewidth of the inter-subband transition:

(i) The finite lifetime τ_{sub} of the upper subband gives a spectral width proportional to $1/\tau_{sub}$. A minimum lifetime of $\tau_{sub} \simeq 500$ fs is estimated if the total linewidth is attributed to this contribution.

(ii) Alloy and well thickness fluctuations modify the energy position of the subbands and result in a distribution of transition energies (inhomogeneous contribution to the line width).

(iii) The different in-plane effective masses and nonparabolicities of the optically coupled subbands result in a k-dependent energy of the intersubband transition. At higher temperatures where the n=1 states populated by electrons cover a wide range in k-space, the overall absorption line represents a superposition of components with different transition energies.

In general, the three effects contribute to the linewidth observed in the steady state spectrum, i.e. a stationary measurement gives no detailed information on the relative strength of the different mechanisms. In the following, we discuss picosecond infrared studies which permit a quantitative characterization of the subband lifetimes and of the broadening due to the subband dispersion.

4.1.2 Picosecond studies of inter-subband scattering

Inter-subband scattering in $Ga_{0.47}In_{0.53}As/Al_{0.48}In_{0.52}As$ MQW's was investigated in experiments with picosecond infrared pulses applying several different pump-probe schemes (Bäuerle 1988). In the first type of measurement, electrons are excited from the n=1 to the n=2 subband by an intense infrared pulse resonant to the inter-subband absorption plotted in Figure 10. The probe pulses which are tunable in the near infrared between 0.92 an 1.05 eV monitor changes of the interband absorption from the n=1 or n=2 valence bands to the respective conduction bands. The different transitions are readily seen in Figure 11 where the steady-state interband absorption of our sample in the near-infrared is plotted (lattice temperature T_L=10 K). Two distinct steps of comparable strength are observed which are, respectively, due to transitions between the n=1 and n=2 valence and conduction subbands.

Excitation of electrons to the n=2 subband by the infrared pump pulse results in a change of the electron distribution function which is monitored via changes of the

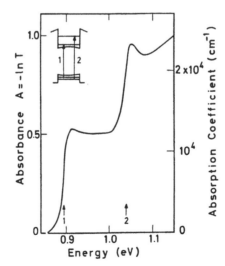

Fig. 11   Interband absorption spectrum of the n-doped $Ga_{0.47}In_{0.53}As/Al_{0.48}In_{0.52}As$ MQW sample. The transitions are depicted schematically in the inset.

interband absorption. A decrease of absorption, i.e. bleaching, is expected if the conduction band states probed by the interband transition are transiently populated. A measurement of this absorption change as a function of delay time between pump and probe gives direct insight in the scattering kinetics of the electrons. The carrier density is constant in this pump-probe scheme, i.e. the excitation pulses exclusively redistribute the electrons present by doping.

In a first experiment, the photon energy of the near-infrared pulses has a value of 1.033 eV, i.e. states near the minimum of the n=2 conduction band and high-lying states in the n=1 subband are probed (transition 2 in Figure 11). Even for excitation of more than 50 percent of the electrons by the pump pulse at 0.148 eV, a transient decrease of absorption due to accumulation of electrons in the n=2 subband could not be detected. This finding suggests a very fast depopulation of the higher subband. Considering the pump intensity and the sensitivity of the experiment, we estimate an upper limit of the lifetime of 3 ps.

In a second measurement, the probe pulses at 0.92 eV monitor changes of the absorption from the n=1 valence to conduction band (transition 1 in Figure 11). The conduction band states probed at 0.92 eV lie above the initial Fermi level of the electrons, i.e. are unoccupied in the non-excited sample held at a lattice temperature of T_L=10 K.

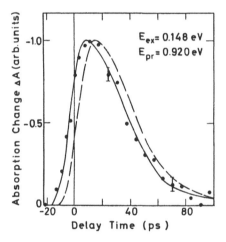

Fig. 12 Time dependent absorption change $\Delta A = -\ln(T/T_0)$ of the MQW sample at a probe energy of E_{pr}=0.92 eV, after excitation of electrons to the n=2 subband by pump pulses at E_{ex}=0.148 eV (points; T, T_0 : transmission of the sample with and without excitation). The solid and the dahed line are calculated from a theoretical model.

In Figure 12, the measured decrease of absorption ΔA is plotted as a function of the time delay between pump and probe pulses. A bleaching of the sample is found which rises with the excitation pulse and decays subsequently on a time scale of 80 ps. Approximately 50 percent of the n=1 electrons are excited by the pump pulse at 0.148 eV, resulting in a maximum absorption decrease of $\Delta A \simeq -0.1$. The very rapid rise of the bleaching demonstrates that the electrons excited to the n=2 subband are scattered back to the n=1 band within the time resolution of our experiment of 3 ps. After this initial relaxation, the electrons thermalize to a hot Fermi distribution by carrier-carrier and carrier-LO phonon scattering on a subpicosecond time scale (Knox 1986, Elsaesser 1991b). Each back-scattered electron supplies an excess energy close to the photon energy of the pump pulse of 0.148 eV to the total energy of the electron distribution. As a result, a Fermi distribution of a temperature of several hundred Kelvin is established in the n=1 subband. The electronic states probed at 0.92 eV are located on the high-energy tail of the distribution function. The population of these states results in a decrease of absorption since the final states of the interband transition are partly blocked. Cooling of the electron distribution leads to the depopulation of those states and a concommittant decay of the absorption change.

The solid and the dashed line in Figure 12 were calculated from a theoretical model of carrier scattering. In the first case, we assume inter-subband scattering of electrons by LO phonon emission with a time constant of approximately 1 ps, i.e. shorter than the time resolution of the experiment. The rise of the calculated signal is in very good agreement with the instantaneous onset of the data points. At later delay times, intraband cooling of the n=1 electrons by emission of LO phonons reduces the calculated absorption change. A more detailed analysis of the latter process which is discussed in (Bäuerle 1988) reveals reduced energy loss rates of the hot electrons in the n=1 subband, most probably due to the generation of an excess population of ('hot') LO phonons (Lobentanzer 1989a, Leo 1988). The dashed line in Figure 12 which does not account for the experimental results represents the absorption change calculated with a intersubband scattering time of 5 ps. In particular, the onset of the theoretical curve is delayed relative to the rise of the data points. This fact demonstrates that the subband lifetime in our $Ga_{0.47}In_{0.53}As/Al_{0.48}In_{0.52}As$ is considerably shorter than 5 ps.

Information on inter-subband scattering of electrons is also gained from a different type of experiment where excess carriers are created by interband excitation (Elsaesser 1989). Here we compare two different experimental situations :

(i) Hot electron-hole pairs in the n=1 conduction and valence subbands are excited by a picosecond near-infrared pulse at 0.99 eV and the resulting change of interband absorption is monitored by probe pulses tunable between 0.92 eV and 1.05 eV. In Figure 13 a, the change of absorption ΔA at a delay time of 6 ps is plotted versus the photon energy of the probe pulses for an excitation density of $(1.5 \pm 0.5) \times 10^{12}$ cm^{-2}. The absolute signal around 0.92 eV has a value of $\Delta A=0.4$, corresponding to a decrease of absorption by $\Delta\alpha = 10000$ cm^{-1}, i.e. the interband absorption with $\alpha_0 \simeq 12000$ cm^{-1} bleaches nearly completely. At higher photon energies, the absorption change decreases due to the smaller population of the higher lying states in the n=1 subbands.

(ii) Tuning the excitation pulses to a higher photon energy of $E_{ex}=1.05$ eV, one simultaneously creates electron-hole pairs in the n=1 and n=2 subbands. The transition from the n=2 valence to n=2 conduction band contributes roughly 50 percent to the total absorption at this photon energy and thus approximately half of the excited electrons are promoted to the n=2 conduction band (total excitation density 2×10^{12} cm^{-2}). The transient spectrum of the sample recorded at a delay time of 6 ps, i.e. at the end of the excitation pulse, is plotted in Figure 13 b. It is important to note that the data points give a negligible absorption change around the bandgap of the n=2 transition at 1.03 eV, i.e. there is no measurable accumulation of electrons in the higher conduction subband. This observation gives independent evidence of the very rapid depopulation of these states by intersubband scattering within the time resolution of the experiment of 3 ps.

The shape of the absorption spectrum plotted in Figure 13 b is very similar to that of Figure 13 a where only n=1 carriers were excited. In both cases, the excited electron-hole pairs relax to a hot Fermi distribution on a subpicosecond time scale, i.e. within the time resolution of the experiment. In the simplest approximation of parabolic subbands, the absorption change $\Delta A(E)$ is given by

$$\Delta A(E) = -\alpha_0(E)nd(f_e[M_e(E - E_G) - E_{Fe}] + f_h[M_h(E - E_G) - E_{Fh}]) \qquad (1)$$

where $\alpha_0(E)$ is the absorption coefficient of the unexcited MQW structure at the photon energy E (E_G: bandgap between the n=1 valence and conduction subband; n=50: number of QW's; d: QW thickness); E_{Fe} and E_{Fh} are the quasi-Fermi levels of electrons and holes, respectively. The reduced mass factors M_e and M_h are given by $M_{e,h} = m_{h,e}/(m_e + m_h)$ with the effective masses $m_e = 0.04m_0$ and $m_h = 0.38m_0$

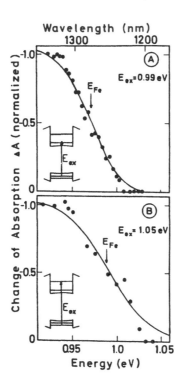

Figure 13   Transient bleaching of the interband absorption for two excitation conditions (points). The change of absorption ΔA after a delay time of 6 ps is plotted vs the photon energy of the probe pulses. (a) Hot carriers are generated in the n=1 valence and conduction bands by excitation at E_{ex}=0.99 eV (excitation density 1.5×10^{12} cm^{-2}). Solid line : calculated spectrum for a carrier temperature of 200 K. (b) Transient spectrum after excitation at 1.05 eV where approximately 50 percent of the carriers are excited in to the n=2 subbbands (total excitation density 2×10^{12} cm^{-2}). An absorption decrease due to an accumulation of electrons in the n=2 conduction band is not observed. Solid line : calculated spectrum for a temperature of the n=1 carriers of 300 K (lattice temperature 10 K).

of electrons and holes (m_0: free-electron mass). The transient spectra in Figure 13 are measured at photon energies between 0.92 and 1.02 eV where α_0 is essentially constant (c.f. Figure 11). Consequently, the shape of the spectra is determined by the sum of the distribution functions of electrons and holes, $f_e + f_h$, depending on the density and temperature of the carriers. The quantitative evaluation of equation (1) shows that bandfilling by holes contributes less than 10 percent to the total absorption change in the range of photon energies plotted in Figure 13. The measured spectra are governed by the high energy tail of the electron distribution in the n=1 conduction subband. The solid lines in Figure 13 a and b which account well for the experimental data, represent spectra calculated from equation (1) for electron temperatures of 200 K and 300 K, respectively.

The cooling of hot carriers determines the kinetics of the absorption changes for the first 100 ps, similar to the data presented in Figure 12. For a more detailed discussion of those processes, the reader is referred to (Lobentanzer 1989a, Elsaesser 1989).

In conclusion we may state: Our different experiments give an upper limit of the inter-subband scattering time of electrons in $Ga_{0.47}In_{0.53}As/Al_{0.48}In_{0.52}As$ MQW's of approximately 3 ps. A lower limit of 0.5 ps is estimated from the linewidth of the inter-subband absorption spectrum in Figure 10, as explained above. These numbers compare well with theoretical estimates of the inter-subband scattering times for electrons highly confined to deep potential wells (Ridley 1987, Riddoch 1983). For an energy separation of the subbands larger than the LO phonon energy of 31 meV, emission of LO phonons represents the dominant scattering process, resulting in a lifetime of the n=2 subband of approxiamately 1.5 ps. Recent studies of inter-subband scattering in GaAs/AlGaAs MQW's of similar QW width give an upper limit of 1 ps for the scattering time of electrons which are well confined to the QW's (Tatham 1989, Jain 1989). In modulation-doped GaAs/AlGaAs structures where a transfer of electrons to the potential minima in the barriers can occur, substantially longer time constants of the repopulation of the n=1 conduction subband have been found (Seilmeier 1987).

4.1.3 Transient inter-subband absorption of hot electrons

The results reported in section 4.1.2. demonstrate that the rapid inter-subband scattering in our $Ga_{0.47}In_{0.53}As/Al_{0.48}In_{0.52}As$ MQW structure creates a hot Fermi distribution of electrons in the n=1 conduction subband. The elevated carrier temperature gives rise to drastic changes of the interband absorption (c.f. Fig. 13). Changes of the inter-subband absorption spectrum were observed as well. Those experimental data are discussed next.

The following pump-probe scheme is applied in the picosecond measurements : Hot electron-hole pairs are generated with a density of 2×10^{11} cm^{-2} by interband excitation with pulses at 1.17 eV (from a mode-locked Nd:YAG laser). Within 3 ps, the electrons relax to the n=1 conduction subband, thermalize with the cold electrons present by doping (density 4×10^{11} cm^{-2}), and form a hot distribution of n=1 electrons. The transient (n=1)→(n=2) inter-subband absorption of the hot electrons is monitored by delayed probe pulses which are tunable in the range of the absorption line. The solid line and the points in Figure 14 represent the normalized steady-state inter-subband absorption line recorded on a infrared spectrophotometer and with picosecond pulses, respectively. After excitation of the hot electrons, we find a strongly broadened band with an increase of absorption both above and below the band center and a small shift

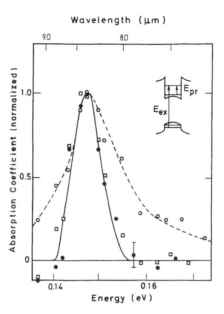

Fig. 14 Inter-subband absorption spectra of the n-doped $Ga_{0.47}In_{0.53}As/Al_{0.48}In_{0.52}As$ MQW structure. The normalized absorption coefficient is plotted as a function of the photon energy E_{pr}. The stationary spectrum of the sample was measured with an infrared spectrophotometer (solid line) and by picosecond pulses (points, lattice temperature T_L=10 K). The shape of the band changes after excitation of hot electrons by a picosecond pulse at E_{ex}=1.17 eV (the circles and the dashed line give results for a delay time of 20 ps). After 100 ps, the spectral broadening has disappeared (squares).

of the line center to slightly higher energies (circles). The halfwidth of the absorption line increases by roughly a factor of two. The broadened band was measured at a delay time of 20 ps, when the excited electrons exclusively populate the n=1 conduction subband. The line broadening disappears during the cooling of the hot carriers on a time scale of approximately 100 ps. The shape of the spectrum observed at that late delay time (squares) agrees with the steady state absorption. (The absolute extinction after 100 ps exceeds the stationary value by approximately 20 percent due to the higher electron density generated by the excitation pulse.)

As pointed out above, the excitation of electron-hole pairs leads to a strong bleaching of the interband absorption close to the band gap between the n=1 valence and conduction bands. The transient absorption spectra in this range give direct information on the momentary temperature of the electrons (c.f. Figure 13). With this technique, we find for the present experiment a transient electron temperature of 350 K at a delay time of 20 ps, where the broadened inter-subband spectrum of Figure 14 was observed.

Two mechanisms have to be considered to explain the reshaping of the inter-subband absorption line : (i) The lifetime of the upper (n=2) subband can change with carrier temperature, resulting in a modified lifetime broadening of the absorption spectrum. (ii) The hot electrons populate a broader interval in k-space where - due to the different effective masses and nonparabolicities of the two conduction subbands - the intersubband transition occurs at a different photon energy. The contribution of sample inhomogeneities to the total linewidth is expected to be independent of the carrier temperature.

To (i) : In our sample, interaction with LO phonons represents the main deactivation mechanism of the n=2 subband, resulting in a lifetime of less than 3 ps. The total electron-LO phonon scattering rate is given by the sum of phonon emission and absorption processes with rates proportional to N_q+1 and N_q, respectively. In equilibrium, the population probability N_q of the LO phonon branch is determined by the lattice temperature T_L. T_L has a constant value of 10 K in our present studies, corresponding to a negligible equilibrium population of the LO phonon states . However, the creation of excess LO phonons by the initial fast inter-subband scattering leads to a time dependent increase of N_q in the relevant q-interval, affecting the inter-subband scattering time. A quantitative estimate of this effect considering the density of excess carriers, the excess energy of the electron-hole pairs, and the q-interval relevant for inter-subband scattering, gives an additional lifetime broadening of the inter-subband absorption that is less than 10 percent of the stationary value of 6 meV at T_L=10 K. Therefore, this process cannot account for the broadening found in our experiments.

To (ii) : Next, the broadening of the inter-subband absorption related to the different k-dispersion of the n=1 and n=2 subbands is discussed. For a quantitative estimate, the quasi-two dimensional bandstructure of the QW's was calculated from a formalism described by (Braun 1985, Malcher 1986, Ekenberg 1987). The carrier motion in the layered system is separated into a first component parallel to the plane of the QW's (wavevector $k_{||}$) and a second quantized part perpendicular to the layers (wavevector k_z). The calculation of the in-plane bandstructure which is relevant for the spectral width of the inter-subband absorption, gives a higher effective mass for the n=2 subband than for the n=1 subband, i.e. the inter-subband energy spacing decreases with increasing wavevector $k_{||}$. The calculation also reveals a substantial nonparabolicity of the two subbands. With this bandstructure, we calculated the inter-subband absorption line by averaging the k-dependent energy of the inter-subband transition over the distribution of hot n=1 electrons. Here the temperature dependent renormalization of the n=1 band was explicitly taken into account. A detailed description of this theoretical model is given in (Bäuerle 1989). For an electron density and temperature of, respectively, 6.2×10^{11} cm^{-2} and 350 K, one finds a half width of the calculated spectrum of 13 meV, in agreement with the experimental data of Figure 14. In comparison to the stationary spectrum, the center of the calculated line is slightly shifted to higher photon energies. This finding is due to the renormalization of the n=1 subband by the excess carriers which are generated by the picosecond excitation pulse. The additional electrons shift the n=1 band to somewhat lower energies, whereas the energy of the n=2 states remains practically unchanged (Levenson 1988). As a result, the energy separation of the states optically coupled by the intersubband transition increases, corresponding to a small shift of the line center to higher energies.

The theoretical analysis demonstrates that the transient broadening of the inter-subband absorption spectrum is mainly due to the different dispersion of the two subbands in k-space. This effect also accounts for part of the temperature dependent broadening found in the steady-state measurements discussed in section 4.1.1. Our time resolved experiments show that the position and the width of inter-subband absorption spectra are a sensitive probe of the band structure and of many-body effects in quasi-two dimesional semiconductors.

4.2 Nonlinear absorption changes due to the n=2 excitons

Optical nonlinearities related to excitons in the lowest (n=1) subband system have been investigated in a number of experiments and theoretical calculations (Schmitt-Rink 1985,1989, Chemla 1985, Wegener 1989). Different new phenomena like the dc and ac quantum confined Stark effect or the screening of the excitonic absorption by free carriers have been observed with n=1 excitons. In the latter case, the magnitude of the absorption change is determined by the density of free carriers which screen the excitons by phase-space filling and exchange interactions (Schmitt-Rink 1985). The absorption change follows in time the population kinetics of the n=1 subbands. Consequently, the original excitonic absorption recovers with the relatively long recombination lifetime of the carriers. A considerably faster response of exciton screening should occur for higher subbands where inter-subband scattering results in a rapid depopulation of the electronic states. In addition, the relaxation processes of hot electrons and holes are expected to be relevant for this type of nonlinearity. In the following, we discuss picosecond infrared studies of the screening of n=2 excitons in an undoped $Ga_{0.47}In_{0.53}As/Al_{0.48}In_{0.52}As$ MQW structure (Lobentanzer 1989b).

The undoped MQW grown by molecular beam epitaxy on an InP substrate consists of 70 $Ga_{0.47}In_{0.53}As$ QW's of 8 nm width alternating with 11.4 nm thick $Al_{0.48}In_{0.52}As$ barriers. At a lattice temperature of $T_L = 10$ K, the n=2 excitonic resonance of this structure is located at a photon energy of 1.07 eV (Stolz 1987b). In the picosecond measurements, electron-hole pairs were excited simultaneously in the n=1 and n=2 subbands by a pumping pulse at $E_{ex}=1.1$ eV. The resulting change of the n=2 excitonic absorption is monitored by probe pulses tunable from 1.0 eV to 1.14 eV. The points in Figure 15 give the n=2 absorption edge of the unexcited sample as measured with the picosecond probe pulses. This spectrum that is identical with the steady-state spectrum

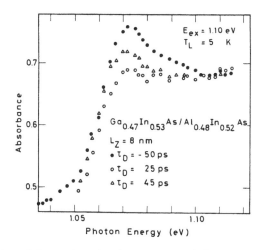

Fig. 15 Bleaching of the n=2 excitonic resonance of the undoped MQW, as measured with picosecond near-infrared pulses. The solid points represent the spectrum found prior to excitation (–50 ps) that is identical with the stationary spectrum. A strong bleaching of the excitonic lines is found 25 ps after excitation (circles). At a delay time of 45 ps, the excitonic absorption has partially recovered (triangles).

clearly shows the excitonic absorption with maximum at 1.07 eV. After excitation of approximately 2×10^{12} electron-hole pairs per cm^2, the excitonic lines bleach strongly. For a delay time of 25 ps (circles), a step-like spectrum is observed. The excitonic absorption recovers within tens of picoseconds as is obvious from the spectrum taken at a delay time of 45 ps (triangles). After 100 ps, we find again the original absorption.

The time dependent absorption change measured at a fixed photon energy of 1.07 eV shows a rapid rise within the time resolution of the experiment. The bleaching decays entirely within 100 ps. This kinetics is very similar to that of carrier cooling in the n=1 subbands discussed previously (Lobentanzer 1989a).

The following findings are relevant to explain the time dependent changes of the excitonic absorption : (i) The lifetime of electrons in the n=2 conduction subband has a value of less than 3 ps, i.e. the depopulation of the electronic n=2 states occurs on a much faster time scale than the recovery of the excitonic absorption. Thus the contribution of excess electrons in n=2 states to the observed bleaching of the excitonic resonances is small. (ii) After inter-subband scattering, the electrons form a hot carrier distribution in the n=1 conduction subband, which lies approximately 150 meV below the n=2 band. The transient carrier temperature reaches a maximum value of 700 K. For an electron density of 2×10^{12} cm^{-2}, the population of electronic n=1 states close to the minimum of the (higher) n=2 subband is very small, i.e. a repopulation of the n=2 band by back-scattering of hot electrons from the n=1 band can be neglected. (iii) The energy separation of the n=1 and n=2 valence subbands of approximately 30 meV is much smaller than for the conduction subbands. In the hot electron-hole plasma, a substantial fraction of holes populates the n=2 valence bands. Thus the bleaching of the n=2 excitons is caused by hot holes which populate states in the n=2 valence band necessary for the formation of the excitons, i.e. the hot holes fill the corresponding phase space. The recovery of the excitonic absorption follows in time the carrier cooling by which the n=2 valence bands are depopulated on a time scale of 100 ps.

In conclusion, the picosecond excitation of electron-hole plasma in the n=2 subband system results in a rapid bleaching of the n=2 excitonic absorption lines. The absorption change recovers within 100 ps, much faster than for the excitonic absorption between the n=1 subbands. The transient bleaching is due to hot holes in the n=2 valence bands and disappears by carrier cooling. The rapid recovery of the excitonic absorption lines may be useful for fast optical switching devices.

References

Akhmanov S.A., Chirkin A.S., Drabovich K.N.. Kovrigin A.I., Khokhlov R.V., Sukhorukov A.P. 1968a *IEEE J. Quant. Electron.* **4** 598

Akhmanov A.G., Akhmanov S.A., Khokhlov R.V., Kovrigin A.I., Piskarskas A.S., Sukhorukov A.P. 1968b *IEEE J. Quant. Electron.* **4** 828

Asai H., Kawamura Y. 1991 *Phys. Rev.* **B 43** 4748

Bäuerle R.J., Elsaesser T., Kaiser W., Lobentanzer H., Stolz W., Ploog K. 1988 *Phys. Rev.* **B 38** 4307

Bäuerle R.J., Elsaesser T., Lobentanzer H., Stolz W., Ploog K. 1989 *Phys. Rev.* **B 40** 10002

Bakker H.J., Kennis J.T.M., Kop H.J., Lagendijk A. 1991 *Opt. Commun.* **86** 58

Bareika B., Dikchyus G., Isyanova E.D., Piskarskas A., Sirutkaitis V. 1981

Sov. Tech. Phys. Lett. **6** 301

Beaud P., Zysset B., Schwarzenbach A.P., Weber H.P. 1986 *Opt. Lett.* **11** 24

Braun M., Rössler U. 1985 *J. Phys.* **C 18** 3365

Chemla D.S., Miller D.A.B. 1985 *J. Opt. Soc.Am.* **B 2** 1155

Edelstein D.C., Wachman E.S., Tang C.L. 1989 *Appl. Phys. Lett.* **54** 1728

Ekenberg U. 1987 *Phys. Rev.* **B 36** 6152

Elsaesser T., Polland H.J., Seilmeier A., Kaiser W. 1984a *IEEE J. Quant.Electron.* **20** 191

Elsaesser T., Seilmeier A., Kaiser W., Koidl P., Brandt G. 1984b *Appl. Phys. Lett.* **44** 383

Elsaesser T., Lobentanzer H., Seilmeier A. 1985 *Opt. Commun.* **53** 355

Elsaesser T., Bäuerle R.J., Kaiser W., Lobentanzer H., Stolz W., Ploog K. 1989 *Appl. Phys. Lett.* **54** 256

Elsaesser T., Nuss M.C. 1991a *Opt. Lett.* **16**, 411

Elsaesser T., Shah J., Rota L., Lugli P. 1991b *Phys. Rev. Lett.* **66** 1757

Fawcett W. 1965 *Proc. Phys. Soc.* **85** 931

Fejer M.M., Yoo S.J.B., Byer R.L., Harwit A., Harris, Jr. J.S. 1989 *Phys. Rev. Lett.* **62** 1041

Göbel E.O. 1990 *Adv. Solid State Phys.* **30** 269 and references therein

Hebling J., Kuhl J. 1989 *Opt. Lett.* **14** 278

Ippen E.P., Haus H.A., Liu L.Y. 1989 *J. Opt. Soc. Am.* **B 7** 1221

Islam M.N., Sunderman E.R., Soccolich C.E., Bar-Joseph I., Sauer N., Chang T.Y., Miller B.I. 1989 *IEEE J. Quant. Electron.* **25** 2454

Jain J.K., DasSarma S. 1989 *Phys. Rev. Lett.* **62** 2305

Jedju T.M., Rothberg L. 1988 *Opt. Lett.* **13** 961

Kahn A.H. 1955 *Phys. Rev.* **97** 1647

Kaiser W., Collins R.J., Fan H.Y. 1953 *Phys. Rev.* **91** 1380

Kane E.O. 1956 *J. Phys. Chem. Solids* **1** 82

Kash J.A., Tsang J.C. 1991 *Light Scattering in Solids VI*, Topics in Appl. Phys. Vol. 68, ed. M. Cardona, G. Güntherodt (Berlin : Springer) pp. 423-518

Knox W.H., Hirlimann C., Miller D.A.B., Shah J., Chemla D.S., Shank C.V. 1986 *Phys. Rev. Lett.* **56** 1191

Laenen R., Graener H., Laubereau A. 1990 *Opt. Lett.* **15** 971

Laubereau A., Greiter L., Kaiser W. 1974 *Appl. Phys. Lett.* **25** 87

Leo K., Rühle W.W., Ploog K. 1988 *Phys. Rev.* **B 38** 1947

Levenson J.A., Abram I., Raj R., Dolique G., Oudar J.L., Alexandre F. 1988 *Phys. Rev.* **B 38** 13443

Lobentanzer H., König W., Stolz W., Ploog K., Elsaesser T., Bäuerle R.J. 1988 *Appl. Phys. Lett.* **53** 571

Lobentanzer H., Stolz W., Nagle J., Ploog K. 1989a *Phys. Rev.* **B 39** 5234

Lobentanzer H., Stolz W., Ploog K., Bäuerle R.J., Elsaesser T. 1989b *Solid State Electron.* **32** 1875

Malcher F., Lommer G., Rössler, U. 1986 *Superlatt. Microstruct.* **2** 267

Mollenauer L.F., Stolen R.H. 1984 *Opt. Lett.* **9** 13

Moore D.S., Schmidt S.C. 1987 *Opt. Lett.* **12** 480

People R., Wecht K.W., Alavi K., Cho A.Y. 1983 *Appl. Phys. Lett.* **43** 118

Polland H.J., Elsaesser T., Seilmeier A., Kaiser W., Kussler M., Marx N.J., Sens B., Drexhage K.H. 1983 *Appl. Phys. B* **32** 53

Reggiani L. 1976 *J. Phys. Chem. Solids* **37** 293

Riddoch F.A., Ridley B.K. 1983 *J. Phys. C* **16** 6971

Ridley B.K. 1987 *Optical Properties of Narrow − Gap Low Dimensional Structures*, ed. C.M. Sotomayor Torres et al (New York : Plenum) p. 177-182

Roskos H., Opitz S., Seilmeier A., Kaiser W. 1986 *IEEE J. Quant. Electron.* **22** 697

Seilmeier A., Spanner K., Laubereau A., Kaiser W. 1978 *Opt. Commun.* **24** 237

Seilmeier A., Hübner H.J., Abstreiter G., Weimann G., Schlapp W. 1987 *Phys. Rev. Lett.* **59** 1345

Schmitt-Rink S., Chemla D.S., Miller D.A.B. 1985 *Phys. Rev.* **B 32** 6601

Schmitt-Rink S., Chemla D.S., Miller D.A.B 1989 *Adv. Physics* **38** 89

Shah J. 1986 *IEEE J. Quant. Electron.* **22** 1728

Shah J. 1989 *Solid State Electron.* **32** 1051 and references therein

Stolz W., Maan J.C., Altarelli M., Tapfer L., Ploog K. 1987a *Phys. Rev.* **B 36** 4301

Stolz W., Maan J.C., Altarelli M., Tapfer L., Ploog K. 1987b *Phys. Rev.* **B 36** 4310

Tatham M.C., Ryan J.F., Foxon C.T. 1989 *Phys. Rev. Lett.* **63** 1637

Vasil'eva M.A., Vorob'ev L.E., Stafeev V.I. 1967 *Sov. Phys. Semicond.* **1** 21

von Allmen, Berz M., Reinhart F.K., Harbeke G. 1989 *Superlatt. Microstruct.* **5** 259

Wegener M., Bar-Joseph I., Sucha G., Islam M.N., Sauer N., Chang T.Y., Chemla D.S. 1989 *Phys. Rev.* **B 39** 12974

West L.C., Eglash S.J. 1985 *Appl. Phys. Lett.* **46** 1156

Woerner M., Elsaesser T., Kaiser W. 1990 *Phys. Rev.* **B 41** 5463

Woerner M., Elsaesser T., Kaiser W. 1991 *Appl. Phys. Lett.* **59** 2004

Woerner M., Elsaesser T., Kaiser W. 1992 *Phys. Rev.* **B 45**, in press

Zhu X., Sibbett W. 1990 *J. Opt. Soc. Am.* **B 7** 2187

Non-local time-resolved spectroscopy tracking of polariton pulses in crystals

Ch. Flytzanis, G.M. Gale and F. Vallée

Laboratoire d'Optique Quantique du C.N.R.S.
Ecole Polytechnique, 91128 Palaiseau cédex, France

1 INTRODUCTION

The propagation, storage and redistribution of electromagnetic energy in crystals underlie many fundamental processes in ordered media. These processes, which are frequently non-linear in character, may also find important applications in optoelectronic devices. Many interesting and potentially useful phenomena in this domain can be delineated by their polariton-like nature, that is to say they consist of a coherent admixture of a dipole allowed material excitation and an electromagnetic mode (Huang (1951), Hopfield (1958, 1966)). The relevance of this type of excitation is not restricted to crystal optics (e.g. Born and Huang (1954), Knox (1963)) but equally plays a crucial role in other more complex systems and processes encountered in photo-biology and photo-chemistry (e.g. Förster (1948), Knox (1963)). It is thus of fundamental interest to have information concerning the dynamics of these quasiparticles and the processes that condition their coherence and lifetime. This information is essential for improving our understanding of certain irreversible processes and also for achieving the ultimate limits of device operation or energy transfer in molecular complexes.

For the relaxation of excitations at the molecular or atomic level the nonlinear optical techniques (e.g. Levenson (1982)) that exploit the different features of the laser, in particular its selectivity and resolution in the frequency or time domains, have provided a unique source of information and constitute the most powerful tools that we have presently at hand to selectively investigate these processes. In its simplest form the problem here reduces to that of the determination of the coherence and population lifetimes, T_2 and T_1 respectively, of independent localized two level systems ; under certain simplifying assumptions (e.g. Slichter (1980)) these enter the Bloch equations (Bloch (1946), Feynman et al (1957)) of the evolution of the system driven off equilibrium and interacting stochastically with its environment. In this picture the relaxation of the population and coherence occurs within the immediate environment of the molecule. In the case of vibrational motion, for instance, the local excite and probe techniques like the Coherent Anti-Stokes Raman Scattering (CARS) or its extensions and variants (e.g.

Laubereau and Kaiser (1978), Flytzanis (1990)) have provided crucial information about these processes by modelling the exictations as effectively independent localized systems and also introducing interactions among them and the bath.

If the two level systems actually interact and also form a periodic array these processes cannot be described in such simple terms. The eigenstates here are delocalized propagating modes with energies $\hbar\omega_\sigma$ ($\underline{k}$) that analytically depend on the wavevector $\underline{k}$, a continuous variable, and are characterized by a density of states $J = |\partial\omega_\sigma/\partial\underline{k}|$ and group velocity :

$$\underline{v}_g\,(\underline{k}) = \underline{\nabla}_k\omega_\sigma\,(\underline{k}) \tag{1}$$

which also measures the flatness of the dispersion relation ω_σ (k). For small values of v_g one essentially recovers a localized state picture. When the group velocity however attains large values, propagation effects, essentially escape of the excitation from an excited molecule and its resonant transfer to the other identical molecules of the periodic array, may interfere with intrinsic relaxation processes and any spatial disorder may substantially complicate the picture. Such effects are expected to be particularly conspicuous in polariton dynamics because their group velocity can be comparable to that of the light in the transparency range of the crystal.

Here we address precisely this situation and we demonstrate that short polariton pulses can be tracked and their spatio-temporal evolution in crystals can be directly followed with a nonlocal time resolved technique (Gale et al (1986), Vallée et al (1991)). Obviously the study of polariton propagation and relaxation has been considered previously but with indirect or insufficiently selective techniques. The present method and its variants possess a very wide flexibility and applicability in most diverse cases and allows an unambiguous separation of temporal and spatial features in the polariton dynamics and damping. The advantage of the non-local technique that we consider here and its comparison with other methods will be addressed below, after a presentation of the basic principles.

2   POLARITONS

We briefly recall the salients features of polariton modes (e.g. Mills and Burstein (1974)). We concentrate our attention on their dispersion relation and relaxation.

2.1 Dispersion

The propagation characteristics of an electromagnetic field in a crystal of dielectric and magnetic permeability tensors, $\underline{\varepsilon}$ ($\underline{k},\omega$) and $\underline{\mu}$ ($\underline{k},\omega$) respectively, is

$$\underline{k} \left[\underline{\underline{\mu}}^{-1}(\underline{k},\omega)(\underline{k}.\underline{E})\right] + \frac{\omega^2}{c^2} \, \underline{\underline{\epsilon}} \, (\underline{k},\omega) \, \underline{E} = 0 \qquad\qquad (2)$$

for which, in the case of an isotropic medium, the transverse part reduces to :

$$\frac{c^2 k^2}{\omega^2} = \epsilon \, (\underline{k},\omega) \, \mu \, (\underline{k},\omega) \qquad\qquad (3a)$$

In the frequency regions where both $\epsilon \, (\underline{k},\omega)$ and $\mu \, (\underline{k},\omega)$ are $\underline{k}$- and $\underline{\omega}$- independent equation (3a) reverts to the simple form :

$$\omega = ck / (\epsilon\mu)^{1/2} \qquad\qquad (3b)$$

and the phase and group velocity of the wave are equal. The situation changes dramatically when either ϵ or μ exhibit a strong dependence on ω or $\underline{k}$. This is mostly observed in the case of the long wavelength limit $k \to 0$ in a nonmagnetic cubic crystal close to an electric dipole allowed resonance that shows up in $\epsilon(\omega)$, for instance a polar phonon or exciton transition. Let us label by $\underline{u}$ the amplitude of such a material mode which in the presence of the electric field $\underline{E}$ satisfies the equation of motion :

$$\ddot{\underline{u}} + \omega_T^2 \, \underline{u} = \frac{e^*}{m} \underline{E} \qquad\qquad (4)$$

where ω_T is the frequency of the mode which we anticipate to be transverse and m and e* are its reduced mass and effective charge respectively (Szigeti (1949)). Then its contribution to the polarisation is :

$$\underline{P} = Ne^*\underline{u} = \frac{Ne^*}{m} \, \frac{1}{\omega_T^2 - \omega^2} \underline{E} \qquad\qquad (5)$$

where N is the mode number density. Then if ϵ_o is the global contribution to the dielectric constant at frequency ω from the other polarisation modes (only the ones with frequencies higher than ω_T contribute) then,

$$\epsilon \, (\omega) = \epsilon_o + \frac{\Omega_p^2}{\omega_T^2 - \omega^2} \qquad\qquad (6)$$

where $\Omega_p^2 = 4\pi Ne^{*2}/m$ and the dispersion relation becomes :

$$\frac{c^2k^2}{\omega^2} = \varepsilon(\omega) = \varepsilon_o + \frac{\Omega_p^2}{\omega_T^2 - \omega^2} \tag{7}$$

from which one obtains the solutions :

$$\omega_\pi^2(k) = \frac{1}{2}\left(\frac{c^2k^2}{\varepsilon_o} + \omega_T^2 + \Omega_p^2\right) \pm \frac{1}{2}\left[\left(\frac{c^2k^2}{\varepsilon_o} - \omega_T^2 - \Omega_p^2\right)^2 + \frac{4c^2k^2}{\varepsilon_o}\Omega_p^2\right]^{1/2} \tag{8}$$

for the eigenfrequencies of the polariton mode propagation ; thus for each value of k there are two frequencies that fall in the upper (+) and lower (-) polariton dispersion curves which are schematically depicted in Fig.1, where we also included the dispersionless nonpropagating longitudinal mode of frequency ω_L such that $\varepsilon(\omega_L) = 0$ or $\omega_L^2 = \omega_T^2 + \Omega_p^2 = \omega_s^2 \varepsilon_s / \varepsilon_o$, the Lyddane-Sachs-Teller relation ; for $\underline{k} = 0$ the upper polariton frequency coincides with this frequency while for the lower branch goes to zero if no other electric dipole resonances lie below ω_T. The form of the total energy U_T stored in the polariton is also quite instructive (Pelzer (1949)) as it is partly electromagnetic (U_E) and partly mechanical (U_M) :

$$U_T = U_E + U_M \approx \frac{1}{2}Nm(\omega^2 + \omega_T^2)u^2 + (\varepsilon_o + \varepsilon)\frac{E^2}{8\pi} \tag{9}$$

which propagates with group velocity :

$$v_g = v_p\left[1 + \frac{1}{2}\frac{\omega}{\varepsilon(\omega)}\frac{\partial\varepsilon(\omega)}{\partial\omega}\right]^{-1} \tag{10}$$

where $v_p = c/\sqrt{\varepsilon(\omega)}$ is the phase velocity ; in Fig. 2 we depict how the U_E and U_M vary with ω. The above picture suffices for the description of the phonon polaritons because m is large and the phonon dispersion is flat (e.g. Born and Huang (1954)). For excitons (e.g. Knox (1963), Rashba and Sturge (1982))with their much smaller values of m (four orders of magnitude smaller), the situation is more complex because of the intrinsic exciton motion which introduces a nonlocality, also termed spatial dispersion, and is present in both transverse and longitudinal modes and leads in particular to the existence of two polaritons with widely

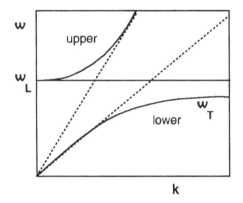

Fig 1. Dispersion curves for the upper and lower branches of a phonon polariton

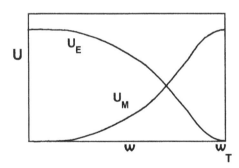

Fig 2. Variation of the electrical (U_E) and mechanical (U_M) components of the polariton energy as a function of frequency

different wavevectors but the same frequency. Polariton effects may also appear with more complex material excitations (bi-excitons, bi-phonons, plasmons, magnons) in interaction with electric dipole excitations to form dressed polaritons (e.g. Loudon and Haynes (1980)) and also in connection with resonances in the magnetic permeability $\mu(\omega)$ that also enters (1) and (2). Finally we wish to point out that there also exist surface polaritons (e.g. Agranovich (1982)) and their dispersion differs from that in three dimensions so that special provisions must be made to phase-match the two.

2.2 Damping

The previous outline overlooks the polariton damping brought in by their stochastic interaction with other degrees of freedom ; this is an intrinsic feature however and in a strict sense the polaritons are not genuine eigenmodes of the crystal. Because of the composite character of the polariton this interaction proceeds through both its electromagnetic and mechanical components and the loss of coherence and energy cannot be reduced within the framework that prevails for localized two-level systems. Since the two parts are coupled coherently any damping mechanism will indiscriminately affect both parts and there is no clear-cut manner to locate the damping in either of them ; however under certain plausible assumptions one may differentiate their origin.

In an ordered crystal the damping of polaritons is mainly due to the finite lifetime of their mechanical component (e.g. Loudon (1970), Barker and Loudon (1972)) which is coupled to other modes through electrical and mechanical anharmonicities (Szigeti (1955), Flytzanis (1972,1975)) ; it is also related to the absorption losses in the reststrahlen region. If this is introduced only through a damping term $\gamma \dot{u}$ in (4) then the dielectric constant becomes :

$$\varepsilon(\omega) = \varepsilon_o + \frac{\Omega_p^2}{\omega_T^2 - \omega^2 + i\omega\gamma} \tag{11}$$

and the polariton damping is given (Loudon (1970), Mills and Burstein (1974)) by :

$$\Gamma_A = S_M \frac{\omega}{\omega_T} \gamma \tag{12}$$

where S_M is the material component strength in the polariton. Actually in a more rigorous description one obtains (Szigeti (1955), Maraddudin and Wallis (1962), Flytzanis (1972)) :

$$\varepsilon\,(\omega)\,=\,\varepsilon_o + \frac{\Omega_p^2}{\omega_T^2 - \omega^2 - 2\omega_T\,\Pi_R\,(\omega) - 2i\omega\,\Pi_I\,(\omega)} \qquad (13)$$

instead of (11) where Π_R and Π_I are the real and imaginary parts of phonon self energy implying non-Lorentzian damping that is strongly frequency dependent but only mildly temperature dependent ; the latter essentially enters through the Bose-Einstein statistics in the density of states.

In imperfect crystals along with the previous damping, which can be termed temporal disorder, an additional damping is introduced through the spatial disorder of the dielectric constant which mostly affects the electromagnetic component of the polariton. Its effect is similar to the one produced on the light scattered by a medium with a spatially random dielectric constant (e.g. Ishimaru (1978), Loudon and Haynes (1980)); light suffers attenuation in the forward propagation mode and a loss of coherence as a consequence to its scattering into other modes. This damping depends on the amplitude of the fluctuations of the random dielectric constant. Neglecting multiple scattering (see section 7 below) to a first approximation we may assume (Gale et al (1986)) that the two damping processes, due to temporal and spatial disorders respectively, are statistically uncorrelated and hence we may write for the total damping rate of the polariton :

$$\Gamma_\Pi \,=\, \Gamma_A + \Gamma_D \qquad (14)$$

where Γ_D depends on the strength of the spatial fluctuations of the dielectric constant and is only mildly frequency dependent as expected from density of states considerations.

It is quite evident that the damping mechanisms play a crucial role in the polariton behavior. Linear optical techniques provide only meager information which is indirectly extracted by fitting the observations with (11) or (12) and the same is true for most incoherent spectroscopic techniques which only provide global information on the energy content and loss in the crystal subsequent to a few scattering events. The most serious drawback of these linear techniques and related incoherent ones is that one is restricted by the optical penetration length in the crystal and extraction of any information about damping and dispersion is complicated by the inherent polariton propagation and transmission at the crystal surfaces which may substantially alter the bulk polariton features. Clearly these difficulties can be only circumvented with nonlinear techniques that take into account the nonlocal character of the polariton and provide high selectivity ; the polariton frequency and wavevector along its dispersion curve can be obtained by nonlinear interaction of optical fields whose frequency and wavevector fall in the transparency region of the crystal.

3   NONLOCAL COHERENT TIME RESOLVED TECHNIQUES

We present now the underlying principle and pattern of a nonlocal nonlinear optical technique (Gale et al (1986), Vallée et al (1991)) that allows one to determine the propagation and damping characteristics of polariton pulses by directly following their evolution in real space and time domain. It employs the following scenario (see also Fig.3).

Excitation stage. An external instantaneous force F_e (t), nonlinear in the electric field amplitudes, resonantly couples to the polariton coordinate and drives it off equilibrium at "instant" t = 0 and at crystal "position" $\underline{X}$ = 0 ; as a consequence the polariton coordinate acquires a phase and an amplitude and its wave vector is fixed by the phase matching condition.

Free precession and propagation. The polariton pulse propagates freely in the direction fixed by the phase matching condition for a time interval t_D after the excitation stage over a distance $\underline{R}_D$ such that :

$$\underline{R}_D = \underline{v}_g \, t_D \qquad\qquad\qquad (15)$$

where $\underline{v}_g$ is its group velocity.

Interrogation stage. At "instant" t_D the polariton pulse is probed at "position" $\underline{X}_D$ with an external weak force F_p (t) of frequency ω_p and the signal generated at the combination frequency $\omega_\pi + \omega_p$ is measured in the matched configuration as a function of t_D.

The exciting and probing forces, $\underline{F}_e$ and $\underline{F}_p$ respectively, are products of electric fields whose order is fixed by the nonlinear interaction one exploits at each stage and to a large extent is fixed by symmetry considerations ; the spatial and temporal overlap of these fields determines the spatial and temporal resolution of the technique respectively. We also stress the fact that in the excitation stage the frequency and wavevector of the driving force F_e is an algebraic combination of those of the fields that constitute this force and one can vary them at will across the polariton dispersion in any direction for both bulk and surface polaritons ; similarly in the interrogation stage the combination frequency can be chosen conveniently to provide best discrimination with the background.

The simplest case for applying this technique is in crystals that lack inversion symmetry in which case the polariton is both single and double photon allowed. Nonlinear coupling in the excitation stage can then be attained with two light fields E_1 and E_2 and is described by the phenomenological energy density (Henry and Garett (1968)) :

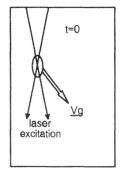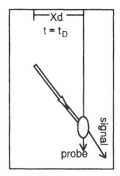

Fig 3. Excitation, propagation and detection of an ultra-short polariton pulse in the bulk crystal

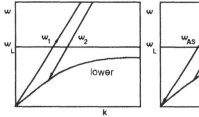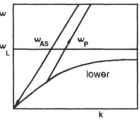

Fig 4. Coherent Raman excitation and anti-Stokes probing of the lower polariton branch

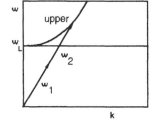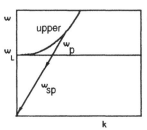

Fig 5. Two-photon excitation and phase-matched parametric probing of the upper polariton branch

$$V = -d_E \underline{E}_1 \underline{E}_2 \underline{E}_\pi - d_M \underline{E}_1 \underline{E}_2 \underline{Q}_\pi \qquad (16)$$

where $\underline{E}_\pi$ and $\underline{Q}_\pi$ represent the electric field and transverse mechanical amplitudes associated with the polariton and d_E and d_M are coupling parameters which vary slowly with the polariton frequency ω_π ; d_E is simply the combination frequency susceptibility or second order susceptibility for sum or difference frequency generation, while d_M is the Raman or two-photon transition amplitudes (Henry and Garett (1968), Flytzanis (1975)). The polariton wavevector is obtained by the phase matching condition :

$$\underline{k}_\pi = \underline{k}_1 \pm \underline{k}_2 \qquad (17)$$

and its frequency from energy conservation conditions :

$$\omega_\pi = \omega_1 \pm \omega_2 \qquad (17b)$$

where the choice of plus (+) or minus (-) sign is fixed by which of the two polariton branches is investigated. Indeed the polariton wavevector changes dramatically close to the resonance ω_T in contrast to $\underline{k}_1$ and $\underline{k}_2$ since the frequencies ω_1 and ω_2 are in the transparency region of the crystal and their vector sum and difference cannot indefinitely follow the polariton dispersion curve. These problems have been extensively exploited in frequency resolved nonlinear spectroscopy of polaritons (Coffinet and de Martini (1969), Fröhlich et al (1971), Haueisen (1971)). Simple vector algebraic considerations indicate that for :

the lower branch, the excitation stage proceeds via coherent Raman excitation and the coupling term (16) is

$$V_- = -d_E E_1 E_2^* E_\pi^* - d_M E_1 E_2^* Q_\pi^* + c.c. \qquad (18a)$$

while the interrogation stage involves phase matched coherent anti-Stokes Raman scattering ; this technique allows the study of the lower branch of phonon polaritons in cubic crystals (Fig.4) : in anisotropic crystals this limitation is circumvented and actually the method can be used to study the upper polariton branch as well (Auston et al (1984), Auston and Cheung (1985)). Recently the electrooptic Cerenkov technique has been proposed to study polariton propagation and damping effects in non centrosymmetric crystals mainly ferroelectrics, for instance ($LiTaO_3$) ; here one exploits the optical rectification effect (inverse electrooptic effect) to generate an extremely fast infrared electromagnetic transient from a single femtosecond pulse to excite the polaritons and subsequently probe with a second femtosecond pulse the electrooptic birefringence induced in the "Cerenkov" cone of the polarization. The selectivity of this method is however limited and the separation of temporal and spatial features is not straightforward. Furthermore this method is restricted to low frequency polaritons with an upper frequency limit given by the infrared femtosecond pulse bandwith.

The upper branch

The excitation stage involves two photon absorption and the coupling term (16) is :

$$V_+ = - d_E \, \underline{E}_1 \, \underline{E}_2 \, \underline{E}_\pi - d_M \, E_1 \, E_2 \, Q_\pi + c.c. \tag{18b}$$

while the interrogation stage is followed by phase matched parametric emission. With present-day short pulse laser sources this technique can be used to study the upper branch of exciton polaritons in cubic noncentrosymmetric crystals (see Fig.5) but with the advent of far infrared short pulse laser sources it can be used to study low frequency phonon-polaritons as well. The exciton-polariton dephasing has also been addressed in the time domain using time-resolved four wave-mixing and transient gratings techniques (Eichler (1977), Fayer (1982)). These techniques however proceed via strong linear polariton excitation within the absorption layer, close to the surface layer, and, hence, only give access to the high excitation regime where the polariton dephasing is dominated by polariton-polariton interactions (Masumoto et al (1983)).

For crystals that possess inversion symmetry the polaritons are not accessible by double photon excitation but one can use the next higher order interaction scheme for instance coherent hyper-Raman excitation followed by coherent anti-Stokes hyper-Raman scattering or three photon absorption followed by the corresponding parametric emission process to study the low and upper polariton branches respectively.

4   BARE PHONON-POLARITON PROPAGATION AND DAMPING

4.1 Polariton propagation

An example of the use of the technique to directly follow (Gale et al (1986)) the propagation of a polariton and determine its characteristics is that of the NH_4Cl crystal, a cubic crystal obtained from the NaCl one by replacing Na^+ with NH_4^+ which is an intrinsically noncentrosymetric molecular entity. At low temperature the ammonium ions are all ordered the same way but at higher temperatures their orientation becomes disordered and the crystal undergoes an order/disorder phase transition. The technique was demonstrated and used to study the polariton related to the ν_4 vibrational mode of the NH_4^+ radical with $\nu_{4T} = 1400$ cm^{-1} and $\nu_{4L} = 1418$ cm^{-1} for the transverse and longitudinal modes respectively : their dispersion curves and Raman spectra have been extensively studied.

The experimental system was driven by a mode-locked Nd^{3+}/glass laser system producing a single 5-ps pulse at 1.054 μm which is frequency converted to generate two fixed frequencies (ω_p and ω_S) and a variable one (ω_L) ; the latter is tuned so that $\omega_L - \omega_S = \omega_\pi$ the polariton frequency. The slope of the linear dependence of the optimum spatial displacement X_D on probe delay t_D precisely gives the polariton group velocity (Fig. 6). This was found to vary dramatically with polariton frequency (Fig.7), from $\approx c/2$ at low frequency down to $\approx c/50$ at high frequency, which reflects the change in polariton character from photon like to phonon-like as we sweep the dispersion relation. These directly determined experimental values were found to perfectly agree with the ones calculated from the dispersion law which has been extensively studied by Raman scattering. This is the first direct determination of a polariton group velocity.

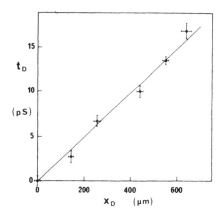

Fig 6. The probe delay at which signal is maximum as a function of lateral probe deplacement

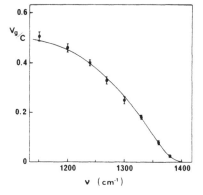

Fig 7. Measured polariton group velocity as a function of polariton frequency

4.2 Polariton damping

If the spatial or temporal spreading (due to velocity dispersion) of the "tracked" polariton packet is small the intensity decrease of its peak coherent anti-Stokes signal with time delay and space displacement directly gives the loss of coherence time constant (T_2) of the polariton packet. Except for extreme cases, for instance when the polariton interacts with the crystal surface, the time decay of the polariton coherence was found to be exponential throughout its dispersion curve and for a wide range of temperatures up to and close to the order disorder transition temperature. The value of T_2 however exhibits a strong resonant behavior with frequency and a characteristically critical behavior with the temperature which we wish to discuss below.

The damping of the non propagating longitudinal component of the v_4-mode at 1418cm^{-1} was also determined and was found to possess a similar behavior.

4.3 Anharmonicity vs Disorder

As previously stated in a perfect crystal the damping of the polariton is mainly due to the finite lifetime of their mechanical component Q_π which is anharmonically coupled to many phonon bands (Loudon (1970)). In ammonium chloride (NH_4Cl) at low temperature this mechanism is the most important one and in particular explains the frequency variation of the polariton damping rate which can be traced to a many-phonon band degenerate with the polariton. This process gives only a weak temperature dependence of the relaxation rate Γ_A, via Bose-Einstein occupation numbers, as experimentally observed between 7 and 130 K.

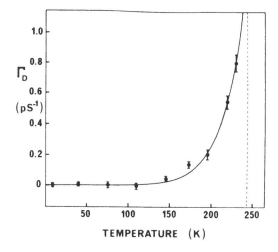

Fig 8. Behavior of the strongly temperature sensitive component Γ_D of the polariton relaxation rate as a function of temperature. The dotted line indicates the order/disorder transition temperature

In the presence of spatial disorder as previously stated an additional damping mechanism is introduced. In the case of NH_4Cl such a spatial disorder can be gradually introduced for instance by increasing the temperature and approaching the order-disorder transition temperature $T_c \approx 243$ K. The effect of the disorder can be globally measured with the order parameter L which can be deduced from thermodynamic data. As can be seen in Fig. 8, the disorder induced part of the damping rate, Γ_D, can be described by the simplest law for a disorder related process : $\Gamma_D = \gamma_D(1-L^2)$. This gives an excellent fit to the experimental data and supports the assumption that dielectric disorder makes a strong contribution to the loss of coherence in polaritons.

5   DRESSED PHONON-POLARITONS

5.1 Polariton Fermi resonance

With this technique it is now possible to investigate some new aspects of the interaction of polaritons with other collective excitations in a crystal. A typical example is the polariton Fermi resonance (Agranovich and Lalov (1971)) : this is produced whenever a polariton branch crosses a many-phonon band, most often a two-phonon band, of the same symmetry as the bare polariton. This situation is frequently encountered in crystals because both polaritons and two-phonon states span large regions in frequency and wave vector space. For a strong anharmonic coupling, the resulting new excitation can be described in terms of a dressed polariton with substantially different characteristics compared to the bare polariton.

The main consequences of polariton Fermi resonance are a partial localization, or slowing down, of the polariton which results from the opening of a new gap in the polariton dispersion curve and a strong, frequency dependent damping of the resulting dressed polariton. These manifestations can be theoretically analyzed by use of the Green's function method, and, in particular, it can be shown that the line shape of the dressed polariton is still lorentzian on the edges of the two-phonon band (where the density of states is relatively low) albeit strongly broadened by the opening of a new direct relaxation channel into the two-phonon continuum .

This lorentzian broadening, directly proportional to the two phonon density of states can be directly evidenced and measured (Vallée et al (1988)) in the time and space resolved CARS experiments by the exponential behavior in time of the polariton-induced anti-Stokes signal. The investigation was performed in ammonium chloride (NH_4Cl) in the frequency region where the polariton is in strong interaction with the polar $2\nu_4$ two-phonon band which extends roughly from 2800 to 2900 cm^{-1} . As expected, an exponential decrease of the coherent signal was observed over several orders of magnitude (up to seven) which demonstrates the

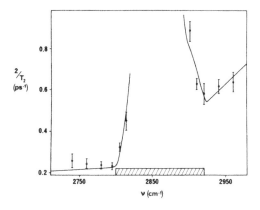

Fig 9. Measured polariton dephasing rate versus polariton frequency under conditions of resonance with a two-phonon band (shaded region)

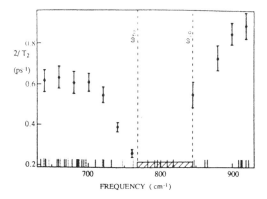

Fig 10. Frequency dependence of the polariton dephasing rate in LiO$_3$ at 80K

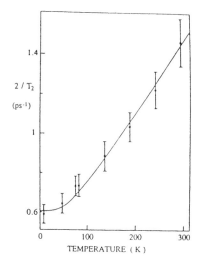

Fig 11. Temperature dependance of the 880 cm^{-1} polariton dephasing rate

lorentzian character of the polariton broadening. The dramatic frequency dependence of the polariton dephasing rate for polariton frequencies close to the edges of the $2v_4$ band is depicted in Fig.9. The rapid increase of the damping rate inside the $2v_4$ band is a clear indication that polariton disintegration into two isoenergetic free phonons constitues a very efficient polariton relaxation channel. Outside the $2v_4$ band indirect phonon assisted processes are operative for the loss of coherence and damping of the polariton by up or down conversion to the $2v_4$ continuum. Taking into account these direct and indirect processes the complete frequency dependence of the measured damping rate can be precisely mapped out. Additional support for the correctness of the chosen mechanisms is provided by the temperature dependence of the damping rate in the range 10-120 K ; at higher temperatures the gradual appearance of disorder leads to a strong increase of the polariton dephasing rate very similar to the one discussed previously.

5.2 Polaritons in uniaxial crystals.

In cubic crystals, the wave-vector restrictions associated with Raman techniques limit the accessible polariton region to a part of its lower branch. This limitation can be circumvented in uniaxial crystals (Anikiev et al (1984)), where the birefringence allows the entire polariton dispersion curve to be observed. By exploiting this fact the time and space resolved CARS technique was extended (Vallée et al (1989)) in uniaxial crystals, like LiO_3, to investigate the upper and lower ordinary E_1 polariton on both sides of the highest frequency reststrahlen band ($\omega_{TO} = 768$ cm^{-1}, $\omega_{LO} = 843$ cm^{-1}).

Here too the group velocity measurements are in good agreement with the indirect value obtained from the dispersion relation. The frequency dependence of the measured polariton dephasing rate is depicted in Fig.10 and shows a minimum in the vicinity of the reststrahlen band. This can be easily interpreted by letting the ω_{TO} - ω_{LO} frequency region play a role similar to the two-phonon band for the indirect mechanism in the case of polariton Fermi resonance (see above), i.e. acting as a reservoir of accessible states for phonon assisted up- or down- conversion of the polariton.

As for the polariton Fermi resonance in NH_4Cl, the temperature dependence of the damping rate is imposed by this interpretation of polariton relaxation, which then can be tested by temperature dependent measurements. This is indeed the case as shown in Fig.11 where the theoretical curve was obtained without fitting any parameters.

6   EXCITON-POLARITONS

The problem of the coherence of exciton-polaritons in polar semiconductors was also addressed with this technique (Vallée et al (1991)) as it allows a direct determination of the dephasing time and is exempt of the drawbacks of previously used techniques : linear excitation within the absorption layer (Masumoto et al (1982)), time resolved luminescence or induced absorption (Askary and Yu (1985), Oka et al (1986)), and four wave mixing in time domain (Masumoto and Shionoya (1982)) and frequency domain (Dagenais and Sharfin (1987)). The demonstration was performed with the measurement of the intrinsic dephasing time of transverse (polariton) and longitudinal components (Vallée et al (1991, 1992)) of the Z_3-exciton in cuprous chloride (CuCl).

The coherent excitation of the exciton-polariton of frequency ω_π and wavevector $k_{e\pi}(\omega_{e\pi})$ is realized in the bulk of the crystal by two photon absorption of two synchronized picosecond pulses with frequencies ω_1 and ω_2 and wave vector $\underline{k}_1$ and $\underline{k}_2$ such that $\omega_{e\pi} = \omega_1 + \omega_2$ and $\underline{k}_{e\pi} = \underline{k}_1 + \underline{k}_2$ (see Fig.5). These conditions introduce certain restrictions in the applicability of the technique since they can only be satisfied for the upper-branch. Close to the exciton resonance and neglecting spatial dispersion, the amplitude of the coherently driven exciton packet is proportional to :

$$d_\lambda = d_{E\lambda}\left(\omega_e^2 - \omega_\lambda^2\right) + d_{M\lambda} \tag{19}$$

where ω_e is the bare exciton frequency at $\underline{k} = 0$ ($\omega_e \sim 3.202eV$ in CuCl) and $\lambda = \pi, L$ labels the exciton polariton (π) and longitudinal (L) exciton respectively. The coupling parameters $d_{M\lambda}$ and $d_{E\lambda}$ are related, respectively, to two-photon absorption and sum frequency generation close to ω_e. The evolution of the exciton coherence, after the local excitation process has terminated, is followed by phase-matched parametric emission at $\omega_d = \omega_e - \omega_p$ stimulated by a third picosecond pulse delayed and spatially separated with respect to the excitation stage. We wish to stress here the fact that the excitation and probing can be done at will anywhere inside the crystal since all involved frequencies are in the transparency range of the medium : the spatial resolution is fixed by the overlap extension of the interacting beams.

The demonstration of this technique was performed (Vallée et al (1991))on two upper-branch polaritons in CuCl with energies $h\omega_{e\pi}^b \approx 3.208eV$ and $\omega_{e\pi}^f = 3.217eV$ corresponding

respectively, to a backward, $\theta = 180°$ and a forward $\theta = 0°$ excitation geometry, and on the longitudinal exciton. As pointed out previously the technique allows the study of the spatiotemporal evolution of the polariton pulse by separating the excitation and probing stages

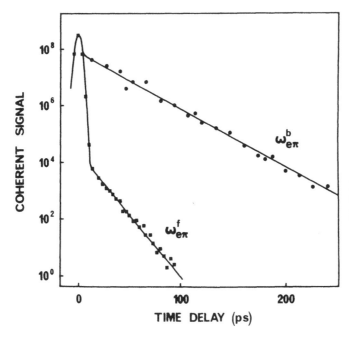

Fig 12. Dephasing rate for the two investigated polaritons in CuCl at 7K

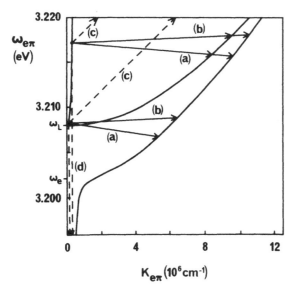

Fig 13. Dispersion curves and relaxation channels for the exciton polariton and longitudinal polariton in CuCl

in time and space. However, in the case of CuCl, relaxation was found to occur much faster than propagation and the polariton wave packet was probed only locally and similarly for the longitudinal exciton since it is not a propagating mode.

The experiments were performed using a passively mode locked Nd^3 glass delivering a single 5-ps pulse in the infrared ($\omega_I = 1.054\mu m$) which is frequency converted to create three independent pulses ω_1, ω_2 and ω_p. The ω_1 frequency is a small part of the initial infrared pulse while ω_2 is tunable around 611μm and the probe beam frequency ω_p, in the infrared, is adapted to phase matching requirements. The experimental procedure is extensively discussed elsewhere. Two samples with different surface orientations were used, one with 110 surface for the study of the transverse (polariton) exciton and another with 111 surface for the study of the longitudinal exciton.

In Fig.12, a measurement is reproduced for the dephasing rate of the two investigated polaritons $\omega_{e\pi}^f$ and $\omega_{e\pi}^b$ at a crystal temperature of 7K. The low intensity ratio of the signals,

$I_f/I_b \approx 10^{-4}$, is a consequence of the destructive interference for $\theta = 0°$ between the material and electric contributions in d_λ in (19). In Fig.13, are reported the measured values of the dephasing rates Γ for different temperatures in the range 7-60K and the calculated ones with an analytical expression of this rate based on the assumption that the dephasing is due to the three main exciton-phonon scattering processes namely the longitudinal optical phonon assisted scattering through the Fröhlich interaction (LO) (e.g. Weisbuch and Ulbrich (1982)), the longitudinal acoustic phonon (LA) scattering mediated by the deformation potential (DP) and the transverse acoustic phonon (TA) one mediated by the pieroelectric effect (PE) (Zook (1964)). The most probable processes for upper-branch polaritons are extraband down- and up-conversion into a lower-branch polariton, with, respectively, emission or absorption of a phonon (Travnikov and Krivolapchuk (1983)). These processes are strongly enhanced compared to intraband ones because of the higher density of final accessible states. At low temperatures scattering off acoustic phonons provide the dominant mechanism while for high temperatures the Fröhlich mechanism becomes dominant. Here, only the up-conversion process needs to be taken into account because of the very low density of accessible states for the down-conversion process. Their compound effect as depicted in Fig.13 satisfactorily reproduces the observed behavior of Γ.

The dephasing rate of the longitudinal exciton was also measured over the same temperature range ; the values are within the same range as those of the transverse (polariton) exciton and the overall temperature dependence is similar implying that the same exciton-phonon mechanisms are at work here too. Thus at low temperature (T<40K) the main relaxation mechanism is scattering by acoustic phonons mediated by the deformation potential (DP) for the LA phonon and by the piezoelectric effect (PE) for the TA phonons ; both up-and

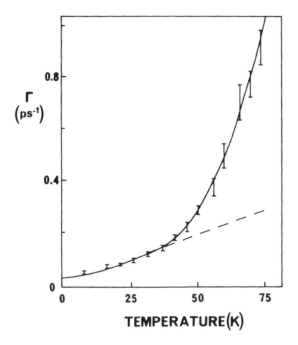

Fig 14. Temperature dependence of the dephasing rate of the longitudinal polariton in CuCl

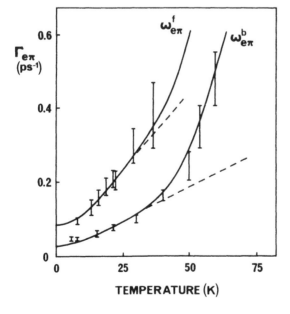

Fig 15. Temperature dependance of the $\omega_{e\pi}^f$ and $\omega_{e\pi}^b$ polariton dephasing rates $\Gamma_{e\pi}$ in CuCl

down- conversion processes are involved (Fig.14). At higher temperatures (T≥40K) the impact of the LO-phonon assisted scattering process mediated by the Fröhlich interaction becomes important and in fact dominates the relaxation as the temperature increases ; here too as in the case of the transverse (polariton) exciton only the up-conversion procession process is relevant because of the very low density of accessible states for the down conversion mechanism. In general the variation of the dephasing rate with frequency both for the transverse and longitudinal excitons can be roughly traced to the variation of the density states over the exciton dispersion curve.

Including all three exciton-phonon mechanisms then one can write :

$$\Gamma = \gamma_{LO}\, n\left(\omega_{LO}\right) + \gamma_{DP}\left(1 + n\left(\omega_{LA}^-\right) + n\left(\omega_{LA}^+\right)\right) + \gamma_{PE}\left(1 + n\left(\omega_{TA}^-\right) + n\left(\omega_{TA}^+\right)\right) \qquad (20)$$

for the intrinsic dephasing rate and obtain a temperature dependance in good agreement with the experimental results. In (20), $n(\omega_i)$ is the occupation number for the ω_i-phonon whose frequency as imposed by energy and wave vector conservation (16) ; γ_{LO}, γ_{DP} and γ_{DP} are frequency-dependent coupling parameters for the three scattering mechanisms LO, DP and PE respectively. These parameters can be estimated by fitting the measured values of the dephasing time with expression (20); they can also be independently estimated (Takaguhara (1985)) using known parameters of CuCl. The fitting of the experimental values of Γ with (20) as depicted in Figs.14 and 15 gives the following table :

	γ_{LO}	γ_{DP}	γ_{PE}
$\omega_{e\pi}^f$	26 μeV	37 μeV	25 μeV
$\omega_{e\pi}^b$	26 μeV	14 μeV	3 μeV
ω_{eL}	26 μeV	14 μeV	3 μeV

The broken lines in Figs.14 and 15 represent the temperature dependance of the dephasing rate if the Fröhlich mechanism was altogether neglected ; as can be seen such a neglect would lead to substantial deviations from the measured values as the temperature increases beyond 40K.

In the case of the longitudinal exciton when the phase matching configuration in the probing stage is relaxed a slower relaxation component was observed (Vallée et al (1992)) at low temperature in addition to the one discussed above. This may originate from other exciton states with slower relaxation rates accessible either through spatial dispersion or through localisation by defects. The precision was not sufficient to make precise statements concerning the origin of this relaxation component.

7   POLARITON TRANSPORT AND LOCALISATION IN A DISORDERED DIELECTRIC

All types of polaritons are composite excitations partly material and partly electromagnetic and as such are expected to share features of both modes. Thus their damping and coherence is affected by both material anharmonicity and spatial disorder. The first has been extensively addressed in the literature both experimentally and theoretically but much less the second which however can be quite critical in certain circumstances as previoulsy shown (Gale et al (1986)). Here we succintly analyse the problem of polariton transport in a crystal with a real refractive index that varies randomly in space in the light of the recent theories that have been developped to take into account effects related to multiple scattering of waves in disordered media (Tsang and Ishimaru (1984)) (Sajeev John (1988), Andersson (1985)).

In the case of pure electromagnetic waves under certain conditions the phase correlations and the interference of multiply scattered waves off the randomly distributed spatial fluctuations of the refractive index can lead to a drastically new behavior in the overall wave propagation, the most prominent feature being the possible localization of the waves (Andersson (1985), Sajeev John (1984, 1985)) : the effective light diffusion constant vanishes at a certain wavelength λ^* which establishes a threshold separating localized from delocalized states of the electromagnetic field. The necessary conditions, however, are very stringent for transparent media (Sajeev John (1984), Genack (1986); it turns out that the corresponding condition is easier to reach in the case of polaritons.

We introduce "dispersive" disorder in the crystal by letting the real background dielectric constant become space dependent $\varepsilon(\underline{r})$ and fluctuate randomly in space around an average value ε_∞ while the oscillator parameters are held constant (e.g. Ishimaru (1973)) ; the fluctuating part $\delta\varepsilon(\underline{r}) = \varepsilon(\underline{r}) - \varepsilon_\infty$ with $<\delta\varepsilon(\underline{r})> = 0$ is assumed to be a gaussian random function with spatial correlation :

$$<\delta\varepsilon(\underline{r})\ \delta\varepsilon(\underline{r}')> = \mu^2\delta(\underline{r} - \underline{r}')$$

where μ is a constant and $< >$ everywhere denotes ensemble averages. For later use we also introduce the renormalized quantities $\tilde{\varepsilon}(\underline{r}) = \varepsilon(\underline{r}) / \varepsilon_\infty$, $\delta\tilde{\varepsilon}(\underline{r}) = \delta\varepsilon(\underline{r}) / \varepsilon_\infty$, $\tilde{\mu} = \mu / \varepsilon_\infty$ and $\tilde{c} = c / \sqrt{\varepsilon_\infty}$.

The polariton electric field E_π and polarization P_d then obey the coupled equations :

$$\left\{ \nabla^2 + \left(\omega/\tilde{c}\right)^2 \left(1 + \delta\tilde{\varepsilon}(\underline{r})\right) \right\} E_\pi = c^{-2} \ddot{P}_c \tag{21a}$$

$$\ddot{P}_d + \omega_T^2 P_d = \omega_p^2 E_\pi \tag{21b}$$

where, anticipating the use of the hydrodynamic regime in deriving the transport coefficients, we have neglected a term $\nabla.\{\delta\varepsilon(\underline{r}) \, \underline{E}_\pi\}$ in (21a). The polariton characteristics can be most conveniently introduced in terms of the Green function $G(\underline{q},\omega)$. Over all disorder configurations one merely obtains an additional isotropic attenuation which can be lumped together with the one due to the material anharmonicity. In the presence of strong dispersive disorder, however, correlation and interference between multiple scattered waves drastically alter the situation. Such effects are suppressed in the short ranged average Green's function $<G(q,\omega)>$ so that the relevant quantity is now the one squared over all disorder configurations (Kane and Stone (1981)) :

$$P(\underline{k},k';\omega,\Omega) = < G_-(\underline{k},\omega + \Omega) \, G_+(\underline{k}',\omega) > \tag{22}$$

where $G_-(k,\omega)$ and $G_+(k,\omega)$ respectively stand for the retarded and advanced one polariton Green's function ; the quantity $P(\underline{k},k';\omega,\Omega)$ in (22) is the averaged intensity propagator and can be expanded in terms of the averaged Green's functions and also averaged over pairs of scattering events which correlate the two Green's functions (Sajeev John (1984), Kirkpatric and Dorfman (1985)). The expansion can be obtained by the Bethe-Salpeter equation along the same lines as for electrons or sound ; to lowest order in $\tilde{\mu}^2$ the expansion of the intensity propagator leads to a diffusion pole while higher order terms yield divergent integrals whose dominant contribution comes from the maximally crossed diagrams.

In the hydrodynamic regime, namely small $\underline{q} = \underline{k} - \underline{k}'$ and Ω, these can be resumed and give :

$$P(\omega;\underline{q},\Omega) \sim (- i \, \Omega + D(\omega,\Omega) \, q^2)^{-1} \tag{23}$$

where the effective diffusion constant $D(\omega,\Omega)$ is given in perturbation theory by

$$D^{-1}(\omega;\Omega) = D_0^{-1}(\omega) + L \int_0^{q_0} dq \left(-i\omega + D_0(\omega) q^2\right)^{-1} \qquad (24)$$

$D_0(\omega)$ is the "Boltzmann" diffusion constant for polariton transport, $L = 1/4\pi^3\rho(\omega)D_0$ where $\rho(\omega)$ is the polariton density of states at ω, and q_0 is a cut off that eliminates high wave vectors that do not correspond to diffusion processes.

The Boltzmann diffusion constant is defined by $D_0 = \ell v_g/3$ where $v_g = \partial\omega/\partial k$ is the polariton group velocity calculated from the dispersion relation (2) and ℓ is the polariton coherence length or mean free path refined by $\ell = v_g\tau$; the elastic scattering time τ is related to the elastic linewidth $\Gamma(\omega) = 1/\tau(\omega)$ of the averaged polariton propagator $<G_+(\underline{k},\omega)>$ and can be calculated from Fermi's rule (e.g. Ginzburg (1985))

$$\frac{1}{\tau} = \frac{2\pi}{\hbar} \left\langle h_\varepsilon^2 \right\rangle \rho(\omega)$$

where the polariton density of states is calculated from $\rho(\omega)\, d(\hbar\omega) = \rho(k)\, dk = (k/\pi)^2 dk$ or

$$\rho(\omega) = \frac{\varepsilon(\omega)\, \omega^2}{\pi^2 c^2} \frac{1}{v_g} \qquad . \qquad (25)$$

and $h_\varepsilon = -\delta \varepsilon(\underline{r})\, E^2/8\pi$ is the effective interaction hamiltonian density for polariton scattering off the refractive index fluctuations ; one obtains

$$\frac{1}{\tau} = \frac{\tilde{\mu}^2}{2\pi} \left(\frac{\omega}{c}\right)^4 \frac{c^2}{\varepsilon(\omega)} \frac{1}{v_g} \qquad (26)$$

or $\ell = 4\pi c^2 v_g^2 \varepsilon(\omega)/\tilde{\mu}^2\omega^4$. The parameters (25) and (26) completely specify the polariton transport within the single scattering approximation and through (24) the multiple scattering correlation and interference effects as well.

We introduce the self-consistency condition (Kirkpatrick and Dorfman (1985)) by replacing $D_0(\omega)$ with $D(\omega,\Omega)$ under the integral in (24) which extends up to $q_0 = \sigma/\ell$ where σ is an undefined constant. Integrating we find :

$$D(\omega,\Omega) = D_0(\omega) - 4\pi h\, D_0(\omega) \{q_0 - (-i\Omega/D(\omega,\Omega))^{1/2}\arctan q_0\ (-i\Omega/D(\omega,\Omega))^{1/2}\} \quad (27)$$

For $\Omega \to 0$ the condition $D(\omega,\Omega \to 0) = 0$ gives

$$4\pi L q_0 = 1 \qquad (28)$$

which gives the relation for the mobility edge ω* for polariton transport

$$\left(\varepsilon_\infty + \frac{\omega_\rho^2}{\omega_T^2 - \omega^{*2}}\right)^5 = \kappa\omega^{*6}\left\{\varepsilon_\infty + \frac{\omega_T^2\omega_\rho^2}{\left(\omega_T^2 - \omega^{*2}\right)^2}\right\}^4 \tag{29}$$

where $\kappa = 12\ \sigma\mu^4/\pi^2 c^6$.

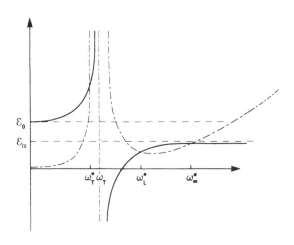

Fig 16. The left (full) and right (dotted) members of eq.29 showing the existance of three mobility edges

   In Fig.16 we give a graphic representation of the right and left members of this relation for a small value of the parameter κ which is related to the disorder. We see that for small κ there are three mobility edges ω_T^*, ω_L^* and ω_∞^* ; with $\omega_T < \omega_L < \omega_\infty$, $\omega_T^* < \omega_T$ and $\omega_L^* > \omega_L$.

The states are localized for $\omega > \omega_\infty$ and within $\{\omega_T^*, \omega_T\}$ and $\{\omega_L, \omega_L^*\}$; thus as a small disorder is introduced the polariton states become localized on either side of the reststrahlen region and in addition above a certain high frequency ω_∞^* as in transparent media. As the disorder increases the localized regions increase and eventually ω_L^* merges with ω_∞^* for a critical value κ_c. Beyond this value there is only one mobility edge ω_T^* and all polariton states with $\omega < \omega_T^*$ are delocalized and above it are localized.

In the region where the polariton is localized one can define a delocalization length $\xi(\omega)$ by

$$D(\omega, \Omega \to 0) = - i \, \Omega \xi^2(\omega) + O \, (\Omega^2)$$

which when inserted in (27) gives

$$\xi_i(\omega) = \frac{2\pi^2 L}{\left(4\pi L q_0 - 1\right)} \tag{30}$$

Since $q_0 \sim 1/\ell$ one easily sees that $\xi(\omega) > \ell$ and close to the mobility edges ξ diverges as $1/|\omega_i^* - \omega|$ namely with a critical exponent of unity.

In the above analysis we ignored the presence of the damping due to the phonon anharmonicity ; its effect is mainly felt close to the reststrahlen region and can be incorporated by introducing a friction term $\gamma \dot{P}_d$ in (21b). which then destroys its time inversion symmetry and diffuses the sharp mobility edges expected from (27). Clearly localization can occur if $\gamma\tau <$ 1 ; otherwise the temporal phase coherence loss due to the anharmonicity washes out the interference of the multiple scattered waves (Golubentsev (1984)).

8   POLARITON OPTICS. CONCLUSIONS

We have presented a nonlocal time-resolved technique that allows one to address and study all aspects of the spatio-temporal evolution of short polariton pulses. This technique opens up new possibilities for the investigation of fundamental problems associated with pulses of collective excitations in crystals regarding their propagation characteristics, dephasing and energy loss processes. In this respect the problem of the polariton pulses is of central interest since it is connected with electromagnetic signal propagation close to a resonance. Many nonlinear optical effects in crystals crucial for optoelectronic devices are greatly enhanced close to electric dipole allowed resonances and must be described in terms of coherent polariton pulse interactions. Of particular interest here are the parametric optical interactions (Akhmanov and Khokhlov (1962)) which are of both fundamental and technological interest : parametric optical amplifiers and oscillators parametric instabilities and chaos...

The space and time resolved CARS technique allows one to "track" a polariton pulse at any "point" inside the crystal and analyse its phase and energy content and assess the impact of temporal and spatial disorder and, in particular, the effect of boundaries. The latter is of particular importance for understanding polariton optics inside a crystal :

reflection and transmission by plane boundaries

polariton Fabry-Perot cavities

polariton total reflection and tunnelling

polariton nonlinear propagation

One can observe reflected polaritons and polaritons transmitted from one crystal to another separated by air (Gale et al (1990)).

A problem of particular fundamental interest is the interaction of polaritons with random spatial disorder. This problem has only been addressed in the case of orientational disorder in NH_4Cl where advantage was taken of the possibility to arbitrarily "tune" the crystal disorder by changing the crystal temperature, which allows a variable degree of disorder to be probed in the same sample. However, measurements in other disordered systems, such as isotopically disordered ones, are of particular interest for assessing the mutual coherence of the electromagnetic and mechanical parts of the polariton ; furthermore, since there is a close connection between polariton and photon scattering by disorder, one can expect to observe polariton localization.

As stated above, the intrinsic limitation of the Raman technique to non centrosymmetric crystals can be circumvented by the use of the hyper-Raman configuration for coherent excitation and probe in crystals like NaCl. Here the decrease in efficiency of the hyper-Raman processes can be counterbalanced by using the very high peak power delivered by picosecond lasers and the much higher damage threshold in shortening the light pulses in such ionic crystals. Furthermore, one could also address the problem of propagation and relaxation of vibrational polaritons in highly disordered media such as glasses or liquids .

REFERENCES

Agranovich V M and Lalov I I 1971, Zh. Eksp. Teor. Fiz. **61** 656 (transl. 1972 Sov. Phys. JETP **34** 350)

Agranovich V M 1982 *Surface Polaritons* North Holland Amsterdam

Akhmanov S A and Khokhlov R V 1962 Zh. Eksp i Teor. Fiz **43** 352 (transl. Soviet Phys. JETP **16** 252)

Andersson P W 1985 Phil Mag. **52** 505

Anikliev A A, Reznik L C, Umarov B S and Scott J F 1984 J. Raman Spectr. **15** 60

Askary F and Yu P Y 1985 Phys. Rev. **B31** 6643

Auston D H, Cheung K P, Valdmanis J A and Kleinman D A 1984 Phys. Rev. Lett. **53** 1555

Auston D H and Cheung K P 1985 J. Opt. Soc. Am. **B2** 606

Barker Jr. A S and Loudon R 1972 Rev. Mod. Phys. **44** 18

Bloch F 1946 Phys. Rev. **70** 460

Born M and Huang K 1954 *Dynamical Theory of Crystal Lattices* Clarendon Press, Oxford

Coffinet J P and de Martini F 1971 Phys. Rev. Lett. **27** 1506

Dagenais M and Sharfin W F 1987 Phys. Rev. Lett. **58** 1776

Eichler H J 1 977 Opt. Acta **24** 631

Fayer M D 1982 Am. Rev. Phys. Chem. **33** 63

Feynman R P Vernon F C and Helwarth R W 1957 J. Appl. Phys. **28** 49

Flytzanis C 1990 in *Applied Laser Spectroscopy* Eds Demtröder W and Inguscio M NATO
    ASI Series, Plenum Press, N.Y.

Flytzanis C 1972 Phys. Rev. Lett. **29** 772

Flytzanis C 1975 in *Quantum Electronics, A Treatise, Vol I*, Eds Tang C L and Rabin H
    Acad. Press

Fröhlich D Möhler I and Wiesner P 1971 Phys. Rev. Lett. **31** 369

Förster Th. 1948 Ann. Phys. (Leipzig) **2** 55

Gale G.M., Vallée F and Flytzanis C 1986 Phys. Rev. Lett. **57** 1867

Genrack A 1987 Phys. Rev. Lett. **58** 2043

Ginzburg V 1975 *Physique Théorique et Astrophysique*, Editions Mir, Moscou

Golubentsev A H 1984 Zh. Eksp. Teor. Fiz. **86** 47 (transl. Sov. Phys. JETP **59** 26)

Haueisen D C and Mahr H 1971 Phys. Rev. Lett. **26** 838

Henry C H and Garett C G B 1968 Phys. Rev. **171** 1058

Hopfield J J 1958 Phys. Rev. **112** 1555

Hopfield J J 1966 J. Phys. Soc. Japan Suppl **21** 77

Huang K 1951 Proc. Roy. Soc. **A208** 352

Ishimaru A 1978 *Wave Propagation and Scattering in Random Media* Academic Press, New
    York

Kane A J and Stone M 1981 Annals Phys. **131** 36

Kirkpatrick T R and Dorfman I R 1985 *in Fundamental Problems in Statistical Mechanics* VI,
    Ed. Cohen E G D Elsevier Science, p. 365

Knox R S 1963 *Theory of Excitons* in Solid State Physics Suppl. 3, Eds Seitz F and Turnbull
    D, Academic Press New York

Laubereau A and Kaiser W 1978 Rev. Mod. Phys. **50** 607

Leung Tsang and Ishimaru A 1984 J. Opt. Soc. Am. **A1** 836

Levenson M 1982 *Introduction to Nonlinear Laser Spectroscopy* Academic Press, New York

Loudon R 1970 J. Phys. **A3** 233

Loudon R and Haynes J 1980 *Light Scattering in Solids*

Maradudin A and Wallis R F 1962 Phys. Rev. **125** 1277

Masumoto Y and Shinoya S 1982 J. Phys. Soc. Jpn **51** 181

Masumoto Y, Shionoya S and Takagahara T 1983 Phys. Rev. Lett. **51** 923

Mills D L and Burstein E 1974 Reps. Prog. Phys. **37** 817

Oka Y, Nakamura K and Fujisaki H 1986 Phys. Rev. Lett. **57** 2857

Pelzer H 1945 Trans. Farad. Soc. **42A** 164

Rashba E I and Sturge M D 1982 *Excitons* North Holland, Amsterdam

Sajeev John 1984 Phys. Rev. Lett. **53** 2169

Sajeev John 1988 Comm. in Cond. Matter **14** 193

Slichter C P 1980 *Principles of Magnetic Resonance* Springer Verlag, Berlin

Szigeti B 1949 Trans. Farad. Soc. **45** 155

Szigeti B 1955 Proc. Roy. Soc. **A252** 217 and **A258** 577

Takagahara T 1985 Phys. Rev. **B31** 8171

Travnikov V V and Krivolapchuk V V 1983 Zh. Eksp. Teor. Fiz. **85** 2087 (transl. Soviet Phys. JETP **58** 1210)

Vallée F, Gale G M and Flytzanis C 1988 Phys. Rev. Lett. **61** 2102

Vallée F, Gale G M and Flytzanis C 1989

Vallée F, Bogani F and Flytzanis C 1991, Phys. Rev. Lett. **66** 1509

Vallée F, Bogani F and Flytzanis C 1992, Proceedings of VII th Interna. Symp. on *Ultrafast Processes in Spectroscopy* Ed. Laubereau A, Adam Hilger, Bristol

Vollhardt D and Wölfle P 1980 Phys. Rev. **22** 4666

Weisbuch C and Ulbrich R G 1982 in *Light Scattering in Solids III* edited by Cardona M and Guntherodt G (Springer-Verlag, Berlin) p. 207

Zook J D 1964 Phys. Rev. **136** A869

Self-trapping of optical beams in photorefractive media

Mordechai Segev, Bruno Crosignani(*) and Amnon Yar

California Institute of Technology,
Pasadena, CA 91125, USA.

We show that an optical beam can be self-trapped in a photorefractive medium, under the combined influence of diffraction and self-scattering (two-wave mixing) of its spatial frequency components. The resulting photorefractive spatial soliton possesses some unique properties, such as independence of the absolute light intensity, and can experience absorption (or gain) with no change in its transverse structure.

(*) Permanent address : Dipartimento di Fisica, Universita' dell'Aquila, L'Aquila, Italy and Fondazione Ugo Bordoni, Roma, Italy

A shortened version of this paper was published as a letter: "Spatial Solitons in Photorefractive Media", published in Physical Review Letters, vol. 68, page 923 (1992).

1.   Introduction

Light solitons in space (spatial solitons) have been the object of intensive theoretical and experimental research during the last three decades. The solitons evolve from nonlinear changes in the refractive index of the material, induced by the light intensity distribution. When the confining effect of the refractive index exactly compensates for the effect of diffraction, the beam becomes self-trapped and is called a spatial soliton. The nonlinear effects which are responsible for soliton formation are in general Kerr-like effects, inducing local index changes proportional to the local light intensities. The index changes needed for spatial solitons require high intensities often exceeding $1MWatt/cm^2$ (see Aitchinson et. al. 1990).

We describe in what follows a new type of spatial solitons, generated by the photorefractive (PR) effect of the medium. The shape of the soliton modulates the refractive index via the PR effect, which results in an exact compensation for the effects of diffraction, and causes the light beam to propagate with an unvarying profile. This index modulation is represented in the formalism as a distribution of index gratings, each induced by the interference between two spatial (frequency) plane wave components of the light beam. Since the efficiency of this effect is independent of the absolute light intensity, these new solitons can be generated at very moderate light intensities. Moreover, a given soliton waveform can propagate unchanged in the medium, at very high or very low light intensities (and at all levels in between).

The PR solitons correspond to steady state solutions of the nonlinear wave equation which describes beam propagation in PR media, and accounts for both diffraction and the mutual interaction between each pair of spatial components of the soliton beam. Since the key to this nonlinear scattering process is grating formation by a <u>continuum</u> of Fourier (plane wave) components of the soliton beam, we cannot resort to the two plane waves analysis commonly applied to PR materials. Our general formalism accounts for the transverse beam spatial structure.

2.   Formulation of the Photorefractive Effect

We start by deriving the nonlinear wave equation which describes the propagation of a monochromatic optical beam of a given frequency (ω) and polarization, traveling in the positive direction of an arbitrary axis z. We assume the absence of nonlinear interaction between orthogonal polarizations (anisotropic scattering, see Temple and Warde 1986), so that our problem can be reduced to a scalar formulation. Light propagation in nonlinear media can be conveniently described using coupled-mode theory (Solimeno et. al. 1986), applied to the case of unbounded media for which an appropriate set of spatial modes is the continuum of plane waves (Crosignani and Yariv, 1984). The electric field associated with the light beam propagating primarily along the z direction is written as

$$E(\mathbf{r},z,t) = \frac{1}{2}[e^{i(kz-\omega t)} \int E(\mathbf{q},\mathbf{r}) \, e^{i(\beta_q - k)z} f(\mathbf{q},z) \, d\mathbf{q} + C.C.] \equiv \frac{1}{2}[A(\mathbf{r},z) \, e^{i(kz-\omega t)} + C.C.] \tag{1}$$

where, in the paraxial approximation: $(k-\beta_q) << k$, and

$$E(\mathbf{q},\mathbf{r}) = \frac{1}{2\pi} \sqrt{\frac{\mu_0}{\varepsilon_0 n_1}} \; e^{i\mathbf{q}\cdot\mathbf{r}} \tag{2}$$

$\mathbf{r} \equiv (x,y)$, $k = \omega n_1/c$ is the light wavenumber, n_1 the unperturbed index of refraction in the medium, and $f(\mathbf{q},z)$ the spatial frequency (angular) distribution of the complex amplitude $A(\mathbf{r},z)$. A spatial mode (plane wave component) is characterized by the projections of its wave vector ($\mathbf{q}$ and β_q) on the transverse ($\mathbf{r}$) and longitudinal (z) directions respectively, with $\beta_q = (k^2-q^2)^{1/2}$ (where q is restricted to $0 \leq q \leq k$). Assuming negligible absorption, and under the rather general conditions specified by Crosignani and Yariv (1984), it is easy to show that $A(\mathbf{r},z)$ obeys, in the presence of a refractive-index distribution $n(\mathbf{r},z) = n_1 + \delta n(\mathbf{r},z)$, the differential equation:

$$(\frac{\partial}{\partial z} - \frac{i}{2k} \nabla_{\mathbf{r}}^2) \, A(\mathbf{r},z) = \frac{ik}{n_1} \, \delta n(\mathbf{r},z) \, A(\mathbf{r},z) \tag{3}$$

The nonlinear term $\delta n(\mathbf{r},z)$ is obtained by considering the mixing process between two plane waves of the same frequency. When only one pair of such plane waves (spatial modes) q_1 and q_2, of field amplitudes $a_1(z)$ and $a_2(z)$, is present in the medium, it induces an index grating $\delta n(r,z)$ which is proportional to the time averaged interference pattern between the waves. The proportionality coefficient is a complex factor $\widehat{\delta n}(q_1,q_2)$, which represents the PR coupling coefficient between the two plane waves, given the material properties (the orientation of the PR crystalline medium, its traps density P_d, its refractive index n_1, the DC dielectric constant ε_r), the externally applied electric field E_0, and the polarization of the waves. In this simple case, $\delta n(r,z)$ is (see for example White et. al. 1985):

$$\delta n(r,z) = \frac{1}{I_0} \left\{ a_1(z)\ e^{i(q_1 \cdot r\ +\ \beta_{q1}z)}\ a_2^*(z)\ e^{-i(q_2 \cdot r\ +\ \beta_{q2}z)}\ \widehat{\delta n}(q_1,q_2) + C.C. \right\} \qquad (4)$$

where $I_0 = |a_1|^2 + |a_2|^2$ is the absolute light intensity. Since the perturbation in the refractive index, $\delta n(r,z)$, is real (no absorption) we get $\widehat{\delta n}(q_1,q_2) = \widehat{\delta n}^*(q_2,q_1)$. This light induced index change may be viewed as the sum of two grating components: the real and imaginary parts of $\widehat{\delta n}(q_1,q_2)$, which correspond to index gratings that are in phase and shifted by $\pi/2$, respectively, with respect to the light interference grating $a_1 a_2^*/I_0$. The $\pi/2$ shifted gratings are responsible for (light) power transfer between the two waves, a phenomenon which can manifest itself as optical amplification, photorefractive oscillation (White et. al 1985), and stimulated noise scattering (Fanning, see Segev et. al 1990a). The in phase gratings are responsible for nonlinear phase coupling. The real / imaginary parts of $\widehat{\delta n}(q_1,q_2)$ are even / odd (respectively) under the exchange $(q_1,q_2) \to (q_2,q_1)$. The index grating $\widehat{\delta n}(q_1,q_2)$, multiplied by "i", is often called γ, and yields an "intensity" coupling coefficient $\Gamma = 2Re\{\gamma\} = -2Im\{\widehat{\delta n}(q_1,q_2)\}$. The index grating $\widehat{\delta n}(q_1,q_2)$ is given by:

$$\widehat{\delta n}(q_1,q_2) = \frac{-\omega}{2c}\ n_1^3\ r_{eff}(q_1,q_2)\ E_m(q_1,q_2)\ (e_1 \cdot e_2^*) \qquad (5)$$

where e_1, e_2 are the polarization vectors of the two interfering plane waves, $r_{eff}(q_1,q_2)$ is a scalar product of the material electroptic tensor (and depends on the orientation of the PR crystalline medium), and $E_m(q_1, q_2)$ is the coefficient of the induced space charge field

$$E_m(q_1, q_2) = E_p(q_1, q_2) \frac{E_d(q_1, q_2) - iE_0}{E_0 + i \, [E_d(q_1, q_2) + E_p(q_1, q_2)]} \tag{6}$$

where E_0 is the externally applied electric field, $E_p = e \, P_d/(\varepsilon_0 \, \varepsilon_r \, K_g)$ the limiting space charge field, and $E_d = k_B T K_g/e$ the diffusion field (e is the electron charge, T the temperatute, and k_B is Boltzman constant). Both E_d and E_p exhibit an angular dependence on the interference grating wave vector $K_g = k_1 - k_2 = (q_1 - q_2, \beta_{q1} - \beta_{q2})$ between the two plane waves. E_p, in some materials, has also a small angular dependance through ε_r (see Segev et. al. 1990b).

Since r_{eff} is always real, the phase shift of $\widehat{\delta n}(q_1, q_2)$ with respect to the interference grating K_g is dictated solely by E_m, which can be split into its real and imaginary parts: $E_m = -(E_1 + iE_2)$, where

$$E_1 = \frac{E_0}{(\frac{E_0}{E_p})^2 + (\frac{E_d}{E_p} + 1)^2} \qquad \text{and} \qquad E_2 = \frac{E_p \, (E_0^2 + E_d^2 + E_d E_p)}{E_0^2 + (E_d + E_p)^2} \tag{7}$$

It is important to note that since K_g is odd under both exchanges $(q_1, q_2) \to (q_2, q_1)$ and $(q_1, q_2) \to (-q_1, -q_2)$ (where $\beta_{q1} - \beta_{q2}$ is very small), E_2 follows the same symmetry, while E_1 is always even. These symmetry properties imply design considerations for a proper choice of photorefractive crystalline medium and direction of propagation, polarization and application of external field, and will be addressed below.

When more than two plane waves are present, $\delta n(r, z)$ involves a summation over all the possible interacting pairs of plane waves. For a given light beam $A(r, z)$, which consists of a continuous spatial-frequency spectrum of plane waves $f(q, z)$, this summation takes the integral form:

$$\iota(\mathbf{r},z)=\frac{1}{|A(\mathbf{r},z)|^2}\int d\mathbf{q}_1\int d\mathbf{q}_2\ f(\mathbf{q}_1,z)\ f^*(\mathbf{q}_2,z)\ E(\mathbf{q}_1,\mathbf{r})\ E^*(\mathbf{q}_2,\mathbf{r})\ e^{i(\beta_{q1}-\beta_{q2})\,z}\ \widehat{\delta n}(\mathbf{q}_1,\mathbf{q}_2)\quad(8)$$

Note that since the PR nonlinearity is independent of the absolute light intensity, δn is normalized by the factor $|A(\mathbf{r},z)|^2$. In addition, a constant factor representing the dark irradiance (see Mamaev and Shkunov 1990) may be added to the light intensity in the denominator of Eq. (8), in order to avoid unrealistic divergence of δn in dark regions. Our expression for δn is based on the assumption that the induced gratings are of low modulation depth, and hence can be linearly superposed. This assumption, however, is not a major restriction when a beam is considered. Moreover, even for an extreme case of two plane wave components of the same amplitude, the modulation depth of their interference grating is given by $a_1 a_2^*/I_0$, where I_0 includes **all** the plane wave component of the beam, and our low modulation depth assumption remains valid. Our representation of the photorefractive light induced index change breaks down only at the margins of the beam, where the local light intensity $|A(\mathbf{r},z)|^2$ is of the order of the dark irradiance, high modulation depth effect become significant (see, for example, Vachss and Hesselink 1988), and the "linearization" assumption is no longer valid. We restrict our analysis to regions where the light intensity is much higher than the dark irradiance, and thus neglect the additional constant in the denominator of Eq. (8). Note, that this model of beam propagation in photorefractive media has proven effective in the interpretation of a variety of wave mixing processes (Segev et. al. 1990a and 1990b) and, in particular, was used for predicting a number of new phenomena (such as incoherent backscattering, see Segev and Yariv 1991).

3. The Photorefractive Soliton Equation

Since diffraction can be viewed as due to a linear phase accumulation in each plane wave component of the light beam, the simplest way to compensate for it, and generate a spatial soliton in a PR medium, is through equal and opposite nonlinear phase delays, as in the Kerr-like solitons. Accordingly, we require the PR coupling coefficient $\widehat{\delta n}(\mathbf{q}_1,\mathbf{q}_2)$ (and hence E_m) to be real and hence to introduce nonlinear phases only (i.e. no "energy transfer" between the pairs of spatial components). Furthermore, since diffraction is essentially a symmetric process, the simplest solution is obtained by requiring a symmetric nonlinear process, i.e. $\delta n(\mathbf{r},z)=\delta n(-\mathbf{r},z)$ [and hence $\widehat{\delta n}(\mathbf{q}_1,\mathbf{q}_2)=\widehat{\delta n}(-\mathbf{q}_1,-\mathbf{q}_2)$], which implies either symmetric or anti-symmetric soliton waveform, i.e. $A(\mathbf{r},z)=A(-\mathbf{r},z)$ or $A(\mathbf{r},z)=-A(-\mathbf{r},z)$. In this paper, we consider even waveform solutions only. The possibility for obtaining PR solitons, induced by a general complex $\widehat{\delta n}(\mathbf{q}_1,\mathbf{q}_2)$, with a soliton waveform [$A(\mathbf{r},z)$] of a non-specific symmetry, will be presented elsewhere (Crosignani et. al. 1992).

In the most general case, $\widehat{\delta n}(\mathbf{q}_1,\mathbf{q}_2)$ can be expressed as:

$$\delta \widehat{n}(\mathbf{q}_1,\mathbf{q}_2) = \iint g(\rho,\rho')\ e^{-i(\mathbf{q}_1\cdot\rho+\mathbf{q}_2\cdot\rho')}\ d\rho\ d\rho' \tag{9}$$

so that, from Eqs. (1) and (8), we get

$$\delta n(\mathbf{r},z) = \frac{1}{|A(\mathbf{r},z)|^2} \iint A(\mathbf{r}-\rho,z)\ A^*(\mathbf{r}+\rho',z)\ g(\rho,\rho')\ d\rho\ d\rho'\ . \tag{10}$$

Note the explicit nonlocal nature of the PR effect, which is brought out by Eq. (10). By inserting Eq. (10) into Eq.(3), the equation of evolution of the electromagnetic field reads

$$(\frac{\partial}{\partial z} - \frac{i}{2k}\nabla_{\mathbf{r}}^2)\ A(\mathbf{r},z) = \frac{ik}{n_1}\frac{1}{A^*(\mathbf{r},z)} \iint A(\mathbf{r}-\rho,z)\ A^*(\mathbf{r}+\rho',z)\ g(\rho,\rho')\ d\rho\ d\rho'\ . \tag{11}$$

where the integral on the right-hand side accounts for the nonlocal nature of the photorefractive effect. In particular, if we look for soliton solutions, we require

$$A(\mathbf{r},z) = U(\mathbf{r})\ e^{i\gamma z} \tag{12}$$

where $U(\mathbf{r})$ and γ are real, γ being the characteristic soliton propagation constant. Eq.(11) becomes then

$$(\gamma - \frac{1}{2k}\nabla_{\mathbf{r}}^2)\ U(\mathbf{r}) = \frac{k}{n_1}\frac{1}{U(\mathbf{r})} \iint U(\mathbf{r}-\rho)\ U(\mathbf{r}+\rho')\ g(\rho,\rho')\ d\rho\ d\rho' \tag{13}$$

The integro-differential equation (13) can be transformed into an ordinary differential equation by using the Taylor expansion of $U(r-\rho)$ around $\rho=0$:

$$U(\mathbf{r}-\rho) = U(\mathbf{r}) - \nabla_\mathbf{r} U(\mathbf{r}) \cdot \rho + \frac{1}{2} [\nabla_\mathbf{r} U(\mathbf{r}) \nabla_\mathbf{r} U(\mathbf{r})] : \rho \rho + \dots \tag{14}$$

along with an analogous expansion for $U(r+\rho')$ around $\rho'=0$, and inserting it into its right-hand side.

Truncating the Taylor expansion to a given order (the second, in our case), requires that the nonlocal influence of the PR effect is restricted to a limited region of a given linear dimension (say d, where d is dictated by the form of $\widehat{\delta n}(q_1, q_2)$) around any position $\mathbf{r}$. It is worthwhile to note, that due to the invariance of $\delta n(\mathbf{r}, z)$ under the exchange $\mathbf{r} \rightarrow -\mathbf{r}$, the odd-order terms of Eq. (14) do not contribute to the right hand side of Eq. (13).

4.   The Two Dimensional Photorefractive Spatial Soliton

For simplicity, we restrict our analysis to a two dimensional case, allowing diffraction in the y direction and looking for self-trapping in the x direction only, or alternatively, examinning self-trapping in a slab PR waveguide, as shown in Fig. 1. A detailed analysis of a three dimensional case will be presented elsewhere (Crosignani et. al 1992). In our two dimensional case $U(\mathbf{r})=U(x)$, and we obtain, by truncating the Taylor expansion after the second term,

$$(\gamma - \frac{1}{2k} \frac{d^2}{dx^2}) U(x) = \frac{km}{n_1} \frac{d^2 U(x)}{dx^2} - \frac{kp}{n_1} \frac{1}{U(x)} \left(\frac{dU(x)}{dx}\right)^2 + \frac{kh}{n_1} \frac{1}{U(x)} \left(\frac{d^2 U(x)}{dx^2}\right)^2 \tag{15}$$

with

$$m = \frac{1}{2} \iint g(\rho, \rho') (\rho^2 + \rho'^2) d\rho\, d\rho' = - \frac{d^2}{dq_1^2} \widehat{\delta n}(q_1, q_2)_{q_1=0, q_2=0} \tag{16}$$

$$p = \iint g(\rho, \rho')\, \rho\ \rho'\, d\rho\, d\rho' = - \frac{d}{dq_1} \frac{d}{dq_2} \widehat{\delta n}(q_1, q_2)_{q_1=0, q_2=0} \tag{17}$$

$$h = \frac{1}{4} \iint g(\rho, \rho')\, \rho^2 \rho'^2\, d\rho\, d\rho' = \frac{1}{2} \frac{d^2}{dq_1^2} \frac{d^2}{dq_2^2} \widehat{\delta n}(q_1, q_2)_{q_1=0, q_2=0} \tag{18}$$

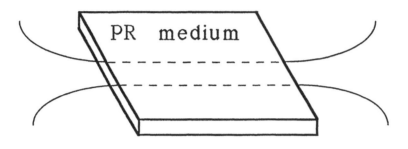

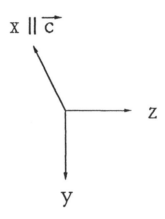

Figure 1- A schematic diagram for observation of a PR spatial soliton in BaTiO$_3$.

where now ρ, ρ' and q_1, q_2 stand for $\rho_x, \rho_{x'}$ and q_{1x}, q_{2x} , and we have also taken advantage of the relations $\widehat{\delta n}(0,0)=0$ [if $\widehat{\delta n}(0,0) \neq 0$ we need to redefine γ by adding a term proportional to $\widehat{\delta n}(0,0)$.] and $\widehat{\delta n}(q_1,q_2) = \widehat{\delta n}(q_2,q_1)$. As will be shown below, h and the higher order terms are smaller than m and p by powers of d, and are therefore neglected. A rearrangement of the terms in Eq.(12) results in

$$a \; U'^2 + b \; UU'' - \gamma \; U^2 = 0 \tag{19}$$

where the prime stands for a derivative with respect to x, and we have set $a = -pk/n_1$ and $b = km/n_1 + 1/2k$. In the special case, where $a = -2b$, Eq. (19) is satisfied by the solution:

$$U(x) = U_0 \; \text{sech}(\alpha x) \tag{20}$$

which is consistent with our requirement of symmetry under the exchange $x \to -x$, and by a proper choice of α with a decay with $|x|$ within a region of linear dimension d. The soliton propagation constant is $\gamma = -b\alpha^2 > 0$, where α has to satisfy the condition $\alpha d \ll 1$ in order to justify the truncation of the Taylor expansion. Note that other solutions for Eq. (19) do exist. A detailed study of the full spectrum of solutions, and their uniqueness properties, will be given elsewhere (Crosignani et. al. 1992).

The nonlinear parameters a and b, and the "effective length" (d) of the PR interaction are determined from $\widehat{\delta n}(q_1,q_2)$. The requirement of a real $\widehat{\delta n}(q_1,q_2)$ for all q_1 and q_2 is equivalent to having a real $E_m(q_1,q_2)$, which in turn implies the application of an external (DC) electric field E_0 to the material. In our case of PR interaction between pairs of plane wave with relatively small angular deviation (spatial components of the same beam, under the paraxial approximation), the limiting space charge field E_p is relatively large and the diffusion field E_d is small. Application of an appropriate external field, such that $|E_d| \ll |E_0| \ll |E_p|$, allows us to neglect the imaginary part of the coupling coefficient, so that

$$E_m(q_1,q_2) \cong E_1(q_1,q_2) \cong \frac{E_0}{(\frac{E_0}{E_p})^2 + 1} \qquad (21)$$

where $E_p = e\, P_d/(\varepsilon_0\, \varepsilon_r\, K_g)$ and, under the paraxial approximation, $K_g = |\mathbf{K_g}| \simeq (q_1 - q_2)$. Note, that E_m now satisfies $E_m(\mathbf{q_1},\mathbf{q_2}) = E_m(-\mathbf{q_1},-\mathbf{q_2})$. In order to maintain this property for $\widehat{\delta n}(q_1,q_2)$ one has to apply this symmetry requirement to $r_{eff}(q_1,q_2)$, and choose the direction of propagation (z) and the polarization of the light beam accordingly. As an example, we consider the coupling coefficient for a field polarized in the x-z plane in $BaTiO_3$ crystal, when the x direction has been adjusted to coincide with the crystalline **c** axis (we neglected the relatively small Pockel's coefficients, other than r_{42}):

$$\widehat{\delta n}(q_1,q_2) \cong \frac{n_1^3}{2}\, r_{42}\, E_0\, \frac{0.5\ (q_1+q_2)^2}{1 + (\frac{E_0\, \varepsilon_0\, \varepsilon_r}{e\, P_d})^2\ (q_1-q_2)^2} \equiv B\, \frac{\frac{(q_1+q_2)^2}{k}}{1 + d^2\ (q_1-q_2)^2} \qquad (22)$$

According to the definition of m, p and h [see Eqs. (16)-(18)] and the expression of $\widehat{\delta n}(q_1,q_2)$, it is a straightforward task to find $m=p=-2B/k^2$ and $h=4Bd^2/k^2$, so that $h=-2d^2m$. We have evaluated the factors B and d by employing the parameters used by Klein (1989), and the condition a=-2b, thus getting B~1 (in dimensionless units) and d 14~μm. The neglect of the "h" term, along with higher order terms, in Eq. (19) is thus justified. It is then possible to show that the condition $\alpha d<1$ is equivalent to $\gamma= -b\alpha^2 = \alpha^2/2k = k(\alpha/k)^2/2 < k$, which can be satisfied by choosing α to be a small fraction of k. Note, that B imposes a positive sign on the externally applied electric field E_0, and a requirement for a large nonlinearity and/or large E_0.

A schematic diagram for observation of a PR spatial soliton in $BaTiO_3$, is shown in Fig. 1. The soliton waveform solution for this configuration, for $\alpha=0.05$ μm^{-1} and at a given instant t_0, is shown in Fig. 2. The vertical axis represents the light wave field amplitude $E(x,z,t=t_0)$ in arbitrary units (a consequence of the PR effect independence of the absolute light intensity $|A|^2$), and the other axes are x and z in μm.

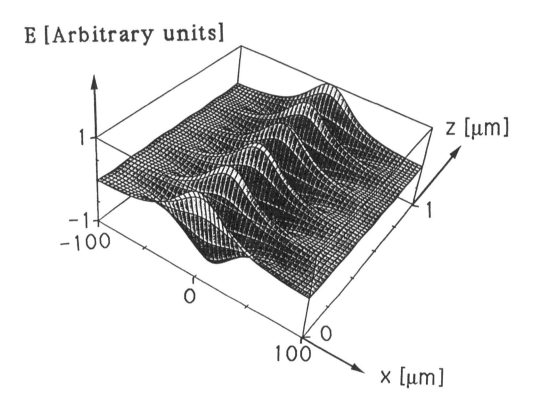

Figure 2- A three dimensional plot of the light electric field E(x,z,t) for a
"frozen" time t=to. The vertical axis gives the amplitude in arbitrary
units, and the horizontal axes are x and z in μm.

5.   The Properties of a Photorefractive Spatial Soliton

Our solution was obtained assuming a real coupling coefficient $\widehat{\delta n}(q_1, q_2)$, which does not allow for energy transfer between plane wave components. As a consequence, the plane wave spectrum of **any** input beam remains unchanged by the medium, and only nonlinear phase changes apply to the individual spatial frequencies. Therefore, an arbitrary input profile cannot evolve into the soliton shape, as in the case with temporal Kerr-like solitons. For that reason, one must start with the correct wave form which compensates exactly for the diffraction, in order to obtain a PR soliton. Once launched, small deviations from the proper solution do neither decay or grow, but maintain both the original deviation and the accompanying diffraction, in a "quasi-stable" situation. A degree of fine tuning of the PR effect (and hence of the soliton width) is allowed by varying the externally applied electric field E_0 (which determines the value of b). Note that for reversal of the polarity of E_0, the linear diffraction and the nonlinear phase are additive and we get "double" the diffraction effect.

The inclusion of absorption (or amplification) in our model results in a soliton which maintains its transverse profile even as the total light intensity increases (gain) or attenuates with propagation. In the absorption case, for example, a linear term $\sigma A(r, z)$ is added to the left hand side of Eq. (8), and the soliton propagation constant γ is allowed to be complex. If we take the imaginary part of γ equal to $-\sigma$, we still get Eq. (10), with γ replaced by its real part γ_r. The resulting soliton is $U(r)\, e^{i\gamma_r z - \sigma z}$, and the transverse structure remains unchanged.

Material considerations are of great importance for the practical realization of a PR soliton. We look for a PR medium in which the diffusion field E_d (responsible for the imaginary part of $\widehat{\delta n}(q_1, q_2)$) is as small as possible, but which still presents a strong PR nonlinearity. Another reason for trying to avoid the energy transfer process is the strong noise amplification mechanism, ("fanning"), which is present in all the PR materials with an imaginary $\widehat{\delta n}(q_1, q_2)$. We expect this effect to be very small in our case, both for the above reason (very small E_d) and because of the small cross-section of interaction with spontaneously scattered noise (Segev et. al 1990a). Recently developed quadratic materials, that belong to the KTN group(See Agranat et. al. 1988 and 1989), can be excellent candidates since they present a very strong nonlinearity with externally applied electric field, and inherently do not support the energy transfer process.

6.   Conclusions

In conclusion, we have presented a new type of a soliton which is based on the photorefractive nonlinearity, discussed the unique properties of this soliton, and considered the conditions necessary for observing it. As an example, we solved for a special two dimentional case, where the light induced index gratings are in phase with the interference gratings (the PR coupling coefficient is real). A study of the general case of a PR medium, and an investigation of the spectrum of the solutions of the soliton equation are currently under way.

References

Agranat A. and Yacoby Y. 1988, J. Opt. Soc. Am. B 5, 1792.

Agranat A., Leyva V. and Yariv A. 1989, Opt. Lett. 14, 1017.

Aitchinson J.S., Weiner A.M., Silberberg Y., Oliver M.K., Jackel J.L., Leaird D.E., Vogel E.M. and Smith P.W. 1990, Opt. Lett. 15, 471.

Crosignani B. and Yariv A. 1984, J. Opt. Soc. Am. A 1, 1034 .

Crosignani B., Segev M., Engin D., Di Porto P. and Yariv A. 1992, to be published.

Klein M.B., Chapter 7, in *Photorefractive Materials and their Applications,* Vol. 1, edited by P.Gunther and J.P.Huignard (Springer-Verlag, Berlin, 1989).

Mamaev A.V. and Shkunov V.V. 1990, Sov. J. Quant. Elect. 19, 1199.

Segev M., Ophir Y. and Fischer B. 1990a, Opt. Comm. 77, 265.

Segev M., Ophir Y. and Fischer B. 1990b, Appl. Phys. Lett. 56, 1086.

Segev M. and Yariv A. 1991, Opt. Lett. 16, 1938.

Solimeno S., Crosignani B. and Di Porto P. 1986, *Guiding, Diffraction and Confinement of Optical Radiation*, Chapter 8 (Academic Press, Orlando).

Temple D.A. and Warde C. 1986, J. Opt. Soc. Am. B 3, 337 .

Vachss F. and Hesselink L. 1988, J. Opt. Soc. Am. A 5, 690.

White J.O., Kwong S.K., Cronin-Golomb M., Fischer B. and Yariv A. 1989, Chapter 4, in *Photorefractive Materials and their Applications,* Vol. 2, edited by GuntherP. and Huignard J.P. (Springer-Verlag, Berlin).

Solitons in quantum nonlinear optics

Raymond Y. Chiao and Ivan H. Deutsch
Department of Physics, University of California, Berkeley, California 94720, U.S.A.

John C. Garrison
Lawrence Livermore National Laboratory, University of California, Livermore, California 94550, U. S. A.

Ewan M. Wright
Optical Sciences Center and Department of Physics, University of Arizona, Tucson, Arizona 85721, U.S.A.

Abstract: Alkali vapors excited by light near resonance are known to possess very large nonlinear optical coefficients n_2, which can have either sign depending on the sign of the detuning. Spatial solitons in a Fabry-Perot resonator filled with such vapors can result from these nonlinearities. The nonlinear Schrödinger equation (NLS) obeyed by these solitons is formally equivalent to a nonrelativistic many-body field theory, in which photons behave like nonrelativistic bosons with pairwise interactions. When n_2 is positive, these interactions are attractive; when n_2 is negative, they are repulsive. With n_2 positive, it is predicted that a self-trapped, 1D spatial soliton (the hyperbolic secant solution to the 1D NLS) should be observable, and we plan to perform soliton-soliton collision experiments with such solitons. At the quantum level, the light inside such a soliton should be self-squeezed, and we plan to observe this using the Shirasaki-Haus method. The possibility of observing the two-photon bound state, the diphoton, will be explored. With n_2 negative, a 1D spatial dark soliton with antibunching of photons should result. Moreover, with n_2 negative, a 2D vortex soliton should form, and we plan to examine the dynamics of these photonic vortices at both classical and quantum levels. The possibility that these vortices obey anyonic statistics will be explored. These optical systems thus constitute an experimentally accessible arena in which many-body phenomena analogous to those in condensed matter physics, especially topological ones, such as the Kosterlitz-Thouless phase transition, superfluidity, and anyonic statistics, may be observed.

TABLE OF CONTENTS

1   Introduction

In memory of S. A. Akhmanov, it is fitting to recall his many contributions to nonlinear optics. He was an inspiring leader in our field, and we shall miss him greatly. One of his last papers was "Controlling transverse-wave interactions in nonlinear optics: generation and interaction of spatio-temporal structures", (Akhmanov et al. 1992). Of these spatio-temporal structures, one was a vortex-like structure, which is similar to one of the solitons to be discussed below.

Solitons have become a central paradigm of nonlinear science. Mathematically, they are important because they are the nonlinear analogs of eigenmodes in linear systems. Physically, they are important because they are stable, particle-like objects, which survive collisions with each other. Not only are these solitons interesting in their own right, but they also possess properties which lend themselves to applications. For example, because of their stable and robust characteristics, they have been proposed as the natural carriers of information in optical fibers. They have also been proposed for optical switching and other information-processing applications (Hasegawa 1989).

Many kinds of solitons are exhibited in nonlinear optical systems. In this review we shall be concentrating mainly on *spatial* soliton solutions of the nonlinear Schrödinger equation, or the Ginzburg-Landau equation, which governs the transverse spatial profile of a light beam propagating through a medium with an intensity-dependent index of refraction $n=n_0+n_2|\mathcal{E}|^2$. Although a complete test to determine which nonlinear wave equations possess soliton solutions has yet to be formulated, some general theorems establish necessary conditions for the existence of solitons (Coleman 1985). The character of the soliton and the conditions for its existence depends strongly on the dimensionality of the configuration space as well as the sign of the nonlinearity. In one spatial dimension solitons exist for either sign of n_2; $n_2>0$ gives rise to self-trapped beams of light with hyperbolic secant profile (Chiao et al. 1964); $n_2<0$ yields "kink" solitons consisting of dark stripes in the light field which are immune to Fresnel diffraction (Swartzlander et al. 1991). In two dimensions, no soliton solutions exist for $n_2>0$. There is a solitary wave solution which is known to be unstable, and can collapse to a self-focusing singularity (Kelley 1965). Wright (1992) has shown that vacuum fluctuations can initiate this collapse, leading to quantum variations in the self-focusing length. Soliton solutions do exist in two dimensions for $n_2<0$; they represent vortices in the optical field (Arrechi et al. 1991, Akhmanov et al. 1992). The practical ease with which we can control both the dimensionality of the system and the sign of the nonlinearity makes nonlinear optics a natural arena for exploring the physics unique to each regime.

When one quantizes the field, these solitons lead to some fascinating new phenomena. One of these is that the soliton undergoes squeezing in the presence of the nonlinearity. We suggest that this can be observed in the 1D spatial soliton by means of the elegant technique of Shirasaki and Haus (1990). The underlying quantum field theory associated with these solitons is equivalent to a nonrelativistic many-body problem for an gas of massive bosons with pairwise interactions (Lieb and Liniger 1963, Kaup 1975, Faddeev & Takhtajan 1987,

Creamer 1980, Sogo et al. 1981). The sign of the nonlinearity determines whether the interactions are attractive or repulsive. These facts mean that the predictions of many-body theories concerning the dilute, interacting Bose gas which have been gleaned over the years can now be tested in actual experiments. For example, in the one-dimensional attractive Bose gas corresponding to the self-trapped light beam, bound-state solutions exist (Chiao et al. 1991). The possibility of observing the two-photon bound state will be discussed here. In the case of the vortex soliton the phenomenon of superfluidity arising from the vortex-binding, or the Kosterlitz-Thouless transition (Kosterlitz & Thouless 1973), and observed in thin films of helium (Bishop & Reppy 1978), should be manifested in the photonic case as well. Perhaps the most intriguing possibility of all for experiments on the many-body system is the detection of identical quasiparticles which obey anyonic statistics. These intermediate statistics interpolate between Bose and Fermi statistics, and are a topological effect possible only in one or two-dimensional many-body physics; they do not exist in three dimensions. Thus, optical solitons provide us with a rich variety of many-body effects. The connections between the various classical solitons and the phenomena associated with their underlying field theories is summarized in Table 1.

Table 1.   Connections between classical solitons and the underlying quantum field theory.

		Classical Soliton	Quantum/Many-Body Effect
1D	$n_2>0$	"Sech" bright soliton (self-trapped beam)	Photonic bound states; Squeezing
	$n_2<0$	"Kink" (dark) topological soliton	Photon antibunching
2D	$n_2>0$	No soliton solutions	Quantum self-focusing singularity
	$n_2<0$	"Vortex" topological soliton	Kosterlitz-Thouless superfluidity; Anyonic statistics

The remainder of the paper is organized as follows. In Sect. 2 we review the classical soliton. We start by considering an experimental realization of our nonlinear equations. This consists of a single longitudinal mode Fabry-Perot resonator filled with an alkali vapor, which can easily be implemented with either sign of the nonlinearity, and in one or two transverse dimensions. The solutions to the nonlinear Schrödinger and Ginzburg-Landau equations are then analyzed. One new result reported here for the first time is that the effective mass of the

1D sech and kink solitons is exactly equal to the effective mass of the photon in the underlying quantum field theory. Another new result is that a gas of optical vortices can spontaneously arise from a *vector* plane wave due to a modulational instability in a self-defocusing medium. In Sect. 3 we consider the quantum field aspects of this problem. First we establish the equivalence between the nonlinear Schrödinger equation and the nonrelativistic interacting Bose gas. We then go on to apply our model to the problems of photonic bound states, the production of squeezed states, and the possibility of Kosterlitz-Thouless superfluidity in the 2D interacting photon gas. Having explored the microscopic theory, we go on to identify the macroscopic, collective coordinates of the soliton, and to "requantize" its motion. This leads to a new result: the macroscopic phenomenon of soliton tunneling in 1D. Also, this leads naturally to the possibility of anyonic statistics for identical vortex solitons.

2   Classical optical solitons

2.1   *Nonlinear system : A Fabry-Perot filled with an alkali-vapor Kerr medium*

Consider a Fabry-Perot resonator filled in its interior by alkali vapor atoms in their ground state. The atoms constitute a nonlinear medium for the light. We shall call this system a "nonlinear Fabry-Perot" (NLFP). Although the nonlinear Fabry-Perot has been previously explored in connection with optical bistability (Szöke et al. 1969, Gibbs 1985), we point out here that there are new features, such as topological solitons in the form of 2D photonic vortices, and possible novel quantum phenomena associated with them, which we believe are worthy of further exploration. For simplicity, let us choose the spacing L between the two parallel mirrors of the Fabry-Perot to be near half a wavelength, so that only the fundamental longitudinal mode will be excited. In this way, the longitudinal degrees of freedom for the photons are frozen out, and the dimensionality of their motion is reduced from three to two. The longitudinal axis of the Fabry-Perot will be called the z-axis, and the two transverse dimensions inside the resonator will be called the x and y axes. Let us excite the fundamental longitudinal mode by means of a laser beam incident along the z axis whose frequency ω is detuned from the resonance fluorescence frequency ω_0 of the alkali atoms by about a Doppler width (for atoms at around room temperature). As a result, there is little absorption, but there is a resonantly enhanced intensity-dependent change in the index of refraction, $\Delta n = n_2|\mathcal{E}|^2$, where the nonlinear, or Kerr effect, coefficient n_2 is given by (Grischkowshky 1970)

$$n_2 = \pi N \mu^4 / [n_0{}^2 \hbar^3 (\omega - \omega_0)^3] \,, \tag{1}$$

in the two-level model of the atom. Here N is the number density of alkali atoms, μ is the electric dipole matrix element for the transition, and n_0 is the linear index of refraction at ω. For example, for rubidium vapor excited by a diode laser tuned one Doppler width away from the D2 resonance at $T=100°C$, where the atomic vapor density is $N=4.6 \times 10^{12}$ cm^{-3} one obtains a nonlinear coefficient $n_2 = 1.6 \times 10^{-3} cm^3/erg$, or $1.3 \times 10^{-5} cm^2/Watt$ in intensity units. This is eight orders of magnitude larger than the n_2 coefficient of liquid carbon disulphide. Therefore we shall see below that it is possible to create solitons with cw diode lasers. The sign of n_2 is *positive*, if the laser is tuned *above* resonance, leading to self-focusing effects (Bjorkholm & Ashkin 1974). The sign of n_2 is *negative*, if the laser is tuned *below* resonance, leading to self-defocusing effects (Swartzlander et al. 1991). The ease with which one can switch the sign of the nonlinearity will be important for the program outlined below.

2.2   *The nonlinear Schrödinger and Ginzburg-Landau Equations: Classical soliton solutions in one and two dimensions*

Here we outline how the nonlinear Schrödinger equation arises from Maxwell's equations. We shall first consider *classical* electromagnetic fields. For the nonlinear Fabry-Perot described above, it is natural to start with the Ansatz

$$E(x,y,z,t) = \tfrac{1}{2}\{\ \mathcal{E}(x,y)\ \sin(\tfrac{\pi z}{L})\ \exp(-i\omega t) + c.c.\ \}, \tag{2}$$

where the complex transverse envelope function $\mathcal{E}(x,y)$ is assumed to vary slowly on the scale of a wavelength, and where infinitely conducting boundary conditions are imposed at the mirrors of the Fabry-Perot, so that $kL = \pi$ for the fundamental mode. We choose the direction of the linearly polarized electric field to be along the x-axis. When this Ansatz is substituted into Maxwell's equations in the form of a wave equation with time and space-averaged nonlinear polarization source terms, we obtain the nonlinear Schrödinger equation (NLS)

$$\nabla_T^2 \mathcal{E} + A\mathcal{E} + B|\mathcal{E}|^2 \mathcal{E} = 0, \tag{3a}$$

$$A = \omega^2 n_0^2/c^2 - \pi^2/L^2, \qquad B = 2\Gamma\omega^2 n_0 n_2/c^2 \tag{3b}$$

where $\nabla_T^2 = \partial^2/\partial x^2 + \partial^2/\partial y^2$ is the transverse Laplacian, and

$$\Gamma \equiv \frac{2}{L}\int_0^L dz\ \sin^4\!\left(\frac{\pi z}{L}\right) = \frac{3}{4}, \tag{3c}$$

is a mode normalization factor. Eq. (3a) is formally identical to the two-dimensional Ginzburg-Landau equation for a complex order parameter Ψ,

$$(-\hbar^2/2m)\nabla_\perp^2 \Psi + \alpha\Psi + \beta|\Psi|^2\Psi = 0 , \tag{4}$$

which applies, for example, to two-dimensional thin films of superfluid helium, if m is the mass of a ^{4}He atom. When the transverse Laplacian is replaced by the minimal-coupling or covariant-derivative form including the electromagnetic vector potential, it applies to type II superconducting thin films. There exist known vortex soliton solutions to the Ginzburg-Landau equation, including the Abrikosov vortex solution (DeGennes 1966). Since Eqs. (3) and (4) are identical, whenever they have the same boundary conditions (see below), they must have the same solutions. In particular, vortex solutions exist in the nonlinear Fabry-Perot when the coefficient n_2 is negative, i.e., for the self-defocusing case. Recently, similar solutions have been observed in ring resonators (Arecchi et al. 1991, Akhmanov et al. 1992). In addition, for this sign of the nonlinearity, one-dimensional "kink" soliton solutions of the Ginzburg-Landau equations are also known to exist (DeGennes 1966) (see below), and these correspond in nonlinear optics to dark solitons recently observed in sodium vapor (Swartzlander et al. 1991).

In connection with superfluid ^{4}He vortices, Fetter has shown that there exist approximate solutions to Eq. (4). Assuming the polar form,

$$\Psi(r,\theta) = \rho(r)^{1/2} \exp(\pm i\theta) , \tag{5}$$

where $r=(x^2 + y^2)^{1/2}$ and $\theta=\arctan(y/x)$, one obtains the approximate solution

$$\rho(r) = \rho(\infty) \, (r/r_0)^2/(1 + (r/r_0)^2) , \tag{6}$$

where r_0 is the vortex core radius (Fetter 1965). This solution satisfies the boundary conditions $\Psi \to$ some constant times $\exp(\pm i\theta)$, as $r\to\infty$. Solutions satisfying these boundary conditions also exist for the photonic case.

There is a general theorem, Derrick's theorem (Coleman 1985), which tells us that soliton-like solutions of Eq. (3), i.e., square-integrable solutions in *two* transverse dimensions, are unstable for the case $n_2 > 0$, i.e., the self-focussing case, whereas they are stable in *one* transverse dimension. For if such 2D solitons existed, they would be unstable against collapse, as a scaling argument shows. However, this theorem does not apply to the above vortex solutions for the case $n_2 < 0$, since these solutions are not square-integrable (for the vortex soliton, the boundary conditions at infinity are such that the density $\rho(r)$ approaches a constant $\rho(\infty)\neq 0$). In light of Derrick's theorem, we shall not consider here any two-dimensional solutions of Eq. (3) for the case $n_2 > 0$.

In the absence of the Fabry-Perot, when the nonlinear coefficient n_2 is positive, there can occur the phenomena of self-focussing and self-trapping. These traveling-wave phenomena were seen in sodium vapor in an early experiment by Bjorkholm and Ashkin (1974). The traveling-wave nonlinear Schrödinger equation, with an additional $i\partial/\partial z$ term obtained in the paraxial approximation, describes the propagation of light in such self-focusing media (Kelley 1965). Traveling-wave solitons, which correspond to self-trapped beams of light in one transverse dimension, are known to exist for this equation in planar geometries (Chiao et al. 1964). Such (1+1) dimensional classical solitons (i.e., those in 1 transverse and 1 longitudinal spatial dimension) have recently been observed in electronic Kerr media with a positive n_2 (Aitchison et al. 1990). The stability of these solitons under soliton-soliton collisions has also been observed (Aitchison et al. 1991).

In the presence of the Fabry-Perot, one might expect analogous standing-wave solitons to exist. However, since the problem is a nonlinear one, one cannot invoke the superposition of two counterpropagating waves to solve the standing-wave soliton problem. Nevertheless, one-dimensional standing-wave solitons, i.e., nonlinear eigenmode solutions of the form given by Eq. (2), do exist. Note that the dependence of the slowly varying envelope function $\mathcal{E}$ on the longitudinal variable z is now eliminated altogether. Assume that $\mathcal{E}(x)$ is a real function of x only, and Eq. (3) becomes

$$d^2\mathcal{E}/dx^2 + A\mathcal{E} + B\mathcal{E}^3 = 0 . \tag{7}$$

For the case $n_2 > 0$, the constant B is positive, and the constant A must be negative, i.e., $A=-|A|$, in order to satisfy the boundary conditions $\mathcal{E}\to0$ as $x\to\pm\infty$, which are required for the square-integrability of the solution. In this case, Eq. (7) possesses an exact solution (Chiao et al. 1964)

$$\mathcal{E}(x) = (2|A|/B)^{1/2} \text{ sech } [(|A|)^{1/2}x] . \tag{8}$$

We shall call this one-dimensional soliton a "sech". The meaning of this solution is that when a nonlinear Fabry-Perot with lossless mirrors is illuminated with an external plane wave with the above transverse profile, it will completely transmit this profile. Hence the nonlinear Fabry-Perot is not only a spectral filter, but also a nonlinear spatial filter with unity transmission coefficient for the above hyperbolic secant profile. For such a Fabry-Perot of finesse 10 filled with rubidium vapor with $n_2= 1.3\text{x}10^{-5}$ cm^2/Watt, a cw diode laser beam with a power of 1 milliwatt, focussed by means of a cylindrical lens to a line focus of 10 μm by 1 mm on its front mirror, will generate such a sech soliton with a 2.4 μm width. An interesting question then arises: Does a coherent state with a hyperbolic secant profile which perfectly matches into the sech solution, remain a coherent state after transmission through the nonlinear Fabry-Perot in its soliton mode, or does it turn itself into a squeezed state? We shall discuss this below.

In the absence of the Fabry-Perot, the ordinary differential equation, Eq. (7), and its sech solution, Eq. (8), also apply to traveling-wave solitons in (1+1) dimensions (Chiao et al. 1964). The fact that $\mathcal{E}(x)$ can be chosen to be real without loss of generality in (1+1) dimensions reflects the underlying global $U(1)$ gauge invariance of the theory. This means that the soliton solutions have uniform phase distributions (i.e., they are "uniphase"). This remarkable property of optical solitons in (1+1) dimensions, that they are characterized by a single global phase, is the basis of all proposed applications of these solitons to all-optical switching.

The above solution can be easily generalized to solutions of the form (Zakharov & Shabat 1972)

$$E(x,y,z,t) = \frac{1}{2}\{ \mathcal{E}(x-vt) \sin(\frac{\pi z}{L}) \exp(i[qx-\omega t]) + \text{c.c.} \}, \tag{9}$$

where q represents a small, x-directed wavevector of a "sliding" soliton. In the slowly varying envelope approximation, it can be shown that the solution, Eq. (8), generalizes to

$$\mathcal{E}(x-vt) = (2|A|/B)^{1/2} \text{ sech } [(|A|)^{1/2}\{x-vt\}] \ , \tag{10}$$

where

$$v = (q/\omega)(c^2/n_0{}^2) \tag{11}$$

represents the x-directed velocity of the sliding motion of the soliton. These solutions permit soliton-soliton collisions in the Fabry-Perot geometry, since we can launch two solitons towards each other, and observe if they survive the collision. If we view the soliton as a particle obeying de Broglie's law, its momentum p in the x-direction is given by $p=\hbar q$. It follows then from Eq. (11) that the sliding soliton behaves like a nonrelativistic particle moving in one dimension with an effective mass

$$M_{eff} = \hbar \omega n_0{}^2/c^2 \ . \tag{12}$$

Maki and Kodama (1986) have shown that it is necessary to quantize the motion of the collective coordinates of the soliton, in particular, its center of mass motion. We shall analyze this point in more detail in Sect. (3.5).

Soliton-soliton collision experiments can be implemented by means of a method similar to that used by Aitchison et al. (1991) in the case of traveling-wave (1+1) dimensional sech solitons. One simply aims two sech-profile laser beams, with a slight inclination angle between them in the x-z plane, at the input face of an alkali vapor cell (without mirrors) such that they cross each other in the middle of the cell. In this way, two colliding sech solitons are launched inside the cell. Observation of their profiles after collision will reveal if the solitons have survived the collision. Also, details of their interaction, such as their repulsion, attraction, and phase shifts accumulated during collision, can be measured. Similarly, collisions of standing-wave solitons in the NLFP can be studied, but the details of how to launch these solitons need to be worked out.

For the self-defocussing case $n_2 < 0$, the constant B is negative, and the constant A must be positive, in order to satisfy the boundary conditions, $|\mathcal{E}(x)|^2 \rightarrow |\mathcal{E}(\infty)|^2 \neq 0$ as $x \rightarrow \pm\infty$, of the dark soliton. In this case, Eq. (7) possesses the solution (DeGennes 1966)

$$\mathcal{E}(x-vt) = (A/|B|)^{1/2} \tanh \left[(A/2)^{1/2} \{x-vt\} \right] \ , \tag{13}$$

where v is given again by Eq. (11). The effective mass given by Eq. (12) also applies to this soliton. Note that the solution possesses a zero, or node, at $x=vt$. This zero corresponds to the center of a dark soliton, which we shall call the "kink". Since the constituent microscopic particles of the soliton (here, photons) are excluded from the zero of the order parameter $\mathcal{E}$, there is a topological hole (i.e., a defect or obstruction) at the kink. This is also true at the core of the vortex soliton, since the x-y plane appears to be punctured at the zero of the order parameter. As far as the constituent particles are concerned, they are excluded from this zero. Hence both the kink and the vortex are classified as "topological solitons", while the sech, which has no such zeros, is classified as a "nontopological soliton" (Coleman 1985).

There are two main reasons why there is so much current interest in topological solitons: First they possess a conserved quantum number known as a "topological charge". For example, in the case of the superfluid helium vortex, its quantized vorticity,

$$\oint \mathbf{v} \cdot d\mathbf{l} = \pm h/m \ , \tag{14}$$

which is the Onsager-Feynman quantization condition, is its topological charge. Note that the vortex can be either positively or negatively "charged". However, the total topological charge of a vortex system, like the total electrical charge, is a conserved quantity. Second, the forces which these vortex exert on each other are long-range, Coulomb-like forces. Hence topological solitons are good models of charged particles in lower dimensional systems.

Many of these solitons arise from the topological features of the underlying many-body problem in *two* spatial dimensions. At the root of these topological phenomena is the fact that in two dimensions, e.g., on a plane, the removal of a single point, i.e., the creation of a hole by puncturing the plane locally, radically changes the topology of the plane globally. Topological quantum numbers, such as the winding number around the hole, can now arise. For example, the Aharonov-Bohm effect can arise from the winding motion on a plane of a charged particle around a thin solenoid which threads through the hole. Now upon puncturing yet more holes into the plane, one further changes its topology with each additional hole. The fundamental topological group for paths on the plane $\mathfrak{R}^2$ punctured by N holes is the braid group $B_N(\mathfrak{R}^2)$, which enumerates the types of braids which can be woven with N strings. As a result, there arises the possibility that identical particles or quasiparticles which are constrained to move on a plane do not obey the usual Bose or Fermi statistics, but obey instead more general statistics, called anyonic (also known as theta, fractional or intermediate) statistics (Leinaas & Myrheim 1977, Goldin et al. 1981, Wilczek 1982, Goldin & Sharp 1983, Wu 1984). We shall discuss this point further in Sect. 3.6.

2.3   *Vortex gas nucleation*
To study the classical and quantum dynamics of two-dimensional optical vortices, it is first

necessary to identify a nonlinear process which can produce the vortices. Ideally we would like to produce a vortex gas of variable density imbedded in a common background, the gas being rare if the mean vortex separation is much larger than the vortex core radius, and dense if the separation approaches the core radius. We propose modulational instability in a self-defocusing nonlinear medium as a mechanism by which a two-dimensional gas of optical vortices can nucleate starting from an initial plane wave electric field. Modulational instability is a nonlinear process that causes an initially homogeneous state to develop spatial structure with a characteristic period (Bespalov & Talanov 1966, Chiao et al. 1966). In the present optical context the initial plane wave plays the role of the condensate in superfluidity since all the energy is concentrated in a single wavevector. The vortex gas will allow us to study the Kosterlitz-Thouless phase transition and superfluidity in the context of nonlinear optics, and also to investigate vortices obeying anyonic statistics in the corresponding quantum theory.

As a model we consider a traveling-wave electric field in a nonlinear medium. For an isotropic medium the third-order nonlinear polarization can be written in the general form (Maker et al. 1964a)

$$\mathcal{P}^{(3)} = 3\chi_{1111}^{(3)}\left[a\left(\mathcal{E}\cdot\mathcal{E}^{*}\right)\mathcal{E} + \frac{1}{2}b\left(\mathcal{E}\cdot\mathcal{E}\right)\mathcal{E}^{*}\right] \tag{15}$$

where $\mathcal{E}$ is the slowly varying vector electric field envelope, and where

$$a = (\chi_{1122}^{(3)} + \chi_{1212}^{(3)})/\chi_{1111}^{(3)} \tag{16a}$$

$$b = 2\chi_{1221}^{(3)}/\chi_{1111}^{(3)} \tag{16b}$$

and $a + b/2 = 1$. For example, for the molecular-orientation Kerr effect in liquids $a = 1/4$, $b = 3/2$, and for an electrostrictive nonlinearity $a = 1$, $b = 0$ (Maker & Terhune 1964b). In alkali vapors the values of a and b depend on the atomic angular momentum (Saikan & Kiguchi 1982). The nonlinear change in refractive-index for a linearly polarized field can be expressed as $\Delta n = n_2|\mathcal{E}|^2$, $n_2 = 3\pi\chi_{1111}^{(3)}/2n_0$, and $n_2 < 0$ for a self-defocusing nonlinear medium. Then using Eq. (15) in Maxwell's equations and employing a circularly polarized basis yields the pair of coupled nonlinear equations for a self-defocusing medium (Maker et al. 1964a)

$$\frac{\partial \mathcal{E}_+}{\partial z} = \frac{i}{2k}\nabla_{\mathrm{T}}^2 \mathcal{E}_+ - ik_0|n_2|\left[a|\mathcal{E}_+|^2 + (a+b)|\mathcal{E}_-|^2\right]\mathcal{E}_+ \tag{17a}$$

$$\frac{\partial \mathcal{E}_-}{\partial z} = \frac{i}{2k}\nabla_{\mathrm{T}}^2 \mathcal{E}_+ - ik_0|n_2|\left[a|\mathcal{E}_-|^2 + (a+b)|\mathcal{E}_+|^2\right]\mathcal{E}_- \tag{17b}$$

where $\mathcal{E}_+$ and $\mathcal{E}_-$ are the envelopes for the right and left circularly polarized field components,

∇^2_T is the transverse Laplacian describing beam diffraction, and $k_0 = n_0 \omega/c$, n_0 being the background refractive index. Here we have assumed that the electric field can be treated as transverse to the propagation (z) axis consistent with the usual paraxial approximation (Lax et al. 1975).

First we consider the vector vortex solutions of Eqs. (17). If we set $\mathcal{E}_- = e^{i\phi}\mathcal{E}_+$ where ϕ is a constant phase, then

$$\frac{\partial \mathcal{E}_+}{\partial z} = \frac{i}{2k}\nabla^2_T\mathcal{E}_+ - 2ik_0\,|n_2|\,|\mathcal{E}_+|^2\,\mathcal{E}_+ \tag{18}$$

and similarly for $\mathcal{E}_-$. These correspond to linearly polarized solutions. For the case of circular polarization with e.g. $\mathcal{E}_- = 0$, the factor 2 in the nonlinear term is replaced by a. We therefore replace the factor of 2 by the general coefficient γ, $\gamma = a$ for circular polarization, $\gamma = 2$ for linear polarization. Equation (18) is now of the same form as Eq. (3) and therefore admits vortex solutions corresponding to both linear and circular polarizations. The vortex core radius is given by (see Eq. (6))

$$r_0^2 = \frac{1}{2\gamma k k_0|n_2|\mathcal{E}_\infty^2}, \tag{19}$$

where $\mathcal{E}_\infty^2$ is the vortex field strength as $r \to \infty$. Initial indications are that the linearly polarized vector vortices are unstable under propagation whereas the circularly polarized vector vortices are stable. This means that the circularly polarized vector vortices will be the dominant excitations which determine the system dynamics. We therefore restrict our attention to these vortices and set $\gamma = a$ hereafter. This means that the relevant core radius is dictated by the a parameter in Eq. (16) as shown in Eq. (19). An estimate of the expected core size can be obtained using an induced index $|n_2|\mathcal{E}_\infty^2 \approx 10^{-4}$, $n_0 \approx 2$, $r_0 \approx 20\lambda_0$. This justifies our use of the paraxial approximation (Lax et al. 1975).

Next we consider the evolution of a linearly polarized condensate with a small perturbation added. Equations (17) have exact linearly polarized plane-wave solutions of the form

$$\mathcal{E}_+(x,y,z) = \mathcal{E}_p(z) = \mathcal{E}_0\,\exp\{-2ik_0\,|n_2|\,|\mathcal{E}_+|^2z\} \tag{20}$$

with $\mathcal{E}_-(x,y,z) = e^{i\phi}\mathcal{E}_p(z)$, and $\mathcal{E}_0$ is the field amplitude. Consider now adding small spatial modulations $\varepsilon_\pm(x,y,z)$ to each circularly polarized component. Then linearizing equations (17) with respect to the perturbations yields the following equations

$$\frac{\partial \varepsilon_+}{\partial z} = \frac{i}{2k}\nabla^2_T\varepsilon_+ - ik_0\,|n_2|\,|\mathcal{E}_0|^2\big[a\,(\varepsilon_+ + \varepsilon_+^*) + (a+b)(\varepsilon_- + \varepsilon_-^*)\big] \tag{21a}$$

$$\frac{\partial \varepsilon_-}{\partial z} = \frac{i}{2k} \nabla_T^2 \varepsilon_- \ - ik_0 |n_2| |\mathcal{E}_0|^2 \left[a\,(\varepsilon_- + \varepsilon_-^*) \ + (a+b)(\varepsilon_+ + \varepsilon_+^*) \right] \tag{21b}$$

If we considered a circularly polarized solution with e.g. $\mathcal{E}_- = 0$, then the equation for the perturbation ε_+ would be the same as Eq. (21a) without the term proportional to $(a+b)$. The resulting equation is identical in form to that for the scalar NLS, and it is well known that the condensate is stable for a self-defocusing nonlinearity (Bespalov & Talanov 1966). The circularly polarized condensates are therefore stable. The reason is that the diffractive term and (linearized) nonlinear term in Eq. (21a) have opposite signs and in this case modulational instability is impossible. However, for the linearly polarized condensate described by Eqs. (17) this can be changed (Agrawal 1990). Here we make the Ansatz

$$\varepsilon_-(x,y,z) = -\varepsilon_+(x,y,z) = -\varepsilon(x,y,z), \tag{22a}$$

so that the two perturbations are π out of phase. Equations (21) then reduce to

$$\frac{\partial \varepsilon}{\partial z} = \frac{i}{2k} \nabla_T^2 \varepsilon \ + \ ibk_0 |n_2| |\mathcal{E}_0|^2 (\varepsilon + \varepsilon^*). \tag{22b}$$

The signs are the same on the two terms, and modulational instability is now possible if $b > 0$. If we now seek a spatially periodic solution of the form

$$\varepsilon(x,y,z) = A_0 e^{gz} \sin(Kx/\sqrt{2}) \sin(Ky/\sqrt{2}) \ , \tag{23}$$

where K is the spatial frequency, then we find for the growth rate g

$$g^2 = \frac{K^2}{2k} \left(2bk_0 |n_2| |\mathcal{E}_0|^2 \ - \frac{K^2}{2k} \right). \tag{24}$$

The solution is then unstable for the range of spatial frequencies $K < K_{max} = \sqrt{4bkk_0 |n_2| |\mathcal{E}_0|^2}$, and the peak gain occurs for $K_p = \sqrt{2bkk_0 |n_2| |\mathcal{E}_0|^2}$. The spatial scale r_{mi} associated with the wave vector of peak gain is then

$$r_{mi}^2 \approx \frac{1}{2bkk_0 |n_2| |\mathcal{E}_0|^2} \ . \tag{25}$$

Note that the spatial scale of the modulational instability is dictated by the b parameter in Eq. (16b).

We now discuss the conditions under which the modulational instability described above can serve to nucleate the formation of a two-dimensional vortex gas from an initial plane wave field which is linearly polarized. Clearly we require the b parameter in Eq. (16b) to be nonzero and positive; otherwise the linearly polarized condensate is stable as in the scalar case (Bespalov & Talanov 1966). We also require that $r_0 > r_{mi}$ when $\mathcal{E}_\infty = \mathcal{E}_0$ is substituted in Eq. (19) with $\gamma = a$. Physically this means that the high spatial frequencies $K_v \approx 2\pi/r_0$ associated with the circularly polarized vortices are within the band of unstable spatial frequencies of the modulational instability $K_p \approx 2\pi/r_{mi}$ (see Eq. (25)). In this way the modulational instability generates the full range of spatial frequencies required for vortex formation. If this condition is not satisfied vortices cannot form and the modulational instability leads to the formation of stripe dark solitons, or roll patterns, with $K \approx K_p$ (Swartzlander et al. 1991). Using Eq. (19) and (25) the condition $r_0 > r_{mi}$ can be expressed in the form

$$b/a > 1, \tag{26}$$

which is a basic material requirement for the proposed mechanism to lead to vortex nucleation. A potential candidate medium is then Kerr liquids for which $a = 1/4$, $b = 3/2$. Alkali vapors are also prime candidates since the values of the a and b parameters can be varied by selecting the total angular momentum values J of the atomic lines (Saikan & Kiguchi 1982).

So far the discussion has been limited to the initial development of the modulational instability described by the linearized analysis given. The rationale for obtaining vortex gas nucleation is as follows: As the modulation instability develops, the initial condensate develops deep modulations which can create zeros in the two-dimensional field distribution. We speculate that this will in turn lead to the generation of vortex-antivortex pairs (since the initial plane-wave field has no net circulation). That this is the case when $b/a > 1$ has been verified in preliminary numerical simulations of the nonlinear propagation equations, Eq. (17), with periodic boundary conditions applied in the transverse dimensions. It remains to investigate whether a finite vortex gas can also be created in a confined region of the transverse plane when a Gaussian input beam is employed.

Finally, we remark that the same basic mechanism may also generate a two-dimensional vortex gas in a nonlinear ring resonator, which has previously been shown to support two-dimensional optical vortices (Arecchi 1991).

3   The quantum properties of optical solitons

3.1   *Quantum field theory of the interacting photon gas*

We turn now from the classical-field solutions of the NLS to its associated underlying quantum field theory. The classical-field Hamiltonian in two dimensions for the Ginzburg-Landau equation, which is formally equivalent to that for NLS, is (DeGennes 1966)

$$H = \int dx\, dy\, \{(\hbar^2/2m)|\nabla_T \Psi|^2 + \alpha\,|\Psi|^2 + \beta\,|\Psi|^4/2\}\,. \tag{27a}$$

where the order parameter Ψ is associated with the slowly varying electric field envelope

$$\mathcal{E}(x,y) = \sqrt{\frac{16\pi\hbar\omega}{n_0^2 L}} \; \Psi(x,y).$$
(27b)

and

$$\alpha = \frac{1}{2}\hbar \left(\frac{\pi^2 c^2}{L^2 n_0^2 \omega} - \omega \right), \qquad \beta = -\frac{16\pi\Gamma(\hbar\omega)^2 n_2}{n_0^3 L}.$$
(27c)

Note that the last term of Eq. (27a) can be written as

$$\frac{1}{2}\int dx\, dy\, dx'\, dy' \; \Psi^*(x',y')\Psi(x',y')V(x'-x,\, y'-y)\, \Psi^*(x,y)\, \Psi(x,y),$$
(28)

where the pairwise interaction potential $V(x'-x,\, y'-y)$ is given by

$$V(x'-x,\, y'-y) = \beta \; \delta^{(2)}(x'-x,\, y'-y) \; .$$
(29)

To quantize the scalar fields $\Psi(x,y)$, we use the equal-time boson commutator relations, which can be derived from the fundamental equal-time commutators for electromagnetic fields in the slowly-varying envelope approximation (Deutsch & Garrison 1991),

$$[\Psi(x',y'),\; \Psi^\dagger(x,y)] = \delta^{(2)}(x'-x,\, y'-y) \; .$$
(30)

The resulting quantum field theory is equivalent to the many-body problem for a gas of nonrelativistic bosons with an effective mass

$$m_{eff} = \hbar k/v_g,$$
(31)

where v_g is the group velocity of the photons (Chiao et al. 1991). Neglecting dispersion, this mass reduces to that given by Eq. (12) for the sech and the kink. Therefore there exists a remarkable coincidence of the effective mass of the underlying *microscopic* constituents of the many-body problem, with that of the *macroscopic* sech and kink solitons.

The constituent bosons (here photons) interact with each other through pairwise delta-function potentials given by Eq. (29). Note that if $n_2 > 0$, then $\beta<0$, implying *attractive* pairwise photon-photon interactions; if $n_2< 0$, then $\beta>0$, implying *repulsive* photon-photon interactions. The origin of the interaction between two photons is their exchange of a virtual excitation in an alkali atom; this is the physics behind Eq. (1). The sign of the interaction can be understood as follows. Consider two photons whose energy is detuned slightly above resonance which participate in a such a virtual exchange. During the virtual absorption of the first photon, whose energy is slightly larger than the two-level energy difference, there is a decrease in the energy of the system. This provides an *attractive* potential well for the second photon which stimulates a virtual emission. The opposite is true for light detuned slightly below resonance. In the approximation that the atom is very small compared with the wavelength of light (the electric dipole approximation), the resulting photon-photon interaction can be approximated by a delta-function potential. The properties associated with this interacting Bose gas can be analyzed for either sign of the potential and for different confinement dimensionalities in the experimentally accessible systems presented here.

3.2   *Quantum effects in one dimension: Photonic bound-states*

One experimental realization of a one-dimensional system whose dynamics are described by the one-dimensional nonlinear Schrödinger equation is the propagation of pulses in a single mode optical fiber, which exhibits both topological (dark-"kink") and nontopological (bright-"sech") soliton solutions (Hasegawa 1989). Yurke and Potasek (1989) used the Gutkin intertwining operator technique to solve the corresponding many-body problem (the one dimensional Bose gas with repulsive interactions), and they predict that photon antibunching statistics should result near the node of the dark soliton. In the case of the bright soliton, Lai and Haus (1989b) solved the associated attractive one-dimensional Bose gas problem, based on the exact Bethe Ansatz solutions (Nohl 1976, Wadati & Sakagami 1933, Thacker 1981). In addition, the Hartree approximation has also been applied to these problems in the large photon number limit (Lai & Haus 1989a, Haus & Lai 1990, Wright 1991). Bright solitons are intimately related to bound-state solutions of this system.

Recently, three of the authors (Chiao et al. 1991) considered the simplest bound state consisting of two photons (the "diphoton") as a possible consequence of an attractive photon-photon delta-function potential, in the context of the NLS for the (1+1) dimensional traveling-wave spatial soliton (one transverse dimension, one longitudinal dimension). The creation of this quasiparticle can be understood from repeated photon-photon scattering mediated by the virtual exchange of the atomic excitation as described above. Experimentally, the signature of the diphoton would be a two-point correlation function with a characteristic profile depending on the relative transverse position of the photons, which is independent of the distance propagated in the nonlinear sample. It is the creation of these bound states which prevents the spreading of the light beam due to diffraction.

To get an order of magnitude estimate of the extent of the diphoton wave function, consider the following model. The soliton arises from a dynamical balance between self-focusing and diffraction. We can obtain an estimate of the width of the classical spatial soliton by setting the self-focusing length (Kelley 1965) equal to the diffraction length (or Rayleigh range) of a beam of a given width and intensity. Thus the (1+1)D soliton width is given by

$$\frac{\lambda}{n_2 |\mathcal{E}|^2} \approx \frac{(\Delta x)^2}{\lambda} \quad \Rightarrow \quad (\Delta x)_{soliton} \approx \frac{\lambda}{\sqrt{n_2 |\mathcal{E}|^2}} \ . \tag{32}$$

For the quantum problem, consider two photons whose relative coordinate wave function is localized in a volume $V = d\,L_y\,L_z$, where L_y is the confinement length of a planar wave guide mode, L_z is the longitudinal localization length, and d is the transverse width of the diphoton wave function; this sets the energy density as $|\mathcal{E}|^2 \approx \hbar\omega/(L_y L_z d)$. Substituting this into Eq. (32), and self-consistently setting $(\Delta x)_{soliton} = d$, one obtains

$$d \approx \frac{\lambda^2 L_y L_z}{n_2 \hbar\omega}. \tag{33}$$

In order to obtain sufficient binding that would make detection of the diphoton possible, we require a sufficiently small localization length L_z for a given nonlinearity.

In our paper (Chiao et al. 1991) we incorrectly assumed that like the classical self-trapped beam, the diphoton would be monochromatic, and identified L_z as the Rayleigh range of the input lens to the nonlinear medium. If the two photons of the traveling wave were truly localized in this short length, then the state would possess a large spectral bandwidth which would be inconsistent with neglecting dispersive effects as we did in our model (for a discussion of quantization in dispersive nonlinear media see Drummond 1990). Thus, the truly localized diphotons in the traveling wave are two dimensional objects (one transverse spatial dimension, one temporal), analogous to the classical light bullets considered by Silverberg (1990) acting under the combined effects of self-focusing, diffraction, self-phase-modulation and anomalous dispersion. The longitudinal localization length of the bound state is set dynamically by $v_g \tau_R$, where τ_R is the relaxation time of the nonlinear response function as was discussed by Blow et al. 1991, in the context of the quantum theory of self-phase modulation. This length is long for systems with large nonlinearities. Thus, from Eq. (33) we see that the diphotons associated with traveling wave solitons are weakly bound.

However, in the geometry of the NLFP, we restrict the interaction to only one longitudinal mode with a harmonic time dependence. The longitudinal localization length is then set by the size of the fundamental mode $L = L_z \approx \lambda/2$, which is quite small. By further restricting the modes to one of the transverse dimensions through the use of a line focus at the input face to the NLFP of length $L_y \gg d$, we arrive at an essentially one-dimensional system for the spatial soliton. This geometry is analogous to that used for optical fibers soltions, whereby single mode operation restricts the number of transverse modes which can participate in the interaction. The diphoton solution of this standing-wave Fabry-Perot configuration can now be analyzed.

The stationary-state Schrödinger equation for the two-photon relative-coordinate wave function determined by the normally order Hamiltonian Eq. (27a) (reduced to one transverse dimension) is

$$\left(-\frac{\hbar^2}{m_{eff}}\frac{d^2}{dx_{rel}^2} - \frac{\beta}{L_y}\delta(x_{rel})\right)u(x_{rel}) = E_{rel}\,u_{rel}(x_{rel})\,, \tag{34}$$

where $x_{rel} = x_1 - x_2$. The sole bound state solution is

$$u_{diphoton}(x_1 - x_2) = \frac{1}{\sqrt{2d}}\,e^{-|x_1 - x_2|/2d}, \tag{35}$$

where d is the diphoton width given by

$$d = \frac{\hbar^2 L_y}{m_{eff}\beta} = \frac{n_0}{48\,\pi^3}\,\frac{\lambda^2 L_y L_z}{\hbar\omega n_2}\,. \tag{36}$$

Note this expression agrees with the result of our simple model given in Eq. (33) modulo numerical factors.

As a test of the feasibility of production and detection of these diphotons consider rubidium vapor at 100°C and λ=780 nm, where n_2=1.6x10^{-5} cm^3/erg, filling a single mode Fabry-Perot, with L_z=0.39 μm. If we use a cylindrical lens to produce a line focus of length L_y= $10^3\lambda$ at the input face of the Fabry-Perot (see Fig. 1), then diphoton width is given by $d \approx 10$ μm. It should be much easier to observe the diphoton associated with this standing-wave mode in experiments than the one associated with the traveling-wave. Detection of the standing-wave diphoton can be achieved by means of coincidence detection of the output light from the nonlinear Fabry-Perot by detectors D1 and D2 in the far field as a function of their transverse separation in the x-direction (see Fig. 1). Rejection of the unbound component of the light is accomplished by means of the slit S1 with a width slightly larger than d after the Fabry-Perot. A Lorentzian-squared two-point correlation function in the coincidence detection by D1 and D2 is the signature of the detection of the diphoton.

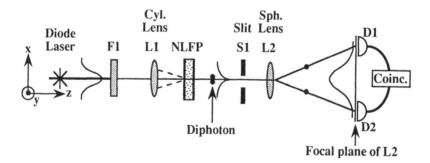

Fig. 1: Schematic for diphoton generation and detection

3.3 Squeezed state generation in NLFP spatial solitons

In the presence of nonlinearities, the unitary evolution does not preserve a coherent state, and thus the photon statistics necessarily become non-Poissonian, or "squeezed". In the case of an optical fiber soliton, it has been shown both theoretically and experimentally that the light inside the soliton experiences squeezing during its propagation (Drummond & Carter 1987, Drummond et al. 1989, Drummond & Carter 1990, Rosenbluh & Shelby 1991). Since the one-dimensional NLS also describes the spatial sech mode of the NLFP, one expects that self-trapping should lead to self-squeezing. It remains to be seen how much squeezing can be realized in this way. One elegant experimental method to measure this is to employ the interference technique of Shirasaki and Haus (1990) which has been carried out by Bergman and Haus (1991). We first give the principle of the method by means of a gedanken experiment (see Fig. 2(a)), and then follow this by a more practical scheme (see Fig. 2(b)).

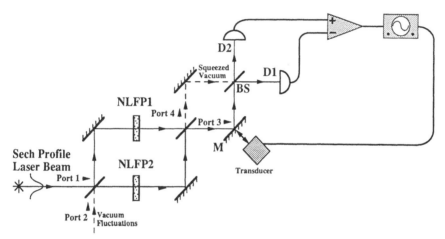

Fig. 2(a): Schematic for detection of squeezing of vacuum fluctuations in two identical nonlinear Fabry-Perots, NLFP1 and NLFP2.

Consider a balanced Mach-Zehnder interferometer with identical nonlinear Fabry-Perots in both arms. A laser beam with a sech profile enters the port 1 of the interferometer, and with such a power level so that it passes through both nonlinear Fabry-Perots with unity transmission coefficient. The arms of the Mach Zehnder are balanced in length so that a classical beam of light would have completely passed out through port 3, and *nothing* would have passed out through port 4. The light leaving port 3 will serve as the local oscillator in the balanced homodyne detector which follows. Vacuum fluctuations enter through the unused entrance port 2, and after passage through the two nonlinear Fabry-Perots along with the laser beam which acts as a pump, experience squeezing. As long as the squeezed vacuum fluctuations are weak compared with the laser beam, they will also pass through the nonlinear

Fabry-Perots with unity transmission along with the laser beams. The squeezed vacuum fluctuations pass out through port 4. The outputs of ports 3 and 4 are then recombined by means of two mirrors on a final beam splitter BS and detected by detectors D1 and D2. The phase of the local oscillator is adjusted by small motions of mirror M, and the noise in the difference signal between the detectors D1 and D2 is measured as a function of this phase in homodyne detection. The balanced detection scheme removes noise due to the fluctuations of the local oscillator (which is also squeezed). The noise in the homodyne current can be compared with the standard quantum limit by blocking the squeezed light leaving port 4.

The use of *two* identical nonlinear Fabry-Perots is cumbersome. A more practical scheme uses a *single* nonlinear Fabry-Perot in a ring interferometer configuration (see Fig. 2(b)). The sech-profile pump laser beam enters port 1 of the interferometer and is split by the input beam splitter B1 into clockwise and counterclockwise senses of propagation around the ring. This sech-profile beam is focused by curved mirrors M1 and M2 in a confocal geometry on the nonlinear Fabry-Perot from both sides, and emerges from it with unity transmission from both sides. The ring interferometer is self-balanced, since the optical path lengths for clockwise and counterclockwise senses of propagation are automatically equal. Hence a classical beam of light would completely emerge out port 1, and *nothing* would emerge out port 2. However, vacuum fluctuations entering through port 2 also exit through port 2 after being squeezed upon passage through the nonlinear Fabry-Perot along with the laser beam. Balanced homodyne detection then follows as before.

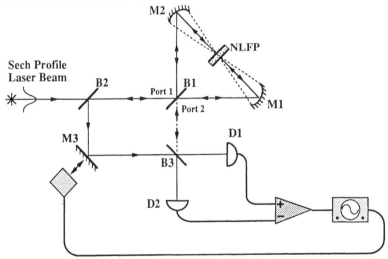

Fig. 2(b): Schematic for detection of squeezed vacuum fluctuations in a single NLFP.

3.4   Photonic thin-film superfluidity

The many-photon problem for the NLFP operating in a single longitudinal mode is the same as the many-body problem of the interacting Bose gas in two dimensions. When n_2 is negative, vortex solutions are possible (see Eq. (6) and the discussion that follows). The special topological nature of two dimensions gives rise to unique many-body phenomena. For example, superfluidity analogous that in thin films of ^{4}He, should appear in the interacting 2D photon gas below its Kosterlitz-Thouless transition (Kosterlitz & Thouless 1973). This phase transition is universal whenever topological charge is present in two dimensions. The transition temperature T_c is given by

$$k_B T_c = \frac{Q_{top}^2}{8\pi} \rho_s(T_c^-),$$ (37)

where Q_{top} is the topological charge (given here by h/m_{eff}) and $\rho_s(T_c^-)$ is the mass per unit area of the superfluid just below T_c. This universal transition applies equally well to both ^{4}He and the photonic case consider here.   In thin films of adsorbed ^{4}He atoms, the Kosterlitz-Thouless transition has been observed (Bishop & Reppy 1978).

The superfluidity which we suggest here differs from that recently suggested based on the BCS photon-pairing model by Cheng (1991). In the BCS model, *attractive* photon-photon interactions are required for photon-pair bound states to form. By contrast, here *repulsive* photon-photon interactions are required, since vortex solutions exists only for this sign of the interaction (see above), and since the existence of vortices is essential for the Kosterlitz-Thouless transition. However, the manifestation of superfluidity is the same in both cases, namely, a suppression of light scattering in the superfluid state. In addition one may be able to monitor the transition to the superfluid state by observing the analog of the Meissner effect. In the superfluid case this corresponds to an expulsion of vorticity.

The meaning here of "temperature" needs further clarification, since the NLFP is a system far from thermodynamic equilibrium. However, previous work on the analogy between lasers near threshold (which are also systems far from thermal equilibrium) and second-order phase transitions (DeGiorgio & Scully 1970, Graham & Haken 1970, Haken 1975) may provide guidance as to the meaning of the Kosterlitz-Thouless transition here.

3.5   Quantum dynamics of solitons

As we have already pointed out earlier, solitons have particle-like properties. For classical nonlinear systems having soliton solutions these properties are exploited in the "effective-particle theories" for perturbed systems (Kaup & Newell 1978). The basic idea is to replace the nonlinear wave equation for a perturbed soliton with a reduced set of equations for the effective particle, i.e., the soliton. The key to doing this is to identify the correct *collective variables* for soliton motion.

The effective particle dynamics for the classical NLS can be deduced from the Hamiltonian, Eq. (27a). First we consider the 1D solitons, i.e., the sech and the kink, for simplicity. In Eq. (27a), the first term is the mean kinetic energy of the nonlinear wave and the last term corresponds to the potential energy U, so the 1D Hamiltonian can be written in the alternative form $H = P^2/2M_{eff} + U$ where

$$P^2/2M_{eff} = (\hbar^2/2m) \int dx |d\Psi/dx|^2 \; , \tag{38}$$

and

$$U = \int dx [\alpha(x)|\Psi|^2 + \frac{1}{2}\beta(x)|\Psi|^4] \; . \tag{39}$$

As mentioned earlier, here the mass m is the effective mass $m_{eff} = \hbar k/v_g$ for the transverse motion of the photon. Note that we have allowed both $\alpha(x)$ and $\beta(x)$ to become functions of the transverse coordinate x. With reference to Eq. (27c) we see that $\alpha(x)$ can become spatially dependent if the refractive index $n_0(x)$ is inhomogeneous, as in a waveguide or at an interface (Aceves et al. 1989a,b). In the case of alkali vapors, such an inhomogeneity can also arise from an inhomogeneous magnetic field applied to the NLFP. Similarly, at a nonlinear interface the value of the nonlinear parameter β can change across the interface giving a spatially inhomogeneous distribution $\beta(x)$. We now substitute the soliton solution in Eq. (39) and assume that during its motion the soliton profile $|\Psi|^2$ remains intact for sufficiently slowly varying functions $\alpha(x)$ and $\beta(x)$ (the soliton is an exact solution only when α and β are constant). Then this profile simply translates under the gradual action of $\alpha(x)$ and $\beta(x)$. This motion can be incorporated into the potential U by replacing $|\Psi(x)|^2$ by $|\Psi(x-X(t))|^2$, where X is the transverse position of the center of the soliton. In this way the effective potential energy becomes a function of the soliton coordinate $U_{eff}(X)$. The key step is now to identify P as the momentum canonically conjugate to X in the Hamiltonian, and this is the essential content of the more elaborate effective particle theory (Kaup & Newell 1978). The reduced effective Hamiltonian for the classical soliton motion is then

$$H_{eff} = P^2/2M_{eff} + U_{eff}(X) \; . \tag{40}$$

where $M_{eff} = m_{eff}$. This explains the coincidence of the effective masses for the soliton and the photon noted earlier. Hamilton's equations then yield Newton's equation for the soliton coordinate

$$M_{eff}\frac{d^2}{dt^2}X = -\frac{d}{dX}U_{eff}(X) \; . \tag{41}$$

Thus the motion of a soliton in a complicated refractive-index structure can be reduced to the solution of Newton's equation, Eq. (41), for the soliton coordinate. The effective particle theory has previously been used to calculate the reflection properties of solitons at a nonlinear interface and also multiple interfaces (Aceves et al 1989). Numerical solutions of the full NLS were found to be in excellent agreement with predictions based on Eq. (41).

So far the discussion has been purely classical. To study the quantum motion of solitons one would ideally start form the quantum version of the Hamiltonian, Eq. (27a), with the boson commutation relations, Eq. (30). It is then conceivable that the appropriate equations of motion for the soliton could be obtained using either Green's function techniques or the time-dependent Hartree approximation. Rather than resorting to these methods we use a notion that is widely applied in nuclear many-body problems, namely we "requantize" the system using the effective Hamiltonian for the collective coordinates of the soliton motion. That is, we start with Eq. (40) as the appropriate quantum Hamiltonian for the soliton center-of-mass coordinate X. This phenomenological quantization scheme for soliton motion has previously been discussed for the case of the NLS for the case of soliton-soliton collisions (Maki & Kodama 1986). The corresponding Schrödinger equation for the wavefunction $\psi(X,t)$ of the soliton as a function of the soliton coordinate X, is then

$$i\hbar \frac{\partial}{\partial t}\psi = -\frac{\hbar^2}{2M_{eff}}\frac{\partial^2}{\partial X^2}\psi + U_{eff}(X)\psi . \tag{42}$$

The implications of Eq. (42) are best illustrated using some special examples. First, consider the case where α and β are constants. In this case U_{eff} is a constant, and Eq. (41) predicts that the classical soliton will travel in a straight line, as a classical free particle would, while maintaining its profile, which is consistent with the "sliding soliton" in Eq. (10). In contrast, according to the Schrödinger equation, Eq. (42), it is only the *average* soliton coordinate which follows a straight line; there must also be quantum fluctuations in this coordinate associated with the Heisenberg uncertainty principle $\Delta P\,\Delta X \geq \hbar/2$. These manifest themselves as the spreading of the wavepacket $\psi(X,t)$ whenever the initial soliton coordinate is known with high precision, and is particularly relevant due to the very small predicted effective mass M_{eff}, Eq. (12), for the soliton. Note that although there is spreading of the wavepacket for the soliton coordinate X, this does *not* mean that the soliton profile itself spreads. Rather the soliton profile stays fixed in size and shape as in the classical case, but we no longer have precise knowledge of the soliton's central position X. Thus when a soliton is detected, it will have a definite profile but uncertain position.

As a second example consider the case of a nonlinear waveguide (Aceves et al. 1989b). Here we can arrange it so that classically the soliton is rejected from a thin film of higher refractive-index than the surrounding media (Wright et al. 1990). The refractive-index step therefore presents a potential barrier for the incident soliton which cannot be overcome classically. However, according to Eq. (42), the soliton can *tunnel* quantum mechanically across this classically forbidden region, a prediction which can potentially be tested experimentally, especially in light of the tiny effective mass M_{eff} of the soliton. The key point is that the quantum theory of soliton motion described by the Schrödinger equation, Eq. (42), allows for classically forbidden processes such as tunneling. The experimental observation of these effects would provide a remarkable *macroscopic* manifestation of the consequences of quantum mechanics, since a huge number of photons inside a single soliton could all simultaneously tunnel together *as a single unit* to the other side of the barrier. This is reminiscent of the macroscopic quantum tunneling observed in SQUIDS (Devoret et al. 1985). Optical solitons thus provide a marvelous testing ground for the relation between classical and quantum field theories as typified by the NLS.

In two dimensions, similar considerations apply to vortex solitons. The collective variables in this case are the coordinates of the center of the vortex, which we shall denote by (X,Y). In the classical limit, instead of obeying Newton's equation of motion, Eq. (41), Creswick and Morrison (1980) have shown, starting from a path-integral approach at the microscopic level, that these coordinates obey Kirchhoff's equations of motion, as suggested earlier by Onsager (1949) on phenomenological grounds (see also Lund 1991),

$$\kappa_i \frac{dX_i}{dt} = +\frac{\partial H}{\partial Y_i} \tag{43a}$$

$$\kappa_i \frac{dY_i}{dt} = -\frac{\partial H}{\partial X_i} \tag{43b}$$

$$H = -\frac{1}{4\pi}\sum_{i<j} \kappa_i \kappa_j \ln\left(\frac{R_{ij}^2}{r_0^2}\right) \tag{43c}$$

$$R_{ij}^2 = (X_i - X_j)^2 + (Y_i - Y_j)^2 , \tag{43d}$$

where $\kappa_i = \pm h/m_{eff}$ is the vorticity (or topological charge) of the ith vortex, (X_i, Y_i) is its position, and r_0 is its core radius. The energy of the system is given here by $Nm_{eff}LH$, where N is the number density of photons in a nonlinear Fabry-Perot of spacing L. Note that these equations are Hamiltonian in form, with X_i and Y_i being canonically conjugate variables. Also note that the effective mass of the vortex soliton is zero in Eq. (43), since these equations of motion have no second derivatives in time. Upon "requantization" (see below), Volovik has shown that this fact leads to the phenomenon of tunneling of vortices over macroscopic distances (Volovik 1972). Recently, however, vortices in Fermi systems have been predicted to have a finite effective mass (Duan & Leggett 1992).

To "requantize" these equations of motion, we invoke the canonical commutation relations (Fetter 1967, Volovik 1972, Chiao et al. 1985, Hansen et al. 1985)

$$[X_i, Y_j] = iC_i \delta_{ij} , \tag{44}$$

where $C_i = \hbar/\kappa_i NL$. It follows from this commutator that there must exist zero-point motion of the vortices due to the uncertainty principle. In the case of a single vortex with $\kappa > 0$,

$$[X,Y] = iC . \tag{45}$$

Therefore the distance squared of this vortex from some arbitrary origin obeys the commutation relations

$$[X^2 + Y^2, X] = -2iCY \qquad (46a)$$
$$[X^2 + Y^2, Y] = +2iCX . \qquad (46b)$$

These have the form of the angular momentum commutators,

$$[\mathcal{L}, X] = +i\,\hbar Y \qquad (47a)$$
$$[\mathcal{L}, Y] = -i\,\hbar X , \qquad (47b)$$

where

$$\mathcal{L} = -\frac{\hbar}{2C}(X^2 + Y^2) \qquad (48)$$

is the generator of rotations, i.e., the angular momentum of the vortex. Since the above algebra is precisely that of the simple harmonic oscillator, the spectrum of $(X^2 + Y^2)$, and hence of the angular momentum $\mathcal{L}$, is shifted from integer values by the *fraction* 1/2, which represents zero-point motion. Recall that zero-point motion is a consequence of the uncertainty principle. The lowest possible angular momentum of a single vortex is therefore $\frac{1}{2}\hbar$. The area $\pi(X^2 + Y^2)$ swept out by the zero-point motion encloses *half* of a constituent boson, and this fraction is a constant of the motion.

3.6   Anyonic statistics in systems of identical photonic vortices

The appearance of *fractional* topological quantum numbers, such as the $\frac{1}{3}e^2/h$ Hall conductance in the fractional quantum Hall effect (Tsui et al. 1982) is believed to be closely connected with the appearance of anyonic statistics. This effect is widely believed to arise from a $\frac{1}{3}e$ charged quasiparticle, which is a vortex-like object in the interacting 2D electron gas in a strong magnetic field (Laughlin 1983). That this quasiparticle may obey fractional statistics has by suggested by Halperin (1984) and by Arovas et al. (1984). Berry's topological phase (Berry 1984, Chiao & Wu 1986, Tomita & Chiao 1986) has also been suggested as a mechanism which could lead to fractional statistics (Arovas et al. 1984). This phase could also give rise to topologically nontrivial persistent currents in mesoscopic systems (Loss et al. 1990, Stern 1992), like those in a vortex .

One of the most intriguing possibilities which arises in the problem of the interacting 2D photon gas is that topological solitons, such as photonic vortices, might obey anyonic statistics. It should be emphasized that such statistics exists only in one or two-dimensional many-body systems, but not in three. Goldin et al. (1982) have pointed out that unusual quantum statistics are possible in one, as well as two, dimensional systems. For photons, as we have pointed out above, two-dimensional systems can be realized by eliminating their z, or longitudinal, degrees of freedom. This could be done by choosing the spacing L of the nonlinear Fabry-Perot to be close to half a wavelength.

There should exist two kinds of photonic vortices: scalar ones, analogous to those in superfluid ^{4}He, and vector ones, analogous to some of the ones in the phases of superfluid ^{3}He (Salomaa & Volovik 1987). Due to the fact that photons have spin 1, photonic vortices are in general similar to those in superfluid ^{3}He, however, with the added constraint that the helicity of the photons must always be ± 1. Both scalar and vector vortices are possible candidates for anyonic statistics, but for simplicity, we shall mainly limit our discussion to scalar vortices here.

As an introduction to anyonic statistics, we shall first consider a specific model for a particle which obeys such statistics, namely the "anyon" model (Wilczek 1982). An anyon is a composite particle consisting of a charge Q attached to a solenoid, or "flux tube", containing flux Φ. This composite particle can be thought of as a bound state of the charge and the flux tube, so that they are constrained to move together as a single unit. Now consider what happens when two identical anyons exchange positions by moving in a circular orbit around each other, on a plane perpendicular to the flux tubes (see Fig. 3). The charge of one particle sees the vector potential arising from the flux inside the other particle during this motion. Therefore upon interchange of the positions of the two identical anyons, the two-anyon system wavefunction picks up the Aharonov-Bohm phase shift,

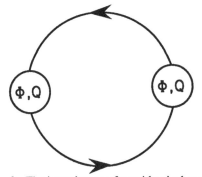

Fig. 3: The interchange of two identical anyons.

$$\Delta\phi = (Q/\hbar c)\oint \mathbf{A}\cdot d\mathbf{l} = 2\pi Q\Phi/hc . \tag{49}$$

This can differ from the zero phase shift picked up by bosons, or the π phase shift picked up by fermions, upon interchange. Since $\Delta\phi$ can in principle be anything, this composite particle was called an "anyon".

Chiao et al. (1985) and Hansen et al. (1985) have suggested that the vortex in 2D superfluid ^{4}He thin films obeys anyonic statistics due to the zero-point motion of the vortex core (a counterargument has been given by Haldane and Wu (1985) based on a compressible thin film superfluid model). This zero-point motion arises from the fact that the X and Y coordinates of the vortex core do not commute with each other (see Eq. (45)). As a result of its motion, the vortex core contains a *fractional* number of atoms. Hence such a vortex can be thought of as a neutral analog of the anyon, that is, a composite particle consisting of an integer number of "flux quanta" (here, phase slips of 2π around its core), and a *fractional* number of "charge quanta" (here, the fractional number of atoms contained in its core). Since vortices in the interacting 2D photon gas obey the same equations of motion as those in the 2D helium films, they should also be neutral anyons. One needs only to replace in the analyses of Chiao et al. (1985) and Hansen et al. (1985) the mass m of the helium atom by the effective mass $m_{eff} = \hbar\omega n_0^2/c^2$ of the photon, to obtain corresponding results.

The presence of anyons implies the appearance of a fractional quantum Hall effect. In order to understand how this topological effect arises, let us consider the following simple gedanken experiment for electromagnetic anyons: Consider a steady flow of anyons in the x direction across a rectangular, anyon-filled, thin-film sample which lies on the x-y plane (see Fig. 4). Each anyon transports both a charge Q and a flux tube Φ. Let us connect an ammeter across the left and right edges of the rectangular sample, so that we can measure the current in the x-direction,

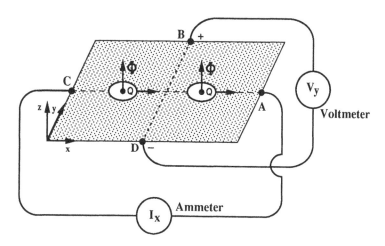

Fig. 4: Hall effect arising from the flow of anyons.

$$I_x = (dN/dt)\, Q ,\qquad\qquad (50)$$

where dN/dt represents the number of anyons crossing per unit time the dashed line BD. Let us also connect a voltmeter across the top and bottom edges of the rectangular sample, which measures the voltage V_y across the y-dimension. The steady flow of anyons across the dashed line BD into the area enclosed by the voltmeter circuit gives rise, by Faraday's law, to a voltage

$$V_y = (dN/dt)\Phi/c \; , \tag{51}$$

where dN/dt is the same as before. Defining the Hall conductance G_{xy} as the ratio I_x/V_y we obtain the result

$$G_{xy} = cQ/\Phi \; . \tag{52}$$

For a 2D electron gas in a strong magnetic field, it is natural to assume that the flux Φ is one flux quantum hc/e (this implies a 2π phase change around the core of the anyon, and therefore *constructive interference* for all electrons circulating around it). Let the charge Q be some fraction f of the electron charge e. Then we obtain the fractional quantum Hall conductance,

$$G_{xy} = fe^2/h \; . \tag{53}$$

The associated phase shift upon interchange of two such identical anyons, which is given by the Aharonov-Bohm phase, Eq. (49), is in this case

$$\Delta\phi = 2\pi Q\Phi/hc = 2\pi f \; . \tag{54}$$

For example, if $f=1/3$, then $\Delta\phi = 2\pi/3$, clearly indicating that the anyon is neither a boson ($\Delta\phi=0$) nor a fermion ($\Delta\phi=\pi$).

Now let us generalize the above results to the case of neutral anyons. Suppose a vortex in a superfluid thin film has trapped at its core, for whatever reason, some fraction f of a constituent particle, for example, a helium atom or a photon. Then each vortex, viewed as a neutral anyon, transports both a phase slip of 2π and a fraction f of this particle. Again, let us consider a rectangular thin-film sample in the geometry of Fig. 4. (In the photonic case, this thin film sample consists of the light in the interior of a thin nonlinear Fabry-Perot.) A flow of vortices dN/dt in the x-direction will then carry a neutral particle current

$$I_x = (dN/dt)f \; . \tag{55}$$

Due to the transport of phase slips by these vortices, there arises a chemical potential difference across the sample in the y-direction, just as in the Josephson effect in superfluid helium (Anderson 1966, Varoquaux et al. 1985, Varoquaux et al. 1990), of an amount

$$\mu_y = (dN/dt)2\pi\hbar \; . \tag{56}$$

Let us define the transverse conductance as $G_{xy} = I_x/\mu_y$. Then the result is

$$G_{xy} = f/h \ . \tag{57}$$

Since Planck's constant h enters inversely into G_{xy}, this quantum-Hall-like effect will be large. Since precise measurements of e^2/h have been made in the quantum Hall effect, one might suppose by analogy that precise measurements of $1/h$ should also be possible here. While this might be true in principle, such measurements will be difficult in practice due to the lack of precise standards for neutral particle current and for chemical potential.

One consequence of anyonic statistics which does not involve Planck's constant explicitly is the Pauli exclusion principle, which can be derived from the following simple argument. Suppose that upon interchange, two identical anyons picked up a phase factor $\exp(i\Delta\phi)$, i.e.,

$$\psi(2,1) = \exp(i\Delta\phi) \ \psi(1,2) \ , \tag{58}$$

where the arguments (1,2) and (2,1) are shorthand for the positions of anyon 1 and 2 before and after interchange, respectively, and ψ represents the wavefunction of the two-anyon system. Now suppose we try to find both anyons at exactly the same place, say, at position 1. Then

$$\psi(1,1) = \exp(i\Delta\phi) \ \psi(1,1) \tag{59}$$

But this implies that

$$(1 - \exp(i\Delta\phi)) \ \psi(1,1) = 0 \ . \tag{60}$$

Hence either the first factor vanishes,

$$(1 - \exp(i\Delta\phi)) = 0 \ , \tag{61}$$

implying Bose statistics ($\Delta\phi=0$), or the second factor vanishes,

$$\psi(1,1) = 0 \ , \tag{62}$$

implying the Pauli exclusion principle. Hence in all cases of anyonic statistics, apart from bosonic statistics, identical particles obey the Pauli exclusion principle, with fermionic statistics as a special case. One consequence of this fact is that a gas of noninteracting nonbosonic anyons should form a Fermi sea at sufficiently low temperatures.

In order to provide a more specific determination of nonbosonic anyonic statistics, which could distinguish them from fermionic statistics, one could utilize the two-vortex solution of Chiao et al. (1985). Due to the Kirchhoffian dynamics of vortices given by Eq. (43), two identical vortices classically orbit around each other in a circular orbit, as in Fig. 3. This motion should be observable for two photonic vortices of the same topological charge in the NLFP in the classical limit. When this motion is quantized, the distance R_{12} between them given by Eq. (43d) is also quantized. The result is

$$R_{12}(n) = \{(4n + 2\Delta\phi/\pi)/(\pi NL)\}^{1/2},\qquad(63)$$

where $n=0,1,2...$ is the radial quantum number for the identical two-vortex system, and N is the number density of photons inside a nonlinear Fabry-Perot of spacing L. Thus a measurement of the *minimum* distance R_{min} between two identical photonic vortices would yield the anyonic phase $\Delta\phi$. For example, for the case $\Delta\phi=\pi/2$, $R_{min}=(\pi NL)^{-1/2}$, or for the case $\Delta\phi=3\pi/2$, $R_{min}=(3/\pi NL)^{1/2}$. For optical electric fields of the order of 0.1 esu inside the NLFP, one obtains a value of R_{min} about 30 µm for the case $\theta=\pi/2$. We hope that this will lead to the first experimental observation of these unusual quantum statistics.

Acknowledgments

R Y C and I H D acknowledge support by ONR under Grant No. N00014-90-J-1259, J C G by the U S Department of Energy at the Lawrence Livermore Laboratory under Contract No. W-7405-Eng-48, and E M W by the Joint Services Optical Program, all agencies of the U S A.

References

Aceves A B, Moloney J V, and Newell A C 1989a *Phys. Rev. A* **39** 1809
Aceves A B, Moloney J V, and Newell A C 1989b *Phys. Rev. A* **39** 1828
Agrawal G P 1991 *J. Opt. Soc. Am. B* **7** 1072
Aitchison J S, Weiner A M, Silberberg Y, Leaird D E, Oliver M K, Jackel J L, and Smith P W E 1991 *Opt. Lett.* **16** 471
Aitchison J S, Weiner A M, Silberberg Y, Oliver M K, Jackel J L, Leaird D E, Vogel E M and Smith P W E 1990 *Opt. Lett.* **15** 471
Akhmanov S A, Vorontsov M A, Ivanov V Yu, Larichev A V, and Zheleznykh N I 1992 *J. Opt. Soc. B* **9** 78
Anderson P W 1966 *Rev. Mod. Phys.* **38** 298
Arecchi F T, Giancomelli G, Ramazza P L, and Residori S 1991 *Phys. Rev. Lett.* **67** 3749
Arovas D, Schrieffer J R and Wilczek F 1984 *Phys. Rev. Lett.* **53** 722
Bergman K and Haus H A 1991 (to be published).

Berry M 1984 *Proc. Roy. Soc. London, Ser. A* **392** 45

Bespalov V I and Talanov V I 1966 *JETP Lett.* **3** 307

Bishop D J and Reppy J D 1978 *Phys. Rev. Lett.* **40** 1727

Bjorkholm J E and Ashkin A 1974 *Phys. Rev. Lett.* **32** 129

Blow K J, Loudon R, and Phoenix S J D 1991, *J. Opt. Soc. Am. B.* **8** 1750.

Cheng Z 1991 *Phys. Rev. Lett.* **67** 2788

Chiao R Y, Deutsch I H, and Garrison J C 1991 *Phys. Rev. Lett.* **67** 1399.

Chiao R Y, Garmire E and Townes C H 1964 *Phys. Rev. Lett.* **13** 479

Chiao R Y, Hansen A, and Moulthrop A M 1985 *Phys. Rev. Lett.* **54** 1339

Chiao R Y, Kelley P L, and Garmire E 1966 *Phys. Rev. Lett.* **17** 1158

Chiao R Y and Wu Y S 1986 *Phys. Rev. Lett.* **57** 933

Coleman S 1985 *Aspects of Symmetry* (Cambridge: Cambridge University Press)

Creamer D B, Thacker H B and Wilkinson D 1980 *Phys Rev. D* **21** 1523

Creswick J and Morrison H L 1980 *Phys. Lett. A* **76** 267

DeGennes P G 1966 *Superconductivity of Metals and Alloys* (New York: W. A. Benjamin)

DeGiorgio V and Scully M O 1970 *Phys. Rev. A* **A2** 1170

Devoret M H, Martinis J M, and Clarke J 1985 *Phys. Rev. Lett.* **55** 1908

Deutsch I H and Garrison J C 1991 *Phys. Rev. A* **43** 2498

Drummond P D and Carter S J 1987 *J. Opt. Soc. Am. B* **4** 1565

Drummond P D, Carter S J and Shelby R M 1989 *Opt. Lett.* **14** 373

Drummond P D 1990 *Phys. Rev. A* **42** 6845

Duan J M and Leggett A J 1992 *Phys. Rev. Lett.* **68** 1216

Faddeev L D and Takhtajan L A 1987 *Hamiltonian Methods in the Theory of Solitons* (Berlin: Springer-Verlag)

Fetter A L 1965 *Phys. Rev.* **138** 429

Fetter A L 1967 *Phys. Rev.* **162** 143

Graham R and Haken H 1970 *Z. f. Phys.* **237** 31

Gibbs H M 1985 *Optical Bistability: Controlling Light by Light* (New York: Academic Press).

Goldin G A, Menikoff R, and Sharp D H 1981 *J. Math. Phys.* **22** 1664

Goldin G A and Sharp D H 1983 *Phys. Rev. D* **28** 830

Grischkowshky D 1970 *Phys. Rev. Lett.* **24** 866

Haken H 1975 *Rev. Mod. Phys.* **47** 67

Haldane F D M and Wu Y S 1985 *Phys. Rev. Lett.* **55** 2887

Halperin B 1984 *Phys. Rev. Lett.* **52** 1583

Hansen A, Moulthrop A M and Chiao R Y 1985 *Phys. Rev. Lett.* **55** 1431

Hasegawa A 1989 *Optical Solitons in Fibers* (Berlin: Springer-Verlag).

Haus H A and Lai Y 1990 *J. Opt. Soc. Am. B* **7** 386

Kaup D J 1975 *J. Math. Phys.* **16** 2036

Kaup D and Newell A C 1978 *Proc. Roy. Soc. London, Ser. A* **361** 413

Kelley P L 1965 *Phys. Rev. Lett.* **15** 1088

Kosterlitz J M and Thouless D J 1973 *J. Phys C* **6** 1181

Lai Y and Haus H A 1989a *Phys. Rev. A* **40** 844

Lai Y and Haus H A 1989b *Phys. Rev. A* **40** 1138
Laughlin R B 1983 *Phys. Rev. Lett.* **50** 1395
Lax M, Louisell W H, and McKnight W B 1975 *Phys. Rev. A* **11** 1365
Leinaas J M and Myrheim M 1977 *Nuovo Cimento B* **37** 1
Lieb E H and Liniger W 1963 *Phys. Rev.* **130** 1605
Lund F 1991 *Phys Lett A* **159** 245
Loss D, Goldbart P, and Balatsky A V 1990 *Phys. Rev. Lett.* **65** 1655
Maker P D, Terhune R W and Savage C M 1964a *Phys. Rev. Lett.* **12** 507
Maker P D and Terhune R W 1964b *Phys. Rev.* **137** 801
Maki J N and Kodama T 1986 *Phys. Rev. Lett.* **57** 2097
Nohl C R 1976 *Ann. Phys. (N.Y.)* **96** 234
Onsager L 1949 *Nuovo Cimento Suppl.* **6** 279
Rosenbluh M and Shelby R M 1991 *Phys. Rev. Lett.* **66** 153
Saikan S and Kiguchi M 1982 *Opt. Lett.* **7** 555
Salomaa M M and Volovik G E 1987 *Rev. Mod. Phys.* **59** 533
Shelby R M, Drummond P D, and Carter S J 1990 *Phys. Rev. A* **42** 2966
Shirasaki M and Haus H A 1990 *J. Opt. Soc. Am. B.* **7** 30
Silverberg Y 1990 *Opt. Lett.* **16** 1282
Sogo K, Uchinami M, Nakamura A and Wadati M 1981 *Prog. Theor. Phys.* **66** 1284
Stern A 1992 *Phys. Rev. Lett.* **68** 1022
Swartzlander Jr. G A, Anderson D R, Regan J J, Yin H, and Kaplan A E 1991 *Phys. Rev. Lett.* **66** 1583
Szöke A, Daneu V, Goldhar J, and Kurnit N A 1969 *Appl. Phys. Lett.* **15** 376
Thacker H B 1981 *Rev. Mod. Phys.* **53** 253
Tomita A and Chiao R Y 1986 *Phys. Rev. Lett.* **57** 937
Tsui D C, Störmer H L, and Gossard A C 1982 *Phys. Rev. Lett.* **48** 1559
Varoquaux E and Avenel O 1985 *Phys. Rev. Lett.* **55** 2704
Varoquaux E, Zimmermann W Jr, and Avenel O 1991 in *Excitations in Two-Dimensional and Three-Dimensional Quantum Fluids* ed A F G Wyatt and H J Lauter (New York: Plenum)
Volovik G E 1972 *Pis'ma Zh. Eksp.Teor. Fiz.* **15** 116 [JETP Lett. **15**, 81]
Wadati M and Sakagami M 1984 *J. Phys. Soc. Jpn.* **53** 1933
Wilczek F 1982 *Phys. Rev. Lett.* **48** 1144
Wilczek F 1982 *Phys. Rev. Lett.* **49** 957
Wright E M, Heatley D R, and Stegeman G I 1990 *Phys. Rep.* **194** 309
Wright E M 1991 *Phys. Rev. A* **43** 3836
Wright E M 1992 (to be published)
Wu Y S 1984 *Phys. Rev. Lett.* **52** 2103
Yurke B and Potasek M J 1989 *J. Opt. Am. B* **6** 1227
Zakharov V E and Shabat A B 1972 *Sov. Phys. JETP* **34** 62

Femtosecond dynamics of third-order nonlinearities in polythiophenes

A.Cybo-Ottone, S.De Silvestri, V.Magni, M.Nisoli, O.Svelto

Centro di Elettronica Quantistica e Strumentazione
Elettronica - CNR
Istituto di Fisica del Politecnico
P.zza L. da Vinci 32 (Milano-Italy)

In this work we review some of the recent results on nonlinear optical properties of thiophene based polymers. The excited state kinetics of these materials is related to the formation of self-localized states (polarons, bipolarons, self trapped excitons) within the band gap. The excited kinetics of thiophene-based polymers has been investigated by picosecond and femtosecond spectroscopic techniques. The transient saturated absorption data of newly synthetized thiophene polymers with low band gap energy transition are also reported. The temporal evolutions may be related to exciton dynamics.

1. Introduction

The growing interest in all optical switching and high speed optoelectronics has promoted the search for new materials with large non-linear optical susceptibilities and fast time response (Kobayashi 1989). Polymers with long conjugated chains are natural candidates and their electrical and optical properties are gaining considerable interest. Such materials can be suitable for the fabrication of waveguides, couplers, switches, modulators etc. The optical properties required for these new materials can be classified in two categories, namely: intrinsic optical properties and bulk mechanical and optical properties. The first category includes a large ($>10^{-9}$ esu) and fast (10^{-12}–10^{-15} s) non-linear optical response, a low energy dissipation and low optical losses in the infrared; the second category a good and long lived chemical stability, easy processability for the fabrication of thin film and finally a remarkable mechanical strength. Considerable efforts have been making in numerous research laboratories to meet the previous requirements. The electronic

excitation properties of conjugated polymers behave quite
differently from the electron-hole pairs formation and decay
in bulk semiconductors (see Etemad et al. 1982 and Orenstein
1986). The presence of a fast lattice relaxation due to a
strong electron-phonon coupling and the quasi one-
dimensional behavior of these compounds result in the
formation of structurally relaxed states (self-trapped
excitations) which include solitons, polarons, charged
bipolarons, and self-trapped excitons (or neutral
bipolarons). The associated electronic levels are located
within the polymer band gap (Friend et al 1987). These
states play an important role in the nonlinear optical
behavior and charge transport in these systems. In
particular very high non-linear optical response may be
related to a redistribution of the oscillator strength due
to the formation of these gap states. In addition these
materials exhibit very large damage threshold and the time
response of the nonlinearities is much shorter than that
observed in semiconductors. Pump and probe and four wave
mixing techniques can be used to determine the amount of
nonlinearity and the time response for these materials.
Transient absorption measurements have been performed to
study the photophysics of trans-polyacetylene (Shank et al.
1982, and Rothberg el al. 1987). In this case the dominant
photoexitation process is represented by the formation of
solitons, but also polarons and bipolarons are further
excitations for these materials. The photogeneration of
solitons is related to the degeneracy of the ground state;
the photogenerated electron-hole pairs created on a single
chain become pairs of oppositely charged solitons S^- and S^+
within 10^{-13}s and separate rapidly (Su et al. 1980 and Su
1981). These solitons can decay or evolve to neutral
solitons (Rothberg et al. 1987 and Heeger et al. 1988). At
longer time polarons may be formed due to interchain
excitation and may subsiquently evolve to charged solitons.
An increasing interest has been developing towards
conjugated polymers whose ground state is nondegenerate. In
this class of materials free solitons can not exist and
confined soliton pairs, polarons or bipolarons are the
fundamental excitations. Cis-polyacetylene, polydiacetylene
and polyphenylene are examples of non-degenerate ground
state polymers. In the literature experiments have been
carried out using ultrafast spectroscopy on cis-polyacetylene (Shank et al. 1982), polydiacetylene (Carter et al.
1986, Huxely et al. 1990 and Kobayashi et al. 1990). An
extension of these studies can provide further insight in
the excited state relaxation mechanisms, whose picture is
much less complete than in degenerate systems. Among the
nondegenerate ground state systems a promising class of
materials is represented by the thiophene-based polymers,

which are characterized by a ring structure in the repeating unit (see Fig.1). The optical properties of these compounds have not yet been fully investigated and a few measurements of excited state relaxation dynamics after photoexcitation have been so far reported (Vardeny et al. 1989, McBranch et al 1990, Kobayashi et al. 1990, Samuel et al. 1991a, Samuel et al 1991b).

In this work we review some of the recent investigations on ultrafast excited state kinetics of thiophene-based polymers using picosecond and femtosecond time resolved spectroscopy. The excited state kinetics of these materials can be related to the formation of self-localized state (polarons, bipolarons, self trapped excitons) within the band gap. The excited kinetics of thiophene-based polymers, has been investigated by picosecond and femtosecond spectroscopic techniques. We report also transient saturated absorption measurements performed in our laboratory on newly shynthetized thiophene polymers with low band gap energy using femtosecond pump and probe technique.

2. Polythiophenes

Polythiophenes possess a simple backbone structure resembling that of cis-polyacetylene. The cis-like structure is stabilized by the sulphur which is known to interact only weakly with the π-electron system of the backbone so that polythiophenes can be considered as a type of pseudopolyenes (Chung et al 1984, Schaffer et al. 1986). Most knowledge of elementary excitation in polythiophenes has been gained from steady-state photoluminance and photoinduced absorption spectroscopies and from light induced electron-spin resonance experiments (Vardeny et al. 1986, Kaneto et al. 1988, Colaneri et al. 1987, Kim et al. 1987, Moraes et al. 1984, Rhue et al. 1990). Electron-hole pairs produced in the same chain by photoexcitation across the band gap may

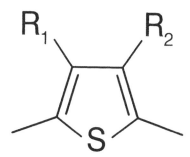

FIG.1: Chemical structure of thiophene monomer; R_1 and R_2 are generic substituents.

remain bound, due to the non degenerate ground state to form
intrachain self-trapped excitons (see Fig.2). These excited
states   seem to be the dominant products of the
photoexcitation at shorter times. In order to form charged
excitations, such as polarons and bipolarons (see Fig.2),
separation of the electron-hole pairs is required. As the
chain distorts around these charges, a simply charged
polaron is formed. After random walk along the polymer chain
two polarons can collide and form a bipolaron. An other
characteristic of these materials is the conformational
flexibility about the quasi single inter-ring C-C bond. The
barrier to rotation has been calculated and measured to be
reasonably low (4 kcal/mole by Lopez Navarrete et al. 1990).
This barrier can be also influenced by the type and position
of the side groups. Moreover the side groups can act as
donor or acceptor of charges as well as may influence the
distortion of the thiophene ring. All these features may
affect the excited state dynamics.

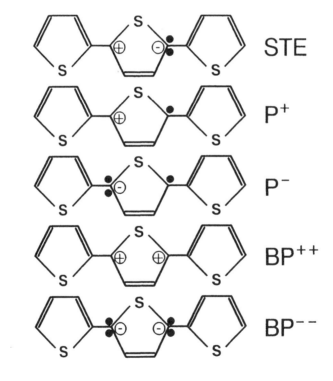

FIG.2: Structures of self-trapped excitons (STE), polarons
(P^+,P^-) and bipolarons (BP^{++}, BP^{--}) in thiophene polymers.

In order to obtain more insight into the elementary processes involved after photoexcitation it is important to investigate the excited-state dynamics by time-resolved methods. Picosecond and femtosecond pump and probe experiments have been performed on a number of thiophene polymers. Hereafter we report the results so far obtained for the different compounds. Polymers of thiophenes, alkylthiophenes and more complicated condensed thiophenes are at present the center of active interest as materials for non-linear optics. In particular polyalchylthiophenes are gaining considerable interest since they are soluble, chemically stable and easily processable.

2.1 Poly(2,5-thienylene)

This compound has been studied by Vardeny et al. 1989. The samples were semitransparent film electrochemically polymerized on a conducting glass substrate. Pump and probe at photon energy of 2 eV with 100 fs time resolution was performed. A photoinduced bleaching has been observed with a decay kinetics rather complex (see Fig.3). An initial exponential decay with a time constant of about 1 ps is observed during the first 0.5 ps then a non-exponential decay develops that lasts for 45 ps, reaching a long-lived plateau. The authors interpret the initial fast decay as geminate recombination of hot photocarriers (probably in form of bound charged solitons) and the change in decay at 0.5 ps as a cross-over to a second state (probably a

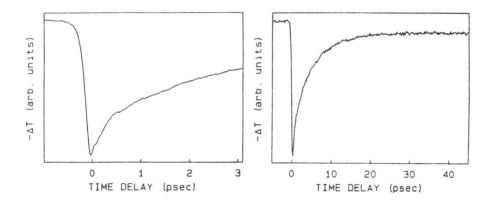

FIG.3: Transient photoinduced transmission changes for cross pump and probe polarizations in a thiophene film measured at room temperature as a function of probe delay (from Vardeny et al. 1989).

bound state of neutral soliton pairs). The power low decay
is due to the recombination of the latter state into the
ground state.

2.2 *Poly(3-alkylthienylenes)*

The polythiophene derivatives with an alkyl substituent
are of great interest since they are soluble and can be
easily prepared as cast film or by spin coating from a
solution. Pump and probe experiments have been performed
with   both picosecond (McBranch et al. 1990) and
femtosecond (Samuel et al. 1991a) resolution. In the
picosecond pump and probe experiments (time resolution 1 ps)
performed on poly(3-hexilthiophene) the initial decay of the
induced photobleaching was observed (see Fig.4a) in about 5
ps down to a long-lived plateau (several hundred
picosecond). By probing at 1.17 eV (1064 nm) with 8 ps time-
compressed pulses a photoinduced absorption was measured
(see Fig.4b), which decays with time constant of 150 ps down
to a long-lived plateau. A similar behavior was observed in
the case of poly(3-octhylthienylene). The data have been
interpreted according to the following model. The fast

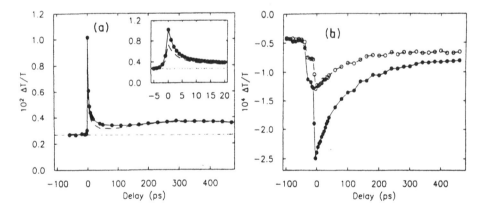

FIG.4:   (a) Photoinduced bleaching at 2.06 eV due to
pumping at 2.06 eV for poly(3-hexylthiophene) as function of
probe delay. The solid curve and the dashed curve are
respectively for pump and probe beam polarized parallel to
each other and at right angle respectively. (b) Photoinduced
absorption observed at 1.17 eV for the same polymer when
pumping at 2.06 eV, as a function of probe delay. The solid
and open circles are respectively for parallel and
perpendicular pump and probe polarizations (from McBranch et
al 1990).

component was attributed to a combination of radiative and non radiative recombination of neutral bipolaron excitons (self-trapped excitons). This channel dominates the photoinduced bleaching. The photoinduced absorption at 1.17 eV was assigned to the formation of gap states with a shift of the oscillator strength from the π-π^* interband transition to transitions involving self-localized states. These structural relaxations may form polarons.

In the femtosecond pump and probe experiment at 2 eV for poly(3-dodecylthienilene) the enhanced time resolution shows a few unusual features (see Fig.5): (i) a photoinduced absorption appears with rise time of 100 fs, then the response changes sign to give a bleaching, which peaks 150 fs after the peak in the absorption and then decays in the following 300 fs leaving a longe lived tail; (ii) oscillations with a period of 120 fs are also observed in the response immediately after the transition from absorption to bleaching. The similar behavior (see Fig.5) has been observed also in the case of poly(3-hexylthienylene). The same results have been obtained for both polymers with pump beam orthogonally polarized to the probe beam. The unusual initial absorption was explained by the authors, at least in part, as due to electromodulation of the band edge produced by separation of electrons and holes. The oscillatory feature are probably due to the coupling between vibrational modes of the polymer chains and the π-electron structure.

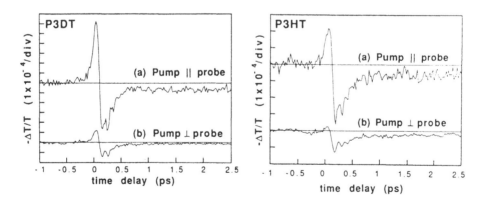

FIG.5: Transient absorption in poly(3-dodecylthienylene) (P3DT) e poly(3-hexylthienylene) (P3HT) for pump and probe pulses at 1.98 eV, as a function of probe delay. The upper and lower curves are respectively for parallel and orthogonal pump and probe polarizations (from Samuel et al 1991a).

2.3 Poly(3-methylthiophene)

This polymer was prepared by electrochemical polymerization on glass substrate. The saturated absorption kinetics was studied by femtosecond pump and probe spectroscopy by Kobayashi et al. 1990. The pump pulses of 100 fs duration were centered at 1.98 eV (628 nm) and the broad band probe pulses were obtained by continuum generation in carbon tetrachloride. The detected photon energy range was 1.3-2.1 eV. The pump and probe data are reported in Fig.6 for a temperature of 10 K. A similar behavior was also observed at room temperature. A clear absorption bleaching is detected near the pump pulse energy. The decay presents two exponential components with time constant of 70 fs and 800 fs. In the energy range 1.35-1.8 eV an induced absorption is present; in particular the rise time of the absorption at 1.86 eV is estimated of the order of the time constant of the initial bleaching decay. The second fast component of the bleaching recovery is also present in the excited state absorption. This complex dynamics is interpreted in terms of a model involving self-trapped excitons formation and decay. After 70 fs the initially free excitons are trapped by the backbone distortion and their excited state absorption at 1.86 eV is the signature. Self-trapped exciton

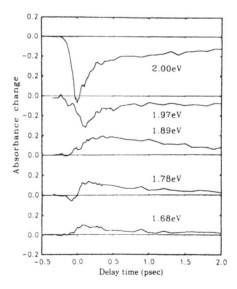

FIG.6: Time dependence of the photoinduced absorption of poly(3-methylthiophene) at 10 K for various energies of the probe pulse, as a function of the probe delay. The pump pulse energy is 1.97 eV (from Kobayashi et al. 1990).

decay occurs within 800 fs by tunnelling to the ground state or formation of long lived species such as triplet excitons, polarons and bipolarons. From absorption bleaching an estimate of the imaginary third order susceptibility can be obtained. The data reported for poly(3-methylthiophene) are of the order of 3.5×10^{-10} esu at 1.97 eV.

3. A new class of polythiophene compounds with low band gap energy.

Here we report recent pump and probe experiments performed in our laboratory on newly synthetized polymers with low band gap namely: poly(3-pentoxythiophene) (PPT), poly(3-decylthienylenevinylene) ($PTVC_{10}$) and poly(3,4-dibutyl- thienylenevinylene) ($PTVC_4$). The chemical structures of these three polymers are shown in Fig.7. The low energy gap in PPT is due the presence of the pentoxy group, which injects electrons. In the case of PTV the vinyle group increasing the conjugation length decreases the band gap. The PTV polymers have been synthetized as described by Galarini et al. 1991. The electronic absorption spectra of PPT and $PTVC_{10}$ are shown in Fig.8a and b respectively; the absorption spectrum of $PTVC_4$ is similar to that of $PTVC_{10}$.

The samples were studied in chloroform solution using the femtosecond pump and probe technique. The excitation source consists of a cavity dumped hybrdly mode-locked Rhodamine 6G dye laser with DODCI as saturable absorber. The use of such femtosecond light source compared to the traditional ones allows to operate with a lower repetition rate and correspondingly higher pulse energy. The advantages are reduced thermal effects and no need of complex amplification systems for single wavelength measurements. The dye laser is synchronously pumped by the second harmonic of a 76 MHz cw mode-locked Nd:YLF laser. The innovative resonator design (Cybo-Ottone et al. 1991) combined to an active system for

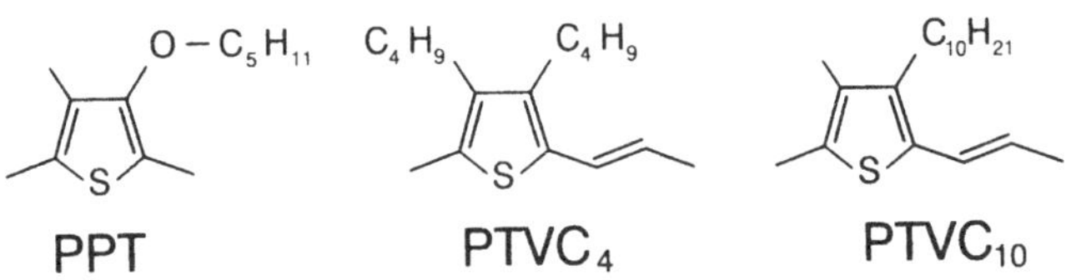

FIG.7: Chemical structures of PPT, $PTVC_{10}$ and $PTVC_4$.

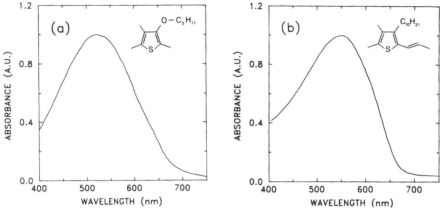

FIG.8:   Electronic absorption spectra of PPT (a) and PTVC$_{10}$
(b). The chemical structures of the repeating units for the
two polymers are reported in the insets.

cavity length stabilization allows to generate pulses of 60
fs duration at 3.8 MHz with an average power of 40 mW and
residual fluctuations of less than one per cent. The peak
wavelength of the pulse spectrum is centred at 627 nm and
falls within the absorption bands of the polymers. In the
femtosecond pump and probe apparatus, the signal to noise
ratio has been improved using the technique of differential
detection, which eliminates the residual fluctuations of the
laser source.
    The transient saturated absorption curves for PPT and
PTVC$_{10}$, as a function of probe delay, are shown respectively
in Fig.9a and b. After subtraction of the coherent coupling

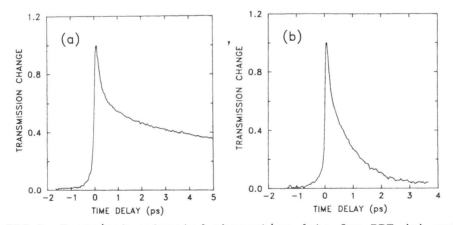

FIG.9: Transient saturated absorption data for PPT (a) and
PTVC$_{10}$ (b) as a function of probe delay.

term the data for PPT can be fitted by a two exponential curve with time constants of 300 fs and 4.8 ps, followed by a long lived plateau. The data for $PTVC_{10}$ can be fitted by a two exponential curve with time constants of 40 fs and 0.9 ps, followed by a plateau of very small amplitude. The pump and probe data for $PTVC_4$ shows a behavior very similar to that of $PTVC_{10}$.

The saturated absorption relaxation in these low band gap energy thiophene polymers occurs on a very fast time scale. This behavior is similar to that observed in the previously mentioned thiophene polymers. The reported results by Kobayashi et al. 1990 on femtosecond transient absorption saturation of a cast film of poly(3-methylthiophene) (see section 2.3) have shown the presence of a two exponential decay, characterized by time constants of 70 fs and 800 fs, which have been attributed respectively to the formation and decay times of self-trapped excitons (induced by strong electron-phonon coupling). The energy of the self-trapped exciton falls in the polymer gap. We believe that the experimental data for PPT, $PTVC_{10}$ and $PTVC_4$ can be interpreted in the framework of this model. Note that in our experiment the wavelength of the pump and probe pulses are the same. The probe pulse transient absorption is thus sensitive both to excited state relaxation and ground state recovery. Therefore, according to the work of Kobayashi et al. 1990, the fast initial decay for PPT and $PTVC_{10}$ can be attributed to the relaxation process of free excitons to self-trapped excitons, while the slow decay can be attributed to the ground state recovery of self-trapped excitons. The remaining plateau may be consistent with the formation of long lived excitations, such as triplet excitons. A single exponential decay was observed in the case of a thin film of poly(2,5-thienylenevinylene) by Samuel et al. 1991b. The absence of a fast initial decay may be attributed to the limit of 100 fs in the time resolution of the experiment.

The different values of the estimated time constants for PPT and $PTVC_{10}$ may be related the different backbone structure. In the case of PPT one has to consider the effects of the electron releasing by the oxygen atom and/or of the conformational flexibility about the inter ring C-C bond. The conformational changes in PTV may be different due to the presence of the vinyle group between the thiophene rings, which makes the structure more rigid. The change of the side groups seems not to be relevant for the relaxation process as demonstrated by the data obtained with $PTVC_4$.

3. Conclusions

In conclusion, we have presented a review of time

resolved transient absorption measurements on thiophene based polymers. The results show very fast excited state kinetics, which occur on a femtosecond and picosecond time scale. The formation dynamics (after photoexcitation) and the identification of self-trapped states, whose energy levels are located in the band gap, are still an open and interesting research area.

Aknowledgments

We would like to thank Prof. R. Tubino of the Istituto di Chimica delle Macromolecole C.N.R. (Milano- Italy) and Prof. G. Zerbi of the Istituto di Chimica Industriale Politecnico (Milano-Italy) for providing us the thiophene polymers used in the femtosecond pump and probe experiments. This work has been partially supported by C.N.R. under the "Progetto Finalizzato" on Telecommunications.

REFERENCES

Carter G M, Hryniewicz J V, Thakur M K, Chen Y J and Meyler S E 1986 Appl. Phys. Lett. **49** 998

Chung T C, Kaufman J H, Heeger J and Wudl F 1984 Phys. Rev. B **30** 702

Colaneri N, Nowak M, Spiegel D, Hotta S and Heeger A J 1987 Phys. Rev. B **36** 7964

Cybo-Ottone A, Magni V and De Silvestri S 1991 Opt. Commun. **82** 137

Etemad S, Heeger A and MacDiarmid A G 1982 Ann. Rev. Phys. Chem. **33** 443

Friend R H, Bradley D D C and Townsend P D 1987 J. Phys. D:Appl. Phys. **20** 1367

Galarini R, Musco A, Pontellini R, Bolognesi A. Destri S, Catellani M, Mascherpa M and Zhuo G 1991 J. Chem. Soc. Chem. Commun. 364

Heeger A J, Kivelson S, Schrieffer J R and Su W P 1988 Re. Modern Phys. **60** 781

Huxley J, Mataloni P, Sheonline R W, Fujimoto J G, Ippen E P and Carter G M 1990 Appl. Phys. Lett. **56** 1600

Lopez Navarrete J T, Tian B and G. Zerbi 1990 Synth. Metals **38** 299

Kaneto K, Hayashi S and Yoshino K J. 1988 Phys. Soc. Jpn **57** 1119

Kim Y H, Hotta S and Heeger A J 1987 Phys. Rev B **36** 7486

Kobayashi T 1989 *Nonlinear Optics of Organics and Semiconductors* (Berlin: Springer-Verlag)

Kobayashi T, Yoshizawa M, Stamm U, Taiji M and Hasegawa M 1990 J. Opt. Soc. Am **B7** 1558

McBranch D, Hays A, Sinclair M, Moses D and Heeger A J 1990
  Phys. Rev. B **42** 3011
Moraes F, Schaffer M, Kobayashi M, Heeger A J and Wudl F 1984
  Phys. Rev. B **30** 2948
Orenstein J 1986 *Handbook of Conducting Polymers* ed. T A
  Skotheim (New York:Dekker) Vol.2
Rothberg L, Jediu T M, Etemand S, Baker G L 1987 Phys. Rev.
  Lett. **36**, 7529
Rhue J, Colaneri N F, Bradley D D C, Friend R H and Wegner G
  1990 J.Phys: Condens Matter **2** 5465
Samuel I D W, Meyer K E, Graham S C, Friend R H, Ruhe J and
  Wegner 1991a   Phys. Rev. B **44** 9773
Samuel I D W, Meyer K E, Bradley D D C, Friend R H, Murata H,
  Tsutsui T and Saito S 1991b Synth. Metals **41-43** 1377
Shaffer H and Heeger A J 1986 Solid State Commun. **59** 415
Shank C V, Yen R, Fork R L, Orenstein J and Baker 1982 J L
  Phys. Rev. Lett. **49**, 1660
Su W P and Schrieffer J R 1980 *Proceedings Nat. Acad Sci.* USA
  77 5626
Su W P 1981 Mol. Cryst. Liq. Cryst **77** 265
Vardeny Z, Grahn H T, Heeger A J and Wudl F 1989 Synth.
  Metals **28**, C299
Vardeny Z, Ehrenfreund E, Brafman O, Nowak M, Schaffer H,
  Heeger J and Wudl F 1986 Phys. Rev. Lett. **56** 671

Nonlinear generation of sub-psec pulses of THz electromagnetic radiation by optoelectronics—applications to time-domain spectroscopy

D. Grischkowsky

IBM Watson Research Center, P.O. Box 218, Yorktown Heights, NY 10598

ABSTRACT

An optoelectronic THz beam system is described, which generates and detects subpsec pulses of freely propagating THz electromagnetic radiation with a time-resolution of 150 fsec and a signal to noise ratio of more than 1000. The generated power of the THz radiation is proportional to the square of the power of the ultrafast laser driving pulses. Some applications of this system to THz time-domain spectroscopy are presented to illustrate its generality and usefulness.

INTRODUCTION

This review will first describe the optoelectronic generation and detection of freely-propagating subpsec pulses of THz electromagnetic radiation, and then it will present applications of an optoelectronic THz beam system to THz time-domain spectroscopy (TDS). Recently, there has been a great deal of work demonstrating the generation of THz radiation (1 THz = 33.3 cm^{-1} = 4.1 meV) via material and electronic excitation by ultrashort laser pulses. Modern integrated circuit techniques have made possible the precise fabrication of micron-sized dipoles, which when photoconductively driven by fsec laser pulses can radiate well into the THz regime, as shown by Auston et al (1984) and by Fattinger and Grischkowsky (1988). An alternative and complimentary approach has been to extend radio and microwave techniques into the THz regime through the use of optoelectronic antennas; Mourou et al (1981), Heidemann et al (1983), DeFonzo et al (1987a,b), Pastol et al (1988, 1990), Smith et al (1988), van Exter et al (1989, 1990) and Dykaar et al (1991). Other sources based on various physical systems and effects include the emission of an electromagnetic shock wave due to a volume dipole distribution moving faster than the phase velocity, i.e., electro-optic Cherenkov radiation discovered by Auston (1983) and developed as a free-space radiation source by Hu et al (1990). A conceptually related radiation source is the electromagnetic shock wave radiated by a surface-dipole distribution propagating faster than the phase velocity, discovered by Grischkowsky et al (1987) and demonstrated as a free-space radiation source by Fattinger and Grischkowsky (1989c). Most recently, radiation has been generated by photoconductively driving the surface field of semiconductors with ultrafast laser pulses, Zhang et al (1990). A new and quite efficient source of broadband THz radiation involves the generation of photocarriers in trap-enhanced electric fields with ultrafast laser pulses, as discovered by Katzenellenbogen and Grischkowsky (1991) and explained by Ralph and Grischkowsky (1991).

Some of these sources are based on an optical type approach whereby a transient point source of THz radiation is located at the focus of a dielectric collimating lens, followed by an additional paraboloidal focusing and collimating mirror, an arrangement introduced by Fattinger and Grischkowsky (1988, 1989a,b) and further developed by van Exter et al (1989a, 1990c). This type of source produces well collimated beams of THz radiation. Matched to an identical receiver, the resulting system has extremely high collection efficiency. With a demonstrated signal-to-noise ratio of 1000, a time resolution of less than 150 fsec and a frequency range from 0.2 THz to more than 5 THz, this optoelectronic THz system is presently the most highly developed and will be the one described in this article. One of the most useful versions of the system is based on repetitive, subpicosecond optical excitation of a Hertzian dipole antenna imbedded in a charged coplanar transmission line structure, first demonstrated by Fattinger and Grischkowsky (1988, 1989a,b) and further developed and characterized by van Exter et al (1989a, 1990c). The burst of radiation emitted by the resulting transient dipole is collimated by a THz optical system into a diffraction limited beam and focused onto a similar receiver structure, where it induces a transient voltage and is detected. The THz optical system gives exceptionally tight coupling between the transmitter and receiver, while the excellent focusing properties preserves the sub-picosecond time dependence of the source.

The combination of THz optics with the synchronously-gated, optoelectronic detection process has exceptional sensitivity for repetitively pulsed beams of THz radiation. Via two stages of collimation a THz beam with a frequency independent divergence is obtained from the THz transmitter. The THz receiver with identical optical properties collects essentially all of this beam. The resulting tightly coupled system of the THz transmitter and receiver gives strong reception of the transmitted pulses of THz radiation after many meters of propagation. Another reason for the exceptional sensitivity is that the THz receiver is gated. The gating window of approximately 0.6 psec is determined by the laser pulsewidth and the carrier lifetime in ion-implanted silicon-on-sapphire SOS. Thus, the noise in the comparatively long time interval (10 nsec) between the repetitive THz pulses is not seen by the receiver. A final important feature of the detection method is that it is a coherent process; the electric field of a repetitive pulse of THz radiation is directly measured. Because a repetitive signal is synchronously detected, the total charge (current) from the signal increases linearly with the number of sampling pulses, while the charge (current) from noise increases only as the square root of the number of pulses.

The powerful technique of time-domain spectroscopy (TDS) has recently been applied to several different systems using a variety of sources, as described and reviewed by Grischkowsky et al (1990). With this technique two electromagnetic pulseshapes are measured, the input pulse and the propagated pulse, which has changed shape due to its passage through the sample under study. Consequently, via Fourier analyses of the input and propagated pulses, the frequency dependent absorption and dispersion of the sample can be obtained. The combination of the TDS technique with THz beams has some powerful advantages compared to traditional c.w. spectroscopy. Firstly, the detection of the THz radiation is extremely sensitive. In terms of average power the sensitivity exceeds that of liquid helium cooled bolometers, by more than 1000 times. Secondly, because of the gated and coherent detection, the thermal background, which plagues traditional measurements in this frequency range, is observationally absent. Comparing time domain spectroscopy with Fourier transform spectroscopy (FTS), it should be clear that the frequency resolution of the two techniques are similar, as they are both based on a scanning delay line, where to first order the frequency resolution is determined by the reciprocal of the time scan. Although for now FTS is superior above 4 THz (133 cm^{-1}), the limited power of the radiation sources and the problems with the thermal background favor TDS below 4 THz.

Several different type measurements will be described that illustrate the generality and usefulness of THz-TDS. An early example of THz time-domain spectroscopy was the characterization of water vapor from 0.25 THz to 1.5 THz, where the cross-sections of the 9 strongest lines were measured with the best accuracy to date by van Exter et al (1989b). Later measurements by Grischkowsky et al (1990), on single crystal sapphire and silicon were motivated by the need to find the best material for the THz lenses in contact with the emitting and detecting chips. The available published data were inadequate for this evaluation. Absorption of the THz radiation by the lens material (initially sapphire) imposed an upper limit on the bandwidth of the entire system. The use of silicon lenses, inspired by the THz-TDS measurements of unusually low absorption and dispersion, immediately increased the system bandwidth from 2 to 3 THz and gave smoother THz pulses with less ringing structure. A THz-TDS measurement of the absorption and dispersion due to carriers in device-grade, doped silicon wafers by van Exter and Grischkowsky (1990a,b), showed that the frequency-dependent properties were completely due to the carriers and not to the host crystal. From these measurements the complex conductance was characterized over the widest frequency range to date. Finally, a experimental and theoretical study of THz coherent transients by Harde et al (1991a,b) will be reviewed. Here, after the excitation of N_2O vapor by a subpsec pulse of THz radiation, the vapor emitted a coherent THz pulse train extending to as long as 1 nsec. The origin of the emitted subpsec THz pulses (commensurate echoes) was a periodic rephasing, during the free-induction decay, of the more than fifty coherently excited rotational lines with commensurate transition frequencies. From the decay and reshaping of the echoes the coherent relaxation time T_2 and the anharmonicity factor for the N_2O molecule were evaluated.

THE OPTOELECTRONIC THz BEAM SYSTEM

A. The Experimental Set-Up.

The setup used to generate and detect beams of short pulses of THz radiation is presented in Fig.1. For this example, the transmitting and receiving antennas are identical, each consisting the antenna imbedded in a coplanar transmission line, shown in Fig. 1a as introduced by van Exter et al (1989a, 1990c). The antenna is fabricated on an ion-implanted silicon-on-sapphire (SOS) wafer. The 20-μm-wide antenna structure is located in the middle of a 20-mm-long coplanar transmission line consisting of two parallel 10-μm-wide, 1-μm-thick, 5 Ω/mm, aluminum lines separated from each other by 30 μm. A colliding-pulse mode-locked (CPM) dye laser, produces 623 nm, 70 fsec pulses at a 100 MHz repetition rate in a beam with 5 mW average power. This beam is focused onto the 5-μm-wide photoconductive silicon gap between the two antenna arms. The 70 fsec laser creation of photocarriers causes subpsec changes in the conductivity of the antenna gap. When a DC bias voltage of typically 10 V is applied to the transmitting antenna, these changes in conductivity result in pulses of electrical current through the antenna, and consequently bursts of electromagnetic radiation are produced. A large fraction of this radiation is emitted into the sapphire substrate in a cone normal to the interface; the radiation pattern is presented in Fattinger and Grischkowsky (1989b). The radiation is then collected and collimated by a dielectric lens attached to the backside (sapphire side) of the SOS wafer, an arrangement introduced by Fattinger and Grischkowsky (1988, 1989a,b). For the work reported here, the dielectric lenses were made of high-resistivity (10 kΩcm) crystalline silicon with a measured absorption of less than 0.05 cm^{-1} in our frequency range. The use of silicon gave significant improvement over the sapphire lenses previously used, although the 10% mismatch in dielectric constant between the silicon lens and the sapphire wafer causes a slight reflection. The center of the truncated 9.5 mm diameter silicon sphere (lens) is 2.0 mm above the ultrafast antenna located at the focus of the lens. As shown in Fig. 1b, after collimation by the silicon lens, the beam propagates and diffracts to a paraboloidal mirror, where the THz radiation is recollimated into a highly directional beam. Although the 70 mm aperture paraboloidal mirrors have a 12 cm focal length, a 16 cm distance was used between the silicon lenses and the mirrors to optimize the response of the system at the peak of the measured spectrum. While the high frequency components of the THz pulses remain reasonably well collimated after leaving the silicon lens, the very low frequency components quickly diffract and illuminate the entire paraboloidal mirror, which presents a solid angle of 0.15 steradians to collect radiation from the source. After recollimation by the paraboloidal mirror, beam diameters (10-70mm) proportional to the wavelength were obtained; thereafter, all of the frequencies propagated with the same 25 mrad divergence. The combination of the paraboloidal mirror and silicon lens (THz optics) and the antenna chip comprise the transmitter, the source of a

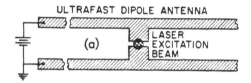

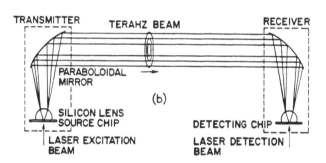

Fig.1 (a) Ultrafast dipolar antenna. (b) THz transmitter and receiver.

highly-directional freely-propagating beam of (sub)-picosecond THz pulses. After a 50 cm propagation distance this THz beam is detected by an identical combination, the THz receiver, where the paraboloidal mirror focuses the beam onto a silicon lens, which focuses it onto a SOS antenna chip, similar to the one used in the emission process. The electric field of the focused incoming THz radiation induces a transient bias voltage across the 5 μm gap between the two arms of this receiving antenna, directly connected to a low-noise current amplifier. The amplitude and time dependence of this transient voltage is obtained by measuring the collected charge (average current) versus the time delay between the THz pulses and the delayed CPM laser pulses in the 5 mW detection beam. These pulses synchronously gate the receiver, by driving the photoconductive switch defined by the 5 μm antenna gap. The detection process with gated integration can be considered as a sub-picosecond boxcar integrator.

B. Measurements of Signal-to-Noise

A typical time-resolved measurement by van Exter and Grischkowsky (1990c) is shown in Fig.2a. The clean pulseshape is a result of the fast action of the photoconductive switch at the antenna gap, the broadband response of the ultrafast antennas, the broadband THz optical transfer function of the lenses and paraboloidal mirrors, and the very low absorption and dispersion of the silicon lenses. The measured pulsewidth of 0.54 psec (FWHM) is only

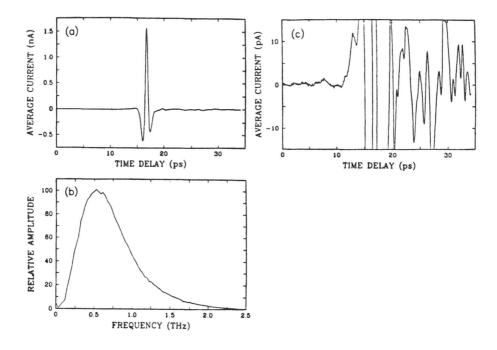

Fig.2 (a) THz pulse measured to 35 psec by scanning the time delay between the optical gating pulses and the incident THz pulses, while monitoring the current induced in the THz receiver. (b) Amplitude spectrum to 2.5 THz of the measured pulse shape. (c) THz pulse on a 100 times expanded vertical scale.

an upper limit to the true pulsewidth, because no deconvolution has been applied to the measurement to take out the response time of the antenna gap. This time-response will be determined in the next section of this article.

In Fig. 2b, the Fourier transform of the measured signal (Fig. 2a) is shown to stretch from about 0.1 to 2.0 THz. This represents only a lower limit to the true extent of the emitted radiation as it contains the frequency response of the receiver. At the low frequency end, the efficiency of both emitter and receiver has been shown to be proportional to the length of the antenna, i.e., proportional to the separation between the two lines of the coplanar transmission line. For extremely low frequencies the size of the paraboloidal mirrors will also limit the efficiency. For the high frequency limit the efficiency of the antenna is strongly reduced when 1/2 the wavelength (in the dielectric) of the emitted radiation is no longer small compared to the antenna length. This frequency for the 30 μm antenna is 1.5 Thz, so that the observed signal and corresponding spectrum is somewhat limited by the antenna response which has dropped by 50% at this frequency. The high frequency part of the spectrum is also limited by the finite risetime of the current transient and the non-ideal imaging properties of the THz optics.

In Fig. 2c, the time-resolved signal is shown on a hundred times expanded vertical scale. The structure observable after the main pulse is reproducible and is due to reflections of the electrical pulse on the transmission line, reflections of the THz pulse from the various dielectric interfaces, and absorption and dispersion of water vapor in the 1 cm path outside the vapor-tight box placed around most of the setup. The observed noise in front of the main pulse is about $1.3x10^{-13}$ A rms for an integration time of 125 ms, corresponding to an integration bandwidth of 1 Hz determined by a 12 dB/octave filter. An identical noise value is obtained when the THz beam is completely blocked. The signal-to-noise ratio in this 4 minute scan is more than 10,000:1. Another 4 minute scan is shown in Fig. 3, for which the intensity of the pump laser beam was reduced from the 6 mW normally used to only 15 μW. This 400-fold reduction in laser power led to a reduction in the transient photocurrent of 320, instead of the expected 400. The discrepancy indicates a slight nonlinearity due to the onset of saturation, related to the fact that the electrical pulses generated on the transmission line are quite strong (almost 1 V in either direction). This 320-fold reduction in photocurrent led to a reduction in the power of the THz beam by the factor $1.0x10^{-5}$. However, despite this enormous reduction in power, the peak amplitude is still more than 30 times larger than the rms noise. Based on previous calculations by van Exter and Grischkowsky (1990c) the average power in the THz beam during this measurement was about 10^{-13} W. If the power of the THz beam were even further reduced, the detection limit of the THz receiver would be reached at $1x10^{-16}$ W, for a signal-to-noise-ratio of unity and a 125 ms integration time. Because the generation and detection of the THz (far-infrared) radiation is coherent, the THz receiver is intrinsically much more sensitive than the incoherent bolometer. The above receiver is approximately 1000 times more sensitive than a helium cooled bolometer described by Johnson et al (1980).

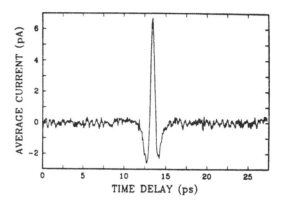

Fig.3 Measured THz pulse to 27.5 psec with a 100,000 times reduction (compared to Fig. 2a) of the THz beam power.

The THz receiver detects, with signal-to-noise ratios of approximately 10,000:1, subpsec, 14 mV pulses coming at a 100 MHz repetition rate in a highly directional beam of THz radiation with an average power of 10 nW. Consequently, the detection limit for these repetitive pulses is about 1.4 μV. However, it has been shown by van Exter and Grischkowsky (1990c) that the sampled voltage, during a single 0.6 psec gating pulse, on the receiving antenna due to the thermal background is about 0.23 mV and due to the vacuum fluctuations is 0.05 mV. Thus, in terms of instantaneous voltages, the receiver can detect 1/160 of the thermal background and 1/35 of the vacuum fluctuations. Beams of THz radiation can be detected with peak powers of only $4x10^{-5}$ that of the incident thermal radiation. This impressive performance is due to the high directionality of the THz receiver and to the fact that the thermal noise is incoherent and adds randomly for successive gating pulses, while the signal propagating in the THz beam is coherent and scales linearly with the number of gating pulses.

C. Measurements of the Time-Dependent Response Function

In this section the experimental study of the THz beam system is extended to smaller 10 μm-long antennas, which have a frequency response extending to 6 THz. From the calculated THz optical transfer function together with the known THz absorption, the limiting bandwidth of the system is extracted following the procedure of Grischkowsky and Katzenellenbogen (1991). Because the transmitter and receiver are identical, identical transmitter and receiver bandwidths are obtained. This result is compared to the calculated radiation spectrum from a Hertzian dipole driven by the current pulse determined by the laser pulsewidth, the current risetime, and the carrier lifetime. From this comparison, the time-domain response function for the antenna current is obtained.

The optoelectronic THz beam system is the same as previously described and as shown in Fig. 1, except that here smaller antennas are used.

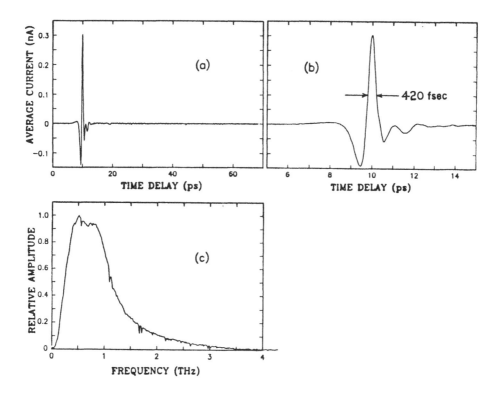

Fig.4 (a) Measured THz pulse to 70 psec. (b) Measured THz pulse on expanded 10 psec time scale. (c) Amplitude spectrum to 4 THz of Fig. 4b.

The antenna structure, again fabricated on ion-implanted SOS, is located in the middle of a 20-mm-long coplanar transmission line consisting of two parallel 5-μm-wide aluminum lines separated from each other by 10 μm. The performance of the colliding-pulse, mode-locked (CPM) dye laser was improved to provide 60 fsec excitation pulses in a beam with an average power of 7 mW on the excitation spot.

For these 10-μm-long antennas the measured transmitted THz pulse is shown in Fig. 4a. This pulse is shown on an expanded time scale in Fig. 4b, where the measured FWHM pulsewidth of 420 fsec (with no deconvolution) is indicated. This pulsewidth is significantly shorter than the 540 fsec pulse (Fig. 2a) obtained from the same experimental arrangement, but with the 30-μm-long antennas as described in the previous section. The use of even still smaller antennas did not significantly shorten the THz pulses. The numerical Fourier transform of Fig. 4a is shown in Fig. 4c, where the amplitude spectrum is seen to extend beyond 3 THz. The sharp spectral features are water lines, from the residual water vapor present in the apparatus.

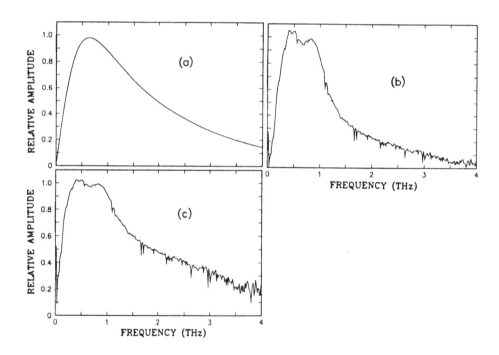

Fig.5 (a) Transmission function to 4 THz. (b) Amplitude spectrum of Fig. 4c divided by transmission function. (c) Amplitude spectral response of transmitter and receiver.

Two effects reduce the spectral extent of the measured pulse (Fig. 4c). These are the frequency-dependent transfer function, explained in detail by Lesurf (1990), of the THz optical system (Fig. 1b) and the THz absorption in the sapphire (SOS) chips The absorption of sapphire has been measured by Russell and Bell (1967), Loewenstein et al (1973) and Grischkowsky et al (1990). The transmission function describing these two effects is presented in Fig. 5a, for our focusing geometry and the SOS chip thickness of 0.46 mm. Dividing the measured spectrum in Fig. 4c by this transmission function we obtain Fig. 5b. Here, the spectral extent is determined only by the product of the receiver response and the transmitted spectrum. Because the transmitter and receiver are identical, by the reciprocity theorem explained by Monteath (1973), the transmitted spectrum is identical to the receiver response, and is given by the square root of Fig.5b shown in Fig. 5c.

D. Calculation of the Semiconductor Time-Dependent Response Function

In the small antenna limit corresponding to the Hertzian dipole, the generated radiation field is proportional to the time-derivative of the current pulse. Based on our study we conclude that the current in the antenna is mainly determined by the intrinsic response of the semiconductor itself. The intrinsic time-domain response function will now be derived for a semiconductor described by the simple Drude formalism. For this case the free carriers are considered as classical point charges subject to random collisions. Here the simplest version of this model is assumed, for which the collision damping is independent of the carrier energy and for which the frequency dependent complex conductivity $\sigma(\omega)$ is given by

$$\sigma(\omega) = \sigma_{dc} \frac{i\Gamma}{\omega + i\Gamma} , \qquad (1)$$

where $\Gamma = 1/\tau$ is the damping rate and τ is the average collision time. The dc conductivity is given by $\sigma_{dc} = e\mu_{dc}N$, where e is the electron charge, μ_{dc} is the dc mobility and N is the carrier density. This relationship is in good agreement with recent time-domain spectroscopy measurements, performed by van Exter and Grischkowsky (1990a,b), on lightly doped silicon from low frequencies to beyond 2 THz. The following procedure is similar to that of Grischkowsky and Katzenellenbogen (1991). It is helpful to recast the formalism into a frequency dependent mobility as

$$\mu(\omega) = \mu_{dc}\frac{i\Gamma}{\omega + i\Gamma}.\tag{2}$$

The dc current density is given by $J_{dc} = \sigma_{dc}E$, or equivalently $J_{dc} = eE\mu_{dc}N$, where E is a constant electric field for the simple case considered here. Because of the linearity of the current in N, for a time dependent carrier density N(t), the time dependent current density can be written as

$$J(t) = eE\int_{-\infty}^{t}\mu(t-t')N(t')dt',\tag{3}$$

where $\mu(t\text{-}t')$ is the time-domain response function for the mobility. This function is determined by the inverse transform of the frequency dependent mobility to be the causal function

$$\mu(t-t') = \mu_{dc}\Gamma e^{-\Gamma(t-t')}\tag{4}$$

which vanishes for negative $(t\text{-}t')$.

In order to facilitate the understanding of the photoconductive switch it is useful to rewrite the basic Eq. (3) in the equivalent form,

$$J(t) = eEA\int_{-\infty}^{t}\mu(t-t')\int_{-\infty}^{t'}R_c(t'-t'')I(t'')dt''dt',\tag{5}$$

where $I(t'')$ is the normalized intensity envelope function of the laser pulse, A is a constant giving the conversion to absorbed photons/volume and R_c is the response function describing the decay of the photogenerated carriers. By defining a new photocurrent response function $j_{pc}(t\text{-}t')$, we can rewrite Eq.(5) in the following way

$$J(t) = \int_{-\infty}^{t}j_{pc}(t-t')I(t')dt',\tag{6}$$

where $j_{pc}(t\text{-}t')$ is obtained by evaluating Eq.(5) with a delta function $\delta(t'')$ laser pulse. Assuming the causal function $R_c(t'\text{-}t'') = \exp\text{-}(t'\text{-}t'')/\tau_c$, describing a simple exponential decay of the carriers with the carrier lifetime τ_c (significantly longer than the average collision time τ) for positive $(t'\text{-}t'')$ and vanishing for negative (t'-t''), and that μ(t-t') is given by the Drude response of Eq. (4), the causal response function j_{pc}(t*) is then evaluated to be

$$j_{pc}(t^*) = \frac{\mu_{dc}eEA\Gamma}{\Gamma - 1/\tau_c}\,(e^{-t^*/\tau_c} - e^{-t^*/\tau}) \qquad (7)$$

for positive t* = (t-t') and shown to vanish for negative t*. In the short pulse limit of the ultrafast excitation pulses, the time dependence of the photocurrent J(t) is approximately equal to that of the photocurrent response function j_{pc}(t*) for positive t*. For a long carrier lifetime, the time dependence of j_{pc}(t*) is described by a simple exponential rise with a risetime of the order of $\tau = 1/\Gamma$, which is equal to 270 fsec and 150 fsec for the electrons and holes, respectively, in lightly doped silicon as measured by van Exter and Grischkowsky (1990a,b). As these results show, the material response can be slow compared to the duration of the ultrafast laser excitation pulses which can be as short as 10 fsec, but is more typically of the order of 60 fsec.

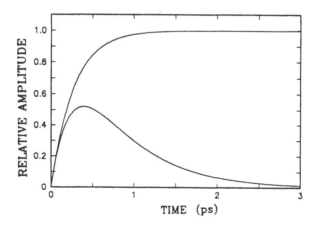

Fig.6   Calculated photoconductive response function to 3 psec with the scattering time τ = 270 fsec and infinite carrier lifetime τ_c (upper curve) and with τ_c = 600 fsec (lower curve).

The time-dependent response function described by Eq. (7) is calculated in Fig. 6 for the two cases; τ = 270 fsec and $\tau_c = \infty$, and τ = 270 fsec and τ_c = 600 fsec. The result for infinite carrier lifetime has the following intuitive interpretation. After the instantaneous creation of carriers, the initial current and mobility is zero. The carriers then accelerate ballistically, as determined by the applied electric field, their charge and effective mass. This acceleration continues for approximately a time equal to the scattering time τ, after which the velocity and current equilibrate to their steady-state value. This discussion will now be shown to accurately describe the mathematical dependence of Eq. (7). With $\tau_c = \infty$, Eq. (7) is equal to

$$j_{pc}(t^*) = \mu_{dc} e E A (1 - e^{-t^*/\tau}),\qquad(8)$$

which for times short compared to τ reduces to

$$j_{pc}(t^*) = \mu_{dc} e E A t^*/\tau.\qquad(9)$$

Remembering that for Drude theory $\mu_{dc} = e/(m^*\Gamma)$, Eq. (9) is equivalent to

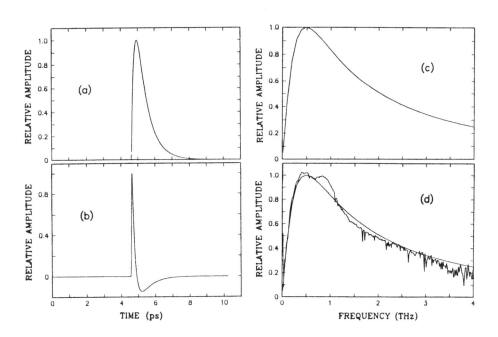

Fig.7 (a) Calculated current pulse (11 psec timescale) in semiconductor and antenna. (b) Time-derivative of current pulse. (c) Amplitude spectrum to 4 THz of Fig. 7b. (d) Comparison of Figs. 5c and 7c.

$$j_{pc}(t^*) = Aet^*(eE/m^*),\tag{10}$$

which describes the ballistic acceleration eE/m^*.

For the photoconductive switches considered here, we assume the time-domain response function $j_{pc}(t^*)$ to be given by Eq. (7). This response function is then convolved with a Gaussian shaped laser pulse with a FWHM of 60 fsec, as prescribed by Eq. (6). The carrier lifetime τ_c has been measured by Doany et al (1987) to be 600 fsec for ion-implanted SOS. As demonstrated in Fig. 7, good agreement with experiment is obtained with the average collision time $\tau = 190$ fsec. With these parameters the calculated shape of the current pulse in the photoconductive switch and the Hertzian dipole antenna is presented in Fig. 7a. The time derivative of this pulse is given in Fig. 7b, where an extremely fast transient, corresponding to the rising edge of the current pulse, is seen. The numerical Fourier transform of Fig. 7b, presented in Fig. 7c, is the predicted amplitude spectrum of the transmitter. In Fig. 7d, this spectrum is compared with the amplitude spectrum of the transmitter from Fig. 5c; the agreement is excellent. Thus, we have determined an experimentally self-consistent time-domain response function describing the current in the Hertzian dipole antenna. For longer antennas for which the radiated pulse is no longer the time-derivative of the current pulse, the calculated current pulse is Fourier analysed and the resulting spectral amplitudes are put into the antenna response to determine the emitted pulse.

In summary, we have shown that the 10-μm-long antenna imbedded in the coplanar transmission line has electrical properties much faster than the semiconductor itself. Consequently, the performance is completely determined (and limited) by the intrinsic response time of the semiconductor. With the 10-μm-long antennas, it is now possible to directly study the dynamical response of free carriers in a variety of semiconductors.

E. A High-Performance Source Utilizing Trap Enhanced Electric Fields.

A different-type, high-performance optoelectronic source chip, first used to generate pulses of freely propagating THz electromagnetic radiation by Katzenellenbogen and Grischkowsky (1991), is shown in Fig. 8d. The simple coplanar transmission line structure consists of two 10-μm-wide metal lines separated by 80 μm fabricated on high-resistivity GaAs. Irradiating the metal-semiconductor interface (edge) of the positively biased line with focused ultrafast laser pulses produces synchronous bursts of THz radiation. This occurs because each laser pulse creates a spot of photocarriers in a region of extremely high electric field, the trap enhanced field described by Ralph and Grischkowsky (1991). The consequent acceleration of the carriers generates the burst of radiation. The CPM dye laser provides 60 fsec excitation pulses with an average power of 5 mW at the $10\mu m$ diameter excitation spot. The major fraction of the laser generated burst of THz radiation is emitted into the GaAs substrate in a cone normal to the interface and is then collected and collimated by a crystalline silicon lens attached to the back side of the chip. This source chip is completely compatible with the previously described optoelectronic THz beam system .

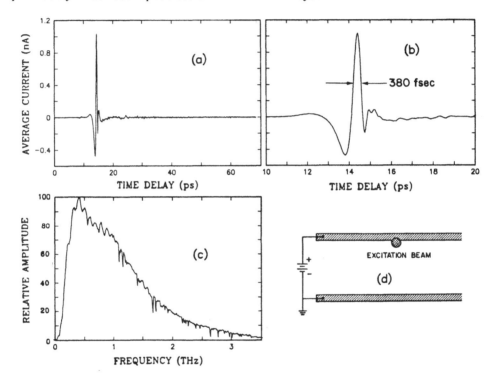

Fig.8 (a) Measured THz pulse to 70 psec. (b) Measured THz pulse on an expanded 10 psec time scale. (c) Amplitude spectrum to 3.5 THz of the THz pulse of Fig. 8a. (d) Source chip configuration used to generate the freely propagating pulses of THz radiation.

The THz radiation detector is an ion-implanted SOS detection chip with the antenna geometry shown in Fig. 1a, but with the faster 10-μm-long antenna.

The measured THz pulse emitted from the laser excited metal-GaAs interface with $+60V$ bias across the transmission line is shown in Fig. 8a, and on an expanded time scale in Fig. 8b. The measured pulsewidth with no deconvolution is seen to be 380 fsec. At the time these results were obtained by Katzenellenbogen and Grischkowsky (1991), they were the shortest directly measured THz pulses; the dip on the falling edge was the sharpest feature ever observed with an ion-implanted detector and indicated a response time faster than 190 fsec. The numerical Fourier transform of the pulse of Fig. 8a, as presented in Fig. 8c, extends to beyond 3 THz; the sharp line structure is due to residual water vapor present in the system.

In work (to be presented at QELS92) using this same type source and detector, but with tighter optical focusing, better matched silicon lenses and with the THz optical system of Fig. 1b optimized to have a unity, frequency-independent, transfer function, ultrashort THz pulses were measured with a ringing structure faster than 160 fsec. Their amplitude spectrum peaked at 1 THz; the relative amplitude at 3 THz was 20%, at 4 THz was 5% and useful radiation extended to 5 THz. These results confirm the response time of the receiver to be faster than 150 fsec, in agreement with the direct characterization of Grischkowsky and Katzenellenbogen (1991) and show that the entire optoelectronic system is competitive with the alternative approach of THz interferometry, Greene et al (1991) and Ralph and Grischkowsky (1992).

F. Other Sources and Receivers

As allowed by the modular nature of the THz beam system, many different combinations of antenna lengths, shapes and other semiconductors, GaAs:As, Warren et al (1990, 1991), and GaAs, Grischkowsky et al (1990), van Exter and Grischkowsky (1990b), Harde et al (1991a,b) and Warren et al (1991), have been used. For the THz source chip the carrier lifetime is of little concern, because the THz radiation is proportional to the time derivative of the current pulse and therefore, is mainly generated on the steep rising edge. Antenna of the same geometry but with longer arms between more widely separated lines produce stronger, slower pulses with lower peak frequencies as shown by van Exter and Grischkowsky (1990a,b). Different detectors can also be used. Although other materials have been tried by Warren et al (1991), ion-implanted SOS material is generally used due to its reproducible properties and short carrier lifetime. The antenna length is chosen to suit the measurement. The previously discussed 10 μm-long antenna has a relatively flat response from low frequencies to 6 THz. Slower more sensitive antennas are obtained simply by increasing the separation between the coplanar lines, while keeping constant the 5 μm photoconductive gap separation between the two arms.

An alternative method of source characterization, which bypasses the problems of receiver bandwidth, is based on far-infrared interferometric techniques using a power detector. This approach was first demonstrated for THz radiation sources by Greene et al (1991), who measured auto-correlation signals with a fullwidth-at-half-maximum (FWHM) of 230 fsec for the THz radiation pulse from laser created carriers accelerated by the surface field of a photoconductive semiconductor. This approach used a single THz radiation source, illuminated by 10 Hz repetition-rate, amplified, 100 fsec pulses from a CPM dye laser, together with a Martin-Puplett interferometer and a liquid helium cooled bolometer. A different interferometric approach, using unamplified, 100 MHz repetition rate, CPM dye laser pulse excitation of a two source interferometer has since been demonstrated by Ralph and Grischkowsky (1992). Via this approach, together with fast, scanning-delay-line averaging, orders of magnitude improvements in the signal-to-noise ratio of the measured interferograms were obtained. Using this method, the THz radiation source of Fig. 8a was shown to produce radiation to 6 THz. In addition, a 230 fsec FWHM auto-correlation signal and an average power of 30 nW for 4 mW of laser excitation power were measured for this source.

THz TIME-DOMAIN-SPECTROSCOPY

The powerful technique of time-domain spectroscopy (TDS) has recently been applied to several different systems using a variety of sources, as described and reviewed by Grischkowsky et al (1990). With this technique two electromagnetic pulseshapes are measured, the input pulse and the propagated pulse, which has changed shape due to its passage through the sample under study. Consequently, via Fourier analyses of the input and propagated pulses, the frequency dependent absorption and dispersion of the sample can be obtained. The useful frequency range of the method is determined by the initial pulse duration and the time resolution of the detection process. Therefore, with each reduction in the generated electromagnetic pulsewidth, and/or the time resolution of detection, there is a corresponding increase in the available frequency range.

The combination of the TDS technique with THz beams has some powerful advantages compared to traditional c.w. spectroscopy. Firstly, the detection of the far-infrared radiation is extremely sensitive. Although the energy per THz pulse is very low (0.1 femtoJoule), the 100 MHz repetition rate and the coherent detection measures the electric field of the propagated pulse with a signal-to-noise ratio of about 10,000 for an integration time of 125 msec, as demonstrated by van Exter and Grischkowsky (1990c). In terms of average power this sensitivity exceeds that of liquid helium cooled bolometers, by more than 1000 times. Secondly, because of the gated and coherent detection, the thermal background, which plagues traditional measurements in this frequency range, is observationally absent. Comparing time domain spectroscopy with Fourier transform spectroscopy (FTS), it should be clear that the frequency resolution of the two techniques are similar, as they are both based on a scanning delay line, where to first order the frequency resolution is determined by the reciprocal of the time scan. However, the fact that TDS scans a delay line with a well-collimated optical beam does present some experimental advantages. Although for now FTS is superior above 4 THz (133 cm^{-1}), the limited power of the radiation sources and the problems with the thermal background favor TDS below 4 THz.

The most serious experimental problem limiting the accuracy of TDS measurements involves the relatively long term changes in the laser pulses and the consequent changes in the input THz pulses. During an experiment, we first measure the input pulse (with no sample in place), then measure the pulse transmitted through the sample, and finally remeasure the input pulse with the sample removed. This sequence is repeated several times to obtain good statistics. Typically, the amplitude spectral ratio of subsequent input pulses varies by +/- 5% over the frequency spectrum. This variation limits the accuracy of an absorption measurement. In the same manner, the relative phase of subsequent input pulses varies by +/- 0.05 radians over the same spectrum, and thereby limits the accuracy of the measurements of the index of refraction. Another experimental consideration involves the frequency distribution across the profile of the THz beam, which consists of a series of overlapping discs with diameters proportional to the wavelength. This feature requires that the sample be uniform and be centered in the beam and that an aperture of the same diameter as the sample blocks any THz radiation from going around the sample. The aperture is kept in place for measurement of the input pulse.

We will now describe several different type measurements that illustrate the generality and usefulness of THz-TDS.

A. Water Vapor

An early example of THz time-domain spectroscopy was the characterization of water vapor from 0.25 THz to 1.5 THz, where the cross-sections of the 9 strongest lines were measured with the best accuracy to date by van Exter et al (1989b).

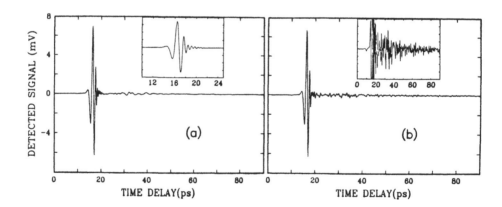

Fig.9 (a) Measured THz pulse (90 psec timescale) propagating in pure nitrogen. Inset shows pulse on an expanded 15 psec time scale. (b) Measured THz pulse with 1.5 Torr of water vapor in the enclosure. Inset shows pulse on a 20x expanded vertical scale.

For these early measurements 30 μm-long source and detector antennas on ion-implanted SOS substrates were used together with MgO lenses instead of the currently used, more optimal high-resistivity silicon. Figure 9a displays the detected THz radiation pulses after propagating through pure nitrogen. This measurement was made in a single 10 minute scan of the 200 psec relative time delay between the excitation and detection pulses. When 1.5 Torr of water vapor, corresponding to 8% humidity at 20.5° C, was added to the enclosure, the transmitted pulse changed to that shown in Fig. 9b. The additional fast oscillations are caused by the combined action of the dispersion and absorption of the water vapor lines. The slower and more erratic variations seen in both Figs. 9a and 9b result from reflections of the main pulse. They are reproducible and divide out in the data analysis. The inset shows the data on a 20X expanded vertical scale. Here, the oscillations are seen to decay approximately exponentially with an average coherent relaxation time T_2.

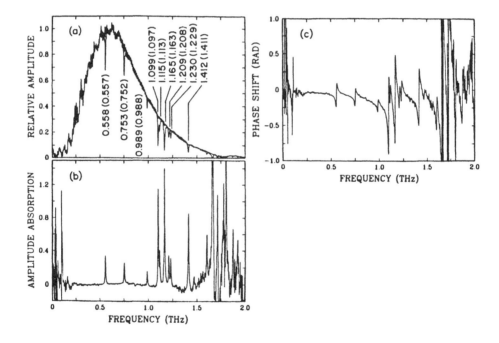

Fig.10 (a) Amplitude spectra to 2 THz of Figs. 9a and 9b. (b) Amplitude absorption coefficient obtained from Fig. 10a. (c) Relative phase of the spectral components of Fig. 10a.

The amplitude spectra of Figs. 9a and 9b are compared in Fig. 10a, where the strong water absorption lines are clearly observable. The additional structure on both spectra is not noise but results from spurious reflections of the main pulse. At each line are indicated the measured frequency with an estimated error of +/- 0.001 THz and in parenthesis the literature value. The corresponding absorption coefficients are displayed in Fig. 10b as the negative of the natural logarithm of the ratio of the two amplitude spectra in Fig. 10a. Because the electric field is directly measured, the relative phaseshift between the two spectra is also obtained as plotted in Fig. 10c, without the need of the Kramers-Kronig relations. As expected for a Lorentzian line, the magnitude of the jump in phase experienced at each resonance equals the peak absorption.

B. Sapphire and Silicon

The following measurements of Grischkowsky et al (1990) on single crystal sapphire and silicon were motivated by the need to find the best material for the THz lenses in contact with the emitting and detecting chips. The available published data were inadequate for this evaluation. Absorption of the THz radiation by the lens material (initially sapphire) imposed an upper limit on the bandwidth of the entire system.

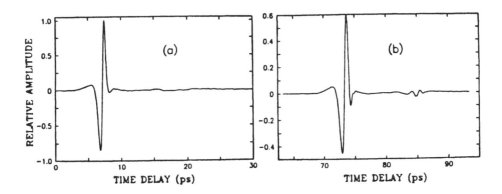

Fig.11 (a) Measured THz pulse to 30 psec. (b) Measured THz pulse (62.5-95 psec) after passage through the sapphire crystal.

The use of silicon lenses, inspired by our measurements of unusually low absorption and dispersion, immediately increased the system bandwidth from 2 to 3 THz and gave smoother THz pulses with less ringing structure.

The single crystal sapphire sample was a polished, 57 mm diameter disc, 9.589 mm thick, and with the C axis in the plane of the disc. A typical THz input pulse incident upon the sample is shown in Fig. 11a, and the output pulse (normalized to the input pulse) after propagation through the sample is shown in Fig. 11b for which the C axis of the crystal was perpendicular to the polarization. The reduction in amplitude is due to the reflective loss at both surfaces and to the absorption suffered during passage through the sapphire. The pulse at 73.4 psec delay is the ordinary pulse, while the smaller pulse at 85.1 psec delay is the extraordinary pulse. The ratio of the peak of the ordinary pulse to that of the extraordinary pulse is approximately 25:1 and gives the polarization sensitivity of our system. In terms of amplitude, the polarization ratio of the generated THz beam is 5:1. The 11.8 psec separation between the two pulses is a measure of the birefringence of sapphire and, neglecting the correction to group velocity due to dispersion, directly gives the difference in the index of refraction between the extraordinary and ordinary ray to be $n_e - n_o = 0.37$ compared to the the value of 0.34 measured by Russell and Bell (1967) and tabulated by Gray (1982). When a full frequency analysis is performed, excellent agreement is obtained with the literature value. The 73.4 psec time delay of the ordinary pulse with respect to the 7.1 psec time delay of the pulse with no sample in place gives the ordinary index of refraction $n_o = 3.07$ in agreement with the literature value. The absorption coefficient vs frequency determined from these pulses is shown in Fig. 12a. Here, we see a monotonic increase in absorption with increasing frequency with the expected quadratic dependence. Due to excessive attenuation caused by the sample being too thick for the weak higher frequency components, the data is considered to be accurate only up to 1.75 THz. Some of the previous work has been indicated on the curve, and shows a rough agreement (within a factor of 2) with the TDS measurement. The earliest work of Russell and Bell (1967) clearly gives too little absorption at low frequencies, where our results are in better agreement with those of Loewenstein et al (1973). The relative phases of the Fourier components determine the index of refraction vs frequency as presented in Fig. 12b, which compares reasonably well with the indicated earlier results.

Crystalline silicon is optically isotropic, eliminating concern about the polarization of the incident THz beam and crystal orientation. Although there is significant literature concerning the far-infrared properties of silicon, below 2 THz there are noteworthy discrepancies among the published data

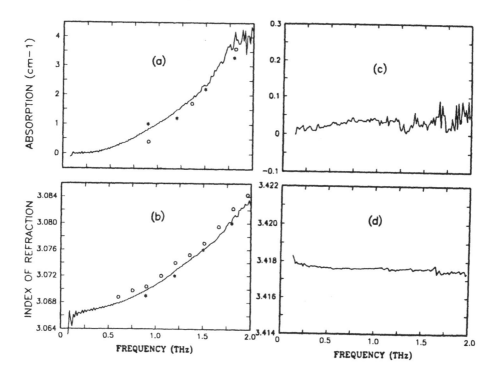

Fig.12 TDS measurements to 2 THz of crystalline sapphire and silicon. The circles are the measurements of Russell and Bell (1967) that are tabulated by Gray (1982); the asterisks are the data of Loewenstein et al (1973). (a) Ordinary ray power absorption coefficient (cm^{-1}) of sapphire. (b) Ordinary ray index of refraction of sapphire. (c) Power absorption coefficient (cm^{-1}) of high-resistivity silicon. (d) Index of refraction of high-resistivity silicon.

with variations in the measured absorption coefficients of up to 10 times. The main reason for this confusion is that below 2 THz the results are extremely sensitive to the presence of carriers. THz-TDS measurements of van Exter and Grischkowsky (1990a,b) show that for 1 Ω-cm, N-type silicon the peak absorption is 100 cm^{-1}, and that for 10 Ω-cm, N-type the absorption is 12 cm^{-1}. Extrapolating these values to 100 Ω-cm, $\alpha = 1$ cm^{-1}; at 1 kΩ-cm, $\alpha = 0.1$ cm^{-1}, and at 10 kΩ-cm, $\alpha = 0.01$ cm^{-1}. Consequently, unless high purity, high-resistivity material is used, what is measured is not the properties of the intrinsic semiconductor but that of the carriers due to residual impurities. This problem is most prevalent in the earlier work on silicon with resistivities of 10 Ω-cm for the sample of Russell and Bell (1967) to 100 Ω-cm for the sample of Loewenstein et al (1973).

The following TDS measurements were made on a 50 mm diameter, 20.046 mm thick single crystal of high-resistivity (greater than 10 kΩ-cm), float-zone silicon. For this material we have measured unprecedented transparency together with a remarkably flat dispersion curve. This is an excellent material for teraHz applications as can be seen from the absorption spectrum presented in Fig. 12c. Throughout the range from low frequencies up to 2 THz the measured absorption coefficient is less than 0.05 cm^{-1}. From the relative phases of the spectral components, the index of refraction vs frequency is obtained as presented in Fig. 12d. Here, the extremely desirable feature of low-dispersion is clearly evident; the index of refraction changes by less than 0.001 over the entire measured spectrum.

C. N-Type and P-Type Silicon

We now describe a THz time-domain-spectroscopy measurement by van Exter and Grischkowsky (1990a,b) of the absorption and dispersion due to carriers in device-grade, doped silicon wafers. The frequency-dependent properties were shown to be completely due to the carriers and not to the host crystal. Consequently, the complex conductance could be characterized over the widest frequency range to date. The samples used were a 283 μm thick wafer of 1.15-Ωcm, N-type and a 258 μm thick wafer of 0.92-Ωcm P-type silicon. The measured absorptions shown in Fig. 13a are more than 2000 times greater than that of the host crystal. Below 0.15 THz the data becomes noisy due to the limited beam power, but the theoretically-predicted drop in absorption at low frequencies is clearly observable. The clear difference between the N and P type material is due to the different dynamic behavior of the electrons and holes. For these measurements the oscillations due to the etalon effects of the sample geometry have been removed numerically. As shown in Fig. 13b, the index of refraction is strongly frequency dependent, having a clear minimum followed by a dramatic increase towards lower frequencies. The agreement with the Drude theory (solid line) is exceptional.

As the THz optical properties of the samples are essentially completely determined by the carrier dynamics, the complex electric conductivity of the doped silicon has also been measured. Independent of Drude theory and relying on only very general assumptions, the electric conductivity is obtained (without any fitting parameters) from the data of Figs. 13a and 13b; the resulting real part of the conductivity is shown

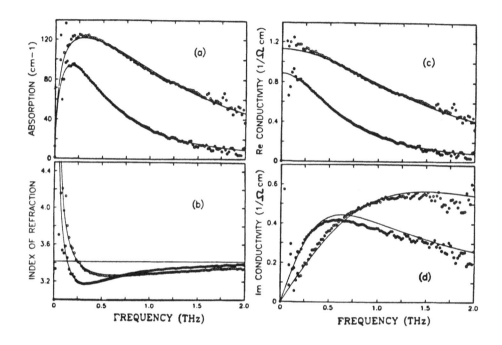

Fig.13 THz-TDS Results to 2 THz for 1.15-Ωcm, N-type (dots, lower curves) and 0.92-Ωcm, P-type (circles, upper curves) silicon. (a) Power absorption. (b) Index of refraction. (c) Real part of the electric conductivity. (d) Imaginary part of the electric conductivity.

in Fig. 13c and the imaginary part in Fig. 13d. The real part is strongly frequency dependent, dropping monotonically from its DC peak to a reduced value at the highest measured frequency of 2 THz. The extrapolated DC conductivities are 0.89 Ωcm for the P-type material and 1.13 Ωcm for the N-type, compared to the directly measured values of 0.92 Ωcm (P-type) and 1.15 Ωcm (N-type). The behavior of the imaginary part is completely different, increasing from zero at low frequencies, peaking at mid-range and then showing a gradual decline. The agreement with the Drude theory (solid line) is quite acceptable.

The two Drude parameters, the plasma angular frequency ω_p and the damping rate Γ, were determined within 5% accuracy from the fits to the data. For 0.92 Ωcm P-type silicon $\omega_p/2\pi = 1.75$ THz and $\Gamma/2\pi = 1.51$ THz, while for 1.15 Ωcm N-type silicon $\omega_p/2\pi = 1.01$ THz and $\Gamma/2\pi = 0.64$ THz. The measured damping rates and the known effective carrier masses, determine mobilities of 1680 cm^2/Vs for the electrons and 500 cm^2/Vs for the holes. The measured plasma frequencies and effective carrier masses determine the carrier densities of $1.4x10^{16}/cm^3$ for the P-type and $3.3x10^{15}/cm^3$ for the N-type silicon. Thus, these device-grade silicon wafers have been electrically characterized from low frequencies to 2 THz by THz-TDS.

D. THz Coherent Transients

An unusual and interesting observation by Harde et al (1991a,b) was that after excitation by a well-collimated beam of subpsec pulses of THz radiation, an N_2O vapor cell emitted a coherent THz pulse train extending to as long as 1 nsec. The N_2O molecule has a permanent electric dipole moment and a large number of strong rotational absorption lines with negligible Doppler broadening within the THz frequency range. The propagation of a 0.5 psec THz pulse through such a molecular vapor excites a multitude of rotational transitions in the impact approximation and causes the molecules to reradiate a free-induction-decay (FID) signal which decays because of relaxation, interference and propagation effects. Due to the experimental system's exceptionally high signal-to-noise ratio, the FID signal can be monitored to delay times as long as 1 nsec. Because the N_2O molecule has an almost constant frequency spacing between the rotational lines, a periodic rephasing and dephasing of the entire ensemble of the more than 50 excited transitions is observed. This situation contrasts to that of water vapor with a multitude of incommensurate resonance lines, studied earlier by THz time-domain spectroscopy by van Exter et al (1989b,c). Consequently, for N_2O vapor after the initial excitation pulse, the sample emits a series of uniformly spaced (39.8 psec) subpsec THz pulses whose amplitude decays with the coherent relaxation time T_2. From these observations T_2 can be directly determined, even under conditions when in the frequency domain the absorption lines would be completely overlapping. From the pulse repetition rate in the train the frequency separation between the rotational lines can be determined, and because of the exceptionally high time resolution of better than 0.5 psec, the anharmonicity in the linespacing can be detected as reshaping of the individual pulses in the train.

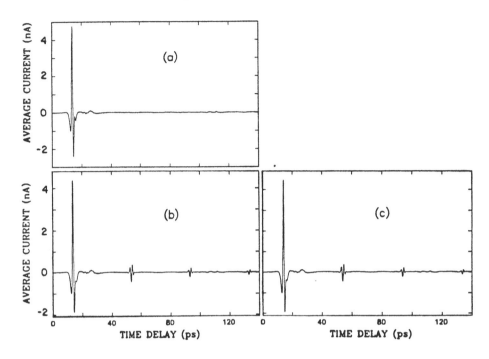

Fig.14 (a) Measured transmitted THz pulse to 140 psec without N_2O in the cell. (b) Measured THz pulse with 600 Torr N_2O vapor. (c) Calculated transmitted THz pulse through 38.7 cm of 600 Torr of N_2O vapor.

The measurements were performed on N_2O vapor within a 38.7 cm-long stainless steel cell having 2-cm-thick, 5-cm-diameter windows of 10 kΩ-cm silicon. For an evacuated cell the output pulse shown in Fig. 14a was obtained. When the cell is filled with 600 Torr of N_2O vapor, the output changes to that shown in Fig. 14b. Here, one sees the transmitted excitation pulse followed by 3 coherent transients emitted by the vapor. The amplitude of the first transient is 1/10 that of the excitation pulse.

Because the THz pulses are in the low-intensity limit of the Maxwell-Bloch equations, the interaction of the electric field with the sample is described in the frequency domain by simply introducing the absorption and dispersion of the vapor. The numerical calculation is performed as follows. The "input pulse" (Fig. 14a) is numerically Fourier analyzed. The resulting frequency components are multiplied by the amplitude absorption and phase change. The inverse numerical Fourier transform is performed giving the predicted output pulse . The fact that the "input pulse" includes reflections, some effects due to residual water vapor and system noise, yields the predicted output pulses shown in Fig. 14c, that are remarkably similar to the actual observation. Every feature seen in the measurement of Fig. 14b, is reproduced in the calculation.

SUMMARY

A optoelectronic THz beam system has been described. The performance of the fastest version of this system is limited by the intrinsic time-response of the semiconductor. Driven by repetitive fsec laser pulses this system can generate and detect subpsec pulses of freely-propagating THz electromagnetic radiation with a signal to noise ratio of 1000:1 and a time resolution of 150 fsec. The importance of this system to THz time-domain-spectroscopy was illustrated by several examples of measurements. For these examples the high performance of the system was responsible for the THz-TDS results exceeding those obtained by the traditional methods of far-infrared (THz) spectroscopy.

ACKNOWLEDGEMENT

This work would not have been possible without the excellent masks and wafer fabrication by Hoi Chan. It has been my pleasure to have worked with the following outstanding scientists in my group for various lengths of time, while we developed the optoelectronic teraHz beam system described in this review and demonstrated its usefulness for THz time-domain-spectroscopy. The individuals are: Christof Fattinger (1987-1989), Martin van Exter (1988-1989), Søren R. Keiding (1989-1990), Hermann Harde (1990-6 months), Nir Katzenellenbogen (1990-, Stephen E. Ralph (1990-. It is my honor to be able to describe their many accomplishments.

REFERENCES

Auston D H 1983 Appl. Phys. Lett., Vol.43, 713.

Auston D H, Cheung K P and Smith P R 1984 Appl. Phys. Lett. Vol. 45, 284.

DeFonzo A P, Jarwala M and Lutz C R 1987a Appl. Phys. Lett., Vol.50, 1155; 1987b Vol.51, 212.

Doany F E, Grischkowsky D and Chi C C 1987 Appl. Phys. Lett. Vol. 50, 460.

Dykaar D R, Greene B I, Federici J F , Levi A F J, Pfeiffer L N and Kopf R F 1991 Appl. Phys. Lett., Vol.59, 262.

Fattinger Ch and Grischkowsky D 1988 Appl. Phys. Lett., Vol.53, 1480 ; 1989a Vol.54, 490; 1989b OSA Proc. on Psec. Elect. and Optoelect., T.C.L. Gerhard Sollner and D.M. Bloom, Eds. (Optical Society of America, Washington, DC), Vol.4.

Fattinger Ch and Grischkowsky D 1989c IEEE J. Quantum Electron., Vol.QE-25, 2608.

Gray D E 1982 "American Institute of Physics Handbook", Third Edition, McGraw-Hill Book Company (New York).

Greene B I, Federici J F, Dykaar D R, Jones R R and Bucksbaum P H 1991 Appl. Phys. Lett., Vol. 59, 893.

Grischkowsky D, Duling,III, I N, Chen J C, and Chi C-C 1987 Phys. Rev. Lett., Vol.59, 1663.

Grischkowsky D, Keiding S, van Exter M and Fattinger Ch 1990 J.Opt.Soc.Am.B, Vol.7, 2006. This paper contains a discussion and an extensive reference list describing the development of time-domain spectroscopy.

Grischkowsky D and Katzenellenbogen N 1991 OSA Proc. on Psec. Elect. and Optoelect., T.C.L. Gerhard Sollner, Jagdeep Shah, eds. (Optical Society of America, Washington, DC), Vol.9.

Harde H, Keiding S and Grischkowsky D 1991a Phys. Rev. Lett., Vol.66, 1834; Harde H and Grischkowsky D 1991b J. Opt. Soc. Am. B, Vol.8, 1642.

Heidemann R, Pfeiffer Th and Jager D 1983 Electronics Lett., Vol.19, 317.

Hu B B, Zhang X-C and Auston D H 1990 Appl. Phys. Lett., Vol.56, 506.

Johnson C, Low F J and Davidson A W 1980 Optical Engr., Vol. 19, 255.

Katzenellenbogen N and Grischkowsky D 1991 Appl. Phys. Lett., Vol. 58, 222.

Lesurf J C G 1990 "Millimetre-Wave Optics, Devices & Systems", (Adam Hilger, Bristol, England).

Loewenstein E V, Smith D R and Morgan R L 1973 Applied Optics, Vol. 12, 398.

Monteath G D 1973 "Applications of the Electromagnetic Reciprocity Principle", (Pergamon Press, Oxford).

Mourou G, Stancampiano C V, Antonetti A and Orszag A 1981 Appl. Phys. Lett., Vol.39, 295.

Pastol Y, Arjavalingam G, Halbout J-M and Kopcsay J V 1988 Electron. Lett., Vol. 24, 1318.

Pastol Y, Arjavalingam G, Halbout J-M 1990 Electron. Lett., Vol. 26, 133.

Ralph S E and Grischkowsky D 1991 Appl. Phys. Lett, Vol. 59, 1972.

Ralph S E and Grischkowsky D 1992 Appl. Phys. Lett., Vol. 60, 1070.

Russell E E and Bell E E 1967 J.Opt.Soc.Am. Vol.57, 543.

Smith P R, Auston D H, and Nuss M C 1988 IEEE J. Quantum Elect. Vol. 24, 255.

van Exter M, Fattinger Ch and Grischkowsky D 1989a Appl. Phys. Lett., Vol.55, 337.

van Exter M, Fattinger Ch and Grischkowsky D 1989b Optics Lett., Vol. 14, 1128; 1989c Laser Spectroscopy IX, Proceedings of the Ninth International Conference on Laser Spectroscopy, Bretton Woods, New Hampshire, June 18-23, 1989, edited by M.S. Feld, J.E. Thomas, and A. Mooradian, (Academic Press, Inc., San Diego).

van Exter M and Grischkowsky D 1990a Appl. Phys. Lett., Vol. 56, 1694; 1990b Phys. Rev. B15, Vol.41, 12,140.

van Exter M and Grischkowsky D 1990c IEEE Trans. Microwave Theory Tech., Vol.38, 1684.

Warren A C, Woodall J M, Freeouf J L, Grischkowsky D, McInturff D T, Melloch M R and Otsuka N, 1990 Appl. Phys. Lett., Vol.57, 1331.

Warren A C, Katzenellenbogen N, Grischkowsky D, Woodall J M, Melloch M R and Otsuka N, 1991 Appl. Phys. Lett., Vol.58, 1512.

Zhang X-C, Hu B B, Darrow J T, and Auston D H 1990 Appl. Phys. Lett., Vol.56, 1011.

Novel non-linear optical techniques for studying chiral molecules of biological importance

Nikolai I Koroteev

International Laser Center and Physics Department, Moscow State University, 119899 Moscow, Russia

1. Introduction

One of the most universal properties of living matter in nature is its exclusive 'chiral purity' namely, the overwhelming abundance of only one of the two possible left or right mirror isomers of biological macro molecules [1-3]. The reason for this fundamental asymmetry of life is still unclear. A further quest for new experimental methods more sensitive to the state of chiral molecules which are part of living cells and organisms or which represent the product of their vital function is obviously needed. From the point of view of a spectroscopist it is essential to work out new schemes of qualitative and quantitative analysis (with the highest possible space and time resolution) of chiral macro-molecules against a background of a racemic mixture. In the simplest case this is a problem of identification of chiral bio-organic molecules in a homogeneous isotropic liquid solution. In the present paper we discuss several non-linear laser spectroscopic techniques which potentially possess selective sensitivity to a mirror asymmetric component of homogeneous isotropic liquid solutions. As shown below these methods provide formerly unavailable physical data on amplitude and frequency dispersion of the optical hyperpolarizabilities of chiral molecules. Appropriate short notes on the same subject were published in [4, 5]. Earlier attempts in the same direction were made more than 25 years ago [6, 7].

A phenomenological classification of all possible linear and non-linear optical processes in non-racemic solutions of chiral molecules up to fourth order of the field amplitude is proposed in section 2. It could be used as a basis of relevant spectroscopic schemes. The most well known and widely used spectroscopy based on frequency dispersion measurements of (linear) optical activity and circular dichroism deals with real and imaginary parts of the linear optical gyration tensor [8] of the medium. However, this spectroscopic method possesses not only many advantages but also some limitations. In particular, linear spectroscopy based on frequency dispersion measurements of optical activity and circular dichroism cannot give any information about the magnitude and dispersion of non-linear optical hyperpolarizabilities of chiral molecules . Non-linear analogs of the above spectroscopy dealing with optical hyperpolarizability of the second and higher orders can provide new additional characteristics. The characteristic feature of homogeneous isotropic non-racemic solutions is their macroscopic asymmetry manifested, in particular, in the presence of dipolar optical susceptibilities of even orders that are absent in the case of optically inactive liquids. It is these susceptibilities that can be used in the most promising non-linear optical schemes based on the measurement of optical rectification (OR) and

Pockel's electro-optic effect (PEOE) using dipolar quadratic susceptibility, a scheme based on the second-harmonic generation (SHG) and five-wave mixing scheme involving fourth-order susceptibilities (section 2). Schemes using non-linear optical activity phenomena (NOA) are also possible; namely analysis of dispersion of the rotation angle of the polarization plane of a weak test wave induced by a considerably more intensive laser pump wave, and a study of new spectral components of the test beam. In the latter scheme a new beam with frequency $\omega_a = 2\omega \pm \omega_L$; $2\omega_L \pm \omega$ (ω, ω_L represent frequencies of the probing and the pumping waves, respectively). In terms of [9], these two schemes are modulation polarization variants of the NOA coherent active spectroscopy. Possible concurrent processes and methods of discriminating useful signals against their background that can provide information about chiral molecules are discussed in section 3. Numerical estimates of chiral molecules non-linear spectroscopy effectiveness are given in section 4.

2. Phenomenological analysis of optical processes in mirror-asymmetric isotopic media

Analysing bulk polarization of a medium induced by optical fields and neglecting magnetodipolar contribution we can single out electrical dipolar (D) and quadrupolar (Q) components, including both linear and non-linear terms with regard to the optical field amplitude:

$$P_i = P_i^D + P_i^Q \tag{1}$$

where

$$P_i^{D,Q} = P_i^{(L)D,Q} + P_i^{(NL)D,Q}$$

while

$$P_i^{(L,NL)Q} = \frac{\partial}{\partial x_j} Q_{ij}^{(L,NL)} \tag{2}$$

$$Q_{ij}^{(L)} = \chi_{ijk}^{(1)Q} E_k \qquad Q_{ij}^{(NL)} = \chi_{ijkl}^{(2)Q} E_k E_l + \chi_{ijklm}^{(3)Q} E_k E_l E_m + \ldots \tag{3}$$

are the components of a unit volume quadrupolar momentum tensor of the medium tested, and

$$\begin{aligned} P_i^{(L)D} &= \chi_{ij}^{(1)} E_j \\ P_i^{(NL)D} &= \chi_{ijk}^{(2)D} E_j E_k + \chi_{ijkl}^{(3)D} E_j E_k E_l + \chi_{ijklm}^{(4)D} E_j E_k E_l E_m + \ldots \end{aligned} \tag{4}$$

are the components of the dipolar momentum vector of a unit volume. The summation is carried out over the repeating Cartesian indices from 1 to 3.

The macroscopic symmetry limits the number of non-zero tensorial components of dipolar (D) and quadrupolar (Q) susceptibilities $\hat{\chi}$.

In contrast to isotropic central symmetrical liquids a non-racemic mixture of chiral biomolecules belonging to a symmetry class $\infty/\infty 2$ is characterized by non-zero odd-rank susceptibility tensors that may represent the most interesting subject in the chiral-molecule spectroscopy. In accordance with general crystallographic 'selection rules' [9], the above tensors of the lowest (third) rank have the following form:

$$\chi_{ijk}^{(1)Q} = \chi^{(1)Q} e_{ijk} \qquad \chi_{ijk}^{(2)D} = \chi^{(2)D} e_{ijk} \tag{5}$$

where e_{ijk} is the Levi–Civita completely antisymmetric unit pseudo-tensor, $\chi^{(1)Q}$, $\chi^{(2)D}$ being (pseudo)scalars. Among 243 tensorial components of $\chi^{(3)Q}_{ijklm}$ and $\chi^{(4)D}_{ijklm}$, there are only 60 non-zero elements (and only 6 of them are independent); all of them have 3 identical indices and the other two are not equal to each other and to the first three (see below, section 2.3). It is rather obvious that the magnitudes of susceptibilities described above are proportional to the difference between the densities of the 'right-hand' and 'left-hand' isomers of chiral molecules, so that they vanish in a racemic solution.

2.1. Linear OA and CD spectroscopy

The linear OA and CD spectroscopy deals with the tensor $\chi^{(1)Q}_{ijk}$ which is in turn connected with the linear gyration tensor g_{lm}:

$$g_{lm} = \frac{\pi}{\lambda} e_{lij} \chi^{(1)Q}_{ijm} = \frac{\pi}{\lambda} g \delta_{lm} = G \delta_{lm} \tag{6}$$

where G is the gyration and δ_{lm} is Kronecker's delta.

Linear optical polarization of chiral medium has the form

$$\boldsymbol{P}^{(L)} = \chi^{(1)D} \boldsymbol{E} + iG[\hat{\boldsymbol{k}}\boldsymbol{E}] \tag{7}$$

where $\hat{\boldsymbol{k}} = \boldsymbol{k}/|\boldsymbol{k}|$, $\boldsymbol{k}$ is the wave vector, $[\hat{\boldsymbol{k}}\,\boldsymbol{E}]_i = e_{ijl}\hat{k}_j E_l$ is a vector product of $\boldsymbol{k}$ and $\boldsymbol{E}$, the natural polarizations of the optically active medium are the right-hand $(+)$ and left-hand $(-)$ circular ones. Proper values of the dielectric constant tensor, i.e. the squares of the indices of refraction for the right- and left-hand circularly polarized waves are

$$n^2_\pm = 1 + 4\pi\chi^{(1)D} \pm 4\pi G$$

so that for the case of non-zero G, the phase velocities $(c^\pm_{\mathrm{ph}} = c/\mathrm{Re}\,n_\pm$, c being the speed of light in vacuum) and the absorption coefficients $(\alpha^\pm = 2k\,\mathrm{Im}\,n_\pm)$ are not the same for the right- and left-hand circular components. These differences are at the foundation of the linear spectroscopy of the natural optical activity and/or the circular dichroism.

2.2. Spectroscopy of optical rectification and electro-optic effect described by the tensor $\chi^{(2)D}_{ijk}(\omega; \omega_1, \omega_2)$ [4]

If $\omega_1 = \omega_2$ and $\omega = 2\omega_1$, all the components of this tensor tend to zero, so that the schemes that use second-harmonic generation and are more easily implemented in all other cases cannot be realized here [6, 11]. All schemes based on mixing non-degenerate frequencies ω_1, ω_2 are unfeasible in the most useful 'collinear' geometry [6, 7] since in this case non-linear polarization has only a longitudinal component at a frequency $\omega = \omega_1 + \omega_2$:

$$\boldsymbol{P}^{(2)D}(\omega) = \chi^{(2)D}(\omega; \omega_1, \omega_2)[\boldsymbol{E}(\omega_1)\,\boldsymbol{E}(\pm\omega_2)]. \tag{8}$$

Consequently it cannot became a source of a free (transversal) electromagnetic wave. Non-collinear schemes are not so effective, owing to a severe infringement of phase-matching conditions [7]. However, schemes based on measuring the dispersion of the

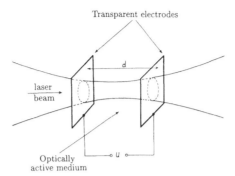

Figure 1. A sketch of a proposed experiment with optical rectification in an optically active liquid

susceptibilities $\chi_{ijk}^{(2)\mathrm{D}}(0; \omega, -\omega)$ (the case of OR) and $\chi_{ijk}^{(2)\mathrm{D}}(\omega; \omega, 0)$ (the case of PEOE) seem to be easily implemented.

Static polarization in the scheme based on OR measurements has the form

$$\boldsymbol{P}^{(2)\mathrm{D}}(0) = \chi^{(2)\mathrm{D}}(0; \omega, -\omega)|A|^2[\boldsymbol{e}\boldsymbol{e}^*] + \mathrm{cc} \tag{9}$$

where $\boldsymbol{e}$ is a complex polarization unit vector of the wave $\boldsymbol{E}(\omega)$ and A is its amplitude. If $\boldsymbol{E}(\omega)$ is an elliptically polarized plane wave with the semi-major (a) and the semi-minor (b) axes arbitrarily oriented in the XY plane, then we have

$$\boldsymbol{P}^{(2)\mathrm{D}}(0) = 2|A|^2 (\sin 2\nu) \operatorname{Im}\left[\chi^{(2)\mathrm{D}}(0; \omega, -\omega)\right] \boldsymbol{e}_z \tag{10}$$

(where $\nu = \pm\arctan(b/a)$), so that the potential difference U is induced between the plates of a capacitor, placed along the direction of plane-polarized light-wave propagation whose intensity is I (see figure 1):

$$U_{\max} \approx \left[64\pi^2 \operatorname{Im}\chi^{(2)\mathrm{D}}/c\varepsilon(0)n(\omega)\right] Id \sin^2\nu \tag{11}$$

where $n(\omega)$ is the index of refraction, $\varepsilon(0)$ is the static dielectric constant of the medium, d is the distance between the capacitor plates. When using a focused laser beam of total power $\mathcal{P}$ and a capacitor with cross-section area S in formula (11), $\mathcal{P}/S$ should be used instead of I. The optical rectification signal is sensitive to the sign of chirality of the medium (determined by the sign of the imaginary part of $\chi^{(2)\mathrm{D}}$) and to the sign of chirality of the light wave (determined by the sign of ν). The signal is non-zero and can be detected when the frequency ω of an optical beam is tuned over the absorption band of chiral molecules.

In the PEOE scheme non-linear polarization has the form

$$\boldsymbol{P}^{(2)\mathrm{D}}(\omega) = \chi^{(2)\mathrm{D}}(\omega; 0, \omega)AE_0[\boldsymbol{e}\boldsymbol{e}_0] \tag{12}$$

where E_0 and $\boldsymbol{e}_0$ are the static field strength and the unit vector along the field, respectively. If $\boldsymbol{e}_0 = \boldsymbol{e}_z$ and $\boldsymbol{e} = \boldsymbol{e}_x$, then $\boldsymbol{P}^{(2)\mathrm{D}}(\omega)$ describes additional (to linear optic activity) rotation of the polarization plane of a light wave, caused by the appearance of the y-component of a light wave $\boldsymbol{E}(\omega)$:

$$E_y = \mathrm{i}[8\pi^2/\lambda n(\omega)]\chi^{(2)\mathrm{D}}AU \tag{13}$$

where $U = E_0 d$ is the potential difference between the plates of the capacitor. Certainly, the effect is a linear one with regard to the amplitude of the light wave and no application of a laser beam is needed.

2.3. Chiral isotropic media spectroscopy based on the five-photon processes

It uses the measurements of dipolar fourth-order non-linearity $\chi_{ijklm}^{(4)D}(\omega_5; \omega_1, \omega_2, \omega_3, \omega_4)$.

The general expression for the vector of the fourth-order non-linear polarization describing a non-degenerate five-photon mixing process in an isotropic homogeneous acentric media, $\omega_5 = \omega_1 + \omega_2 + \omega_3 + \omega_4$, has the following vector form (invariant under the choice of coordinate axes):

$$P^{(4)}(\omega_5) = DA_1A_2A_3A_4\Big\{\chi_{23111}^{(4)D}[e_1e_2](e_3e_4) + \chi_{31121}^{(4)D}[e_2e_3](e_1e_4)$$
$$+ \chi_{31112}^{(4)D}[e_3e_4](e_1e_2) + \chi_{13112}^{(4)D}e_3(e_1[e_2e_4])$$
$$+ \chi_{13121}^{(4)D}e_4(e_1[e_2e_3]) + \chi_{11123}^{(4)D}e_1(e_3[e_4e_2])\Big\}. \tag{14}$$

where ω_5 is the frequency of the new wave generated inside the non-linear medium by the interaction of the four incoming harmonic optical plane waves with frequencies $\omega_1, \ldots, \omega_4$, unit polarization vectors $e_1, \ldots, e_4$ and amplitudes $A_1, \ldots, A_4$, respectively), and $\chi_{23111}^{(4)D}$ etc are six non-zero independent components of the fourth-order non-linear optical susceptibility tensor $\chi_{ijklm}^{(4)D}$; the factor D reflects the possible degeneracy among the frequencies $\omega_1, \ldots, \omega_4$ (in the case when all frequencies are different, $D = 4! = 24$; if only three out of four frequencies are different, than $D = 3! = 6$, and so on). Again, as above, square brackets correspond to the vectorial product and circular brackets, to the scalar product of the vectors they enclose.

The unit vectors $e_1, \ldots, e_4$ ($|e_i|^2 = 1$) are generally complex, describing any possible elliptical polarization states of the incident beams. In the simplest case of non-collinear second-harmonic generation involving non-linearity $\chi^{(4)D}$ (when two intersecting beams with equal frequencies ω interact with each other according to the scheme $2\omega = \omega + \omega + \omega - \omega$) phase-matched generation is possible in media with normal linear dispersion (see figure 2)

$$k_{\mathrm{SH}} = 3k - k'$$

where k and k' are the wave vectors of intersecting beams with frequency ω and k_{SH} is the wave vector of the (free) second-harmonic wave. (In an aqueous solution and with ω corresponding to the Nd:YAG radiation, $\alpha \approx 8 \div 10°$.) The non-linear polarization at the SHG frequency has the form implied by (14):

$$P^{(4)D}(2\omega) = 4\chi_{31112}^{(4)D}A^3A'^*[ee'^*](ee) \tag{14a}$$

where A, A', e and e' are the complex amplitudes and unit polarization vectors of the intersecting beams. As can be seen non-linear polarization (14a) goes to zero if the wave $E(\omega)$ is circularly polarized. The asterisk (*) indicates complex conjugation.

The optimal choice would be the linear polarization of the waves k, k' lying in the plane of interaction; the undesirable geometry is the one in which the vectors e and e' are both directed along the normal to the interaction plane containing the vectors k, k' and k_{SH}, i.e. the one in which $P^{(4)D}$ tends to zero.

The pulse power $\mathcal{P}_{\mathrm{SH}}$ of the second-harmonic signal generated along the phase-matching direction by the intersecting focused Gaussian beams of frequency ω, power $\mathcal{P}$ and waist radius w_0, linearly polarized in the plane of interaction, can easily be

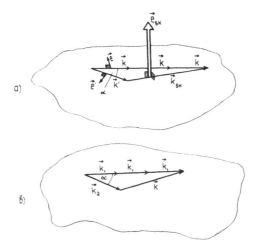

Figure 2. (*a*) Phase-matched geometry of second-harmonic generation in an optically active medium; also shown an optimal polarization configuration of the process; e, e' lie in the plane of interaction, e_{SH} is normal to the plane of k, k'-wave vectors of the incoming beams with frequency ω. (*b*) Phase-matched geometry of the five-photon mixing process: $\omega = 3\omega_1 - \omega_2$, $k = 3k_1 - k_2$

estimated:

$$\mathcal{P}_{\rm SH} \approx \pi w_0^2 I_{\rm SH} = \frac{c w_0^2 n}{8} |E_{\rm SH}|^2 \approx \frac{\pi^2 w_0^2 \omega^2}{2 c^2 n} \left(\sin \alpha \frac{2 w_0}{\tan \alpha} \right)^2$$

$$\times |e_{\rm SH} \boldsymbol{P}^{(4)\rm D}(2\omega)|^2 \approx \frac{2^{15} \pi^4 \left| 4\chi^{(4)\rm D}_{31112} \right|^2 \mathcal{P}^4}{\lambda_\omega^2 c^3 n_{2\omega}^4 w_0^4 n_\omega}. \tag{15}$$

Here n is the index of refraction and we assume that $\alpha \ll 1$, so that $\sin \alpha \approx \tan \alpha$. We can see that the total power of the second harmonic along the phase-matching direction is independent of α, though, of course, this effect is not observable in the exactly collinear geometry ($\alpha = 0$).

Tuning the frequency ω over a selected range and recording the relative intensity of the second harmonic, the researcher can measure the dispersion curve $|\chi^{(4)\rm D}_{31111}|^2$; in this particular case one-, two- and three-photon resonances of the medium under study are observable in the spectrum.

The second-harmonic generation scheme can be generalized over non-degenerate cases, when beams with different frequencies ω_1 and ω_2 interact, so that a five-photon process $\omega = 3\omega_1 - \omega_2 = \omega_1 + \omega_1 + \omega_1 - \omega_2$ with susceptibility $\chi^{(4)\rm D}_{ijklm}(\omega; \omega_1, \omega_1, \omega_1, -\omega_2)$ is detected:

$$\boldsymbol{P}^{(4)\rm D}(\omega) = 4\chi^{(4)\rm D}_{31112}(\omega; \omega_1, \omega_1, \omega_1, -\omega_2) A_1^3 A_2^*[e_1 e_2^*](e_1 e_1). \tag{16}$$

This non-collinear process can be again phase-matched in a medium with normal dispersion, by adjusting an angle between two intersecting beams ω_1 and ω_2. The phase-matching condition is determined as in the earlier case of SHG: $\boldsymbol{k} = 3\boldsymbol{k}_1 - \boldsymbol{k}_2$. To a certain extent, this scheme resembles a widely used CARS scheme, based on the four-photon mixing process of the type $\omega_{\rm a} = \omega_1 + \omega_1 - \omega_2$ [7]. Like CARS, a five-photon process makes it possible to examine the resonances of a substance not only for the harmonics of the frequencies used, but for the difference and combination frequencies, such as $\omega_1 - \omega_2$, $2\omega_1 - \omega_2$ and $3\omega_1 - \omega_2$, and also in the scheme $\omega' = \omega_2 + \omega_2 + \omega_2 - \omega_1$ and at the frequencies $2\omega_2 - \omega_1$ and $3\omega_2 - \omega_1$.

It is convenient in the above-mentioned scheme to choose a circular-polarized wave with tunable frequency (say, ω_2); scanning this frequency over one-photon absorption bands will not then cause the dispersion of the (linear) optical activity; the latter can make the contribution to the five-photon signal dispersion in the case of linear polarization of the wave with tunable frequency.

2.4. Spectroscopy of chiral molecules based on the measurements of non-linear optical activity (NOA)

This deals with cubic quadrupolar third-order susceptibilities in the form $\chi_{ijklm}^{(3)Q}(\omega; \omega_L, \omega, -\omega_L)$ and $\chi_{ijklm}^{(3)Q}(\omega_a; \omega_1, \omega_1, -\omega_2)$; the effect is also critically dependent on the chirality of the medium under study.

The first susceptibility describes the non-linear optical activity effect itself, namely, changing the angle of rotation of a test-beam (ω) polarization plane induced by a strong optical field with frequency ω_L. In fact, non-linear optical activity provides renormalization of the linear gyration rate due to a non-linear addition proportional to the intensity of the pump wave ω_L. The corresponding non-linear quadrupolar polarization is expressed as

$$
\begin{aligned}
\boldsymbol{P}^{(3)Q}(\omega) = \mathrm{i}6|A_L|^2 A \Big\{ & \chi_{23111}^{(3)Q}[\boldsymbol{k}\boldsymbol{e}_L](\boldsymbol{e}\boldsymbol{e}_L^*) + \chi_{31121}^{(3)Q}[\boldsymbol{e}_L\boldsymbol{e}](\boldsymbol{k}\boldsymbol{e}_L^*) \\
& + \chi_{31112}^{(3)Q}[\boldsymbol{e}\boldsymbol{e}_L^*](\boldsymbol{k}\boldsymbol{e}_L) + \chi_{13112}^{(3)Q}\boldsymbol{e}([\boldsymbol{k}[\boldsymbol{e}_L\boldsymbol{e}_L^*]]) \\
& + \chi_{13121}^{(3)Q}\boldsymbol{e}_L^*(\boldsymbol{k}[\boldsymbol{e}_L\boldsymbol{e}]) + \chi_{11123}^{(3)Q}\boldsymbol{k}(\boldsymbol{e}[\boldsymbol{e}_L^*\boldsymbol{e}_L]) \Big\}.
\end{aligned}
\tag{17}
$$

Here $\boldsymbol{k}$ is the wave vector of the test wave, A, A_L and $\boldsymbol{e}$, $\boldsymbol{e}_L$ are the amplitudes and the unit vectors of polarization of the waves ω and ω_L, respectively.

Finally, in addition to frequency dispersion of NOA angle, we can examine the frequency dispersion of a new spectral component $\omega_a = 2\omega_L - \omega$ of a test beam generated by a non-linear source:

$$
\begin{aligned}
\boldsymbol{P}^{(3)Q}(\omega) = \mathrm{i}3A_L^2 A^* \Big\{ & \chi_{23111}^{(3)Q}[\boldsymbol{k}_a\boldsymbol{e}_L](\boldsymbol{e}_L\boldsymbol{e}^*) \\
& + \chi_{31112}^{(3)Q}[\boldsymbol{e}_L\boldsymbol{e}^*](\boldsymbol{k}_a\boldsymbol{e}_L) + \chi_{13112}^{(3)Q}\boldsymbol{e}_L([\boldsymbol{k}_a[\boldsymbol{e}_L\boldsymbol{e}^*]]) \Big\}
\end{aligned}
\tag{18}
$$

where $\boldsymbol{k}_a = 2\boldsymbol{k}_1 - \boldsymbol{k}$ is the wave vector of the anti-Stokes wave under phase-matching conditions.

2.5. Two-photon circular dichroism

In addition to the usual linear optical circular dichroism (different absorption of circularly left- and right-polarized light beams in chiral molecules), its non-linear analogs are also possible, namely, an experimenter can measure the difference in two-photon absorption coefficients for oppositely circularly-polarized waves.

Phenomenologically, the effect is caused by the non-vanishing imaginary part of the chiral susceptibility $\chi_{ijklm}^{(3)Q}(\omega; \omega_L, -\omega_L, \omega)$, where ω is the frequency of a probe beam whose additional absorption is induced by the presence of the pump wave with frequency ω_L. The total rate of two-photon absorption at frequency ω is

$$w_{\text{TPA}}(\omega, \omega_L) = -\overline{E\frac{\partial P^{(3)Q}}{\partial t}} = -\frac{\omega}{2}\,\text{Im}(E^* P^{(3)}(\omega)) \tag{19}$$

where $\boldsymbol{P}^{(3)}(\omega)$ is a non-linear polarization source at a frequency $\omega = \omega_L - \omega_L + \omega$ of third order in the amplitudes of the optical fields; it is expressed by the corresponding terms from (3) and (4).

For left- $(-)$ or right- $(+)$ circularly polarized beams with identical probe and pump photons ($\omega = \omega_L$), the TPA rates, according to (17) and (19), have the form

$$w_{\text{TPA}}^{(\pm)}(\omega, \omega) = -\frac{\omega}{2}|A|^4 \left\{ \left(\chi_{1122}^{(3)D''} + \chi_{1212}^{(3)D''}\right) \pm k\left(\chi_{23111}^{(3)Q''} + \chi_{13112}^{(3)Q''}\right) \right\} \tag{20}$$

where $\chi_{1122}^{(3)D''}$ etc stand for the imaginary parts of the corresponding tensorial components of susceptibilities.

3. Polarization discrimination of signals from chiral molecules against the background of solvent masking effect

Effects connected with non-linear optical response of chiral molecules can develop against the background of non-informative processes involving solvent molecules, windows substance, etc. The most universal masking processes occur as a result of spatial dispersion of optical non-linearity and most often are caused by the quadrupolar non-linearity $\chi_{ijklm}^{(2)Q}$.

3.1. The corresponding masking non-linear polarization in application to optical rectification

This has the following form:

$$P^{(2)Q}(0) = 2|A|^2 \,\text{Im}\left\{ \chi_{1122}^{(2)Q} \right\} \boldsymbol{k}(\boldsymbol{e}\boldsymbol{e}^*).$$

This signal is non-zero in all media including those with central symmetry. However, owing to its quadrupolar character, it is rather weak: in pure optically active liquids it can be three to four orders of magnitude less than that provided by dipolar optical rectification effects. This signal can be further weakened through the use of the standing-wave geometry. In this case the quadrupolar effect will vanish and the dipolar one will double. The same approach can be used for discriminating the dipolar electro-optic effect.

3.2. Quadrupolar polarization in the scheme of non-collinear second-harmonic generation

This can be written as

$$P^{(2)Q}(2\omega) = \text{i}2AA' \left\{ \chi_{1122}^{(2)Q}(\boldsymbol{k} + \boldsymbol{k}')(\boldsymbol{e}\boldsymbol{e}') + \chi_{1212}^{(2)Q}(\boldsymbol{e}'(\boldsymbol{k}'\boldsymbol{e}) + \boldsymbol{e}(\boldsymbol{k}\boldsymbol{e}')) \right\}. \tag{21}$$

However, in a medium with normal dispersion this effect is non-phase-matched; in the direction $(\boldsymbol{k} + \boldsymbol{k}')$ the phase mismatch is minimal. As we can see from (21) the non-linear polarization source along this direction is defined by the normal component of the vector, which takes the right position in the expression following $\chi_{1212}^{(2)Q}$ in (19), and does not depend on the characteristics of a medium. Consequently, this component of a non-linear source can be suppressed through selection of a combination of the polarizer and the retardation plates. The equation defining the appropriate complex unit vector $\boldsymbol{p}$ has the form:

$$(\boldsymbol{p}\boldsymbol{e}')(\boldsymbol{k}'\boldsymbol{e}) + (\boldsymbol{p}\boldsymbol{e})(\boldsymbol{k}\boldsymbol{e}') - \frac{2(\boldsymbol{k}'\boldsymbol{e})(\boldsymbol{k}\boldsymbol{e}')}{\boldsymbol{k} + \boldsymbol{k}')^2}(\boldsymbol{k} + \boldsymbol{k}', \boldsymbol{p}) = 0.$$

The second-harmonic beam transmitted through the analyser–retardation plate will then lack the contribution from quadrupolar second-order non-linearity but will result completely from the contribution of chiral non-linearity $\chi^{(4)D}$.

When a non-degenerate five-photon process is used, one can detect, in addition to all the parasite effects described above, competitive cascade processes involving dipolar quadratic susceptibility $\chi^{(2)D}$ and dipolar cubic susceptibility $\chi^{(3)D}$ (see section 4).

3.3. Common characteristics inherent in the masking of non-linear optical processes

This competes with the effects based on the non-linear optical activity, namely the CARS-type scheme and the optical Kerr effect (OKE) [7] involving dipolar susceptibilities of third order, i.e. four-photon processes $\omega_a = \omega_1 + \omega_1 - \omega_2$ (CARS scheme) and $\omega = \omega + \omega_L - \omega_L$ (OKE scheme).

The difference between the polarization states of non-linear sources remains the feature which makes it feasible to discriminate the specific effect of chiral molecules from the background of much-higher-intensity signals.

It can be shown that there are two polarization configurations, in which the dipolar OKE does not disturb the linear polarization state of a test wave . The non-linear optical activity effect does not vanish in either of these configurations.

Let us set the x-axis along e and the z-axis along k. Then for the above mentioned cases we have

$$(a) \qquad \boldsymbol{e} = \boldsymbol{e}_L = \boldsymbol{e}_x$$
$$P_y^{(3)D}(\omega) = 0$$
$$P_y^{(3)Q}(\omega) = i6|A_L|^2 A \chi_{23111}^{(3)Q} k$$

$$(b) \qquad \boldsymbol{e} = \boldsymbol{e}_x \qquad \boldsymbol{e}_L = \boldsymbol{e}_y$$
$$P_y^{(3)D}(\omega) = 0$$
$$P_y^{(3)Q}(\omega) = i6|A_L|^2 A \chi_{12131}^{(3)Q} k.$$

The polarization suppression of the CARS signal can be realized in three polarization configurations:

$$(a) \qquad \boldsymbol{e}_1 = \boldsymbol{e}_2 = \boldsymbol{e}_x \qquad \boldsymbol{k}_a = k_a \boldsymbol{e}_z$$
$$P_y^{(3)D}(\omega_a) = 0$$
$$P_y^{(3)Q}(\omega_a) = i3A_1^2 A_2^* k_a \chi_{23111}^{(3)Q}$$

$$(b) \qquad \boldsymbol{e}_1 = \boldsymbol{e}_x \qquad \boldsymbol{e}_2 = \boldsymbol{e}_y \qquad \boldsymbol{k}_a = k_a \boldsymbol{e}_z$$
$$P_y^{(3)D}(\omega_a) = 0$$
$$P_y^{(3)Q}(\omega_a) = i3A_1^2 A_2^* k_a \chi_{13112}^{(3)Q}$$

$$(c) \qquad e_1 = (1/\sqrt{2})(e_x \mp \mathrm{i}e_y)$$

$$e_2 = (\cos\alpha\, e_x \pm \mathrm{i}\cos\alpha\, e_y \mp \mathrm{i}\sin\alpha\, e_z)/\sqrt{1+\cos^2\alpha}$$

$$k_1 = k_1 e_z \qquad k_2 = k_2(\sin\alpha\, e_y + \cos\alpha\, e_z)$$

$$k_\mathrm{a} = k_\mathrm{a}(\sin\gamma\, e_y + \cos\gamma\, e_z).$$

(where α and γ are angles between the wave vectors k_1 and k_2, and k_1 and k_a, respectively).

In all the three configurations, the CARS signal is completely suppressed:

$$P_y^{(3)\mathrm{D}}(\omega_\mathrm{a}) = 0.$$

The quadrupolar polarization component orthogonal to k_a has the form

$$P_\perp^{(3)\mathrm{Q}}(\omega_\mathrm{a}) = 3A_1^2 A_2^* k_\mathrm{a} \frac{\sin\alpha\,\sin\gamma}{2\sqrt{1+\cos^2\alpha}} \left(\chi_{31\dot{1}2}^{(3)\mathrm{Q}} + \chi_{13\dot{1}12}^{(3)\mathrm{Q}} \right) \left(e_x \mp \mathrm{i}\cos\gamma\, e_y' \right)$$

where e_y' is a unit vector orthogonal to e_x and k_a.

4. Interference of 'direct' and 'cascade' processes in the non-linear optic scheme. Numerical estimates

In spectroscopic schemes involving non-linearities of higher than second order not only 'direct' processes of signal generation (described in sections 2 and 3) but also two- and three-stage, so-called 'cascade', processes involving non-linearities of lower orders will take place.

Two sequences of processes in a cascade are thus possible in the case of the NOA spectroscopic scheme.

(a) First, an electromagnetic wave with the difference frequency of IR-range $\Omega = \omega_1 - \omega_2$ is generated and then the wave of the sum (difference) frequency is generated: $\omega_\mathrm{a} = \omega_1 + \Omega$ (or $\omega_2 = \omega_1 - \Omega$).

(b) During the first stage the light wave of the sum frequency $\omega' = \omega_1 + \omega_2$ (or $\omega'' = \omega_1 + \omega_2$) is being generated, while during the second stage NOA the spectroscopic signal itself $\omega_\mathrm{a} = \omega' - \omega_2$ (or $\omega_2 = \omega'' - \omega_1$) is generated.

The general scheme and the resulting formula of the problem for the signal cascade generation in the processes of (a) and (b) types are well established in non-local non-linear optics (see [12, 13] and section 4 of [9]).

The amplitude of an 'induced' electromagnetic wave of intermediate frequency (say, ω'), generated at the first stage, can be expressed through the quadratic polarization amplitude $P^{(2)}(\omega')$:

$$E(\omega') = -\frac{4\pi\omega'^2}{c^2(k'^2 - q^2)} \left\{ P^{(2)}(\omega') - \frac{q(qP^{(2)}(\omega'))}{k'^2} \right\}. \tag{22}$$

Here q is the wave vector of the polarization wave and $k' = (\omega'/c)\sqrt{\varepsilon(\omega')}$ is the wave number of a 'free' electromagnetic wave having frequency ω'.

During the second stage, the field $E(\omega')$ interacting with the incoming laser field E_L or E induces in the medium a non-linear polarization at the frequency of the detected signal. Being expressed through the amplitude of the waves A_L and A introduced into the medium, it represents the 'cascade' cubic polarization. The corresponding tensor of fifth (or sixth) rank, that couples the components of the vector amplitudes of the three fields, and of one or two wave vectors, to the vector of the 'cascade' non-linear polarization can be called the tensor of the cubic 'cascade' non-linearity of quadrupolar (or dipolar, or hybrid) type.

Relevant general expressions for the case of an arbitrary mutual orientation of polarization unit vectors and the wave vectors of interfering waves are rather cumbersome and are not represented here.

As an example, let us consider the simplest expression for the cascade polarization of a 'hybrid' type, realized in the NOA spectroscopic scheme at a frequency $\omega_a = 2\omega_1 - \omega_2$ through a polarization configuration in which the normal CARS signal is suppressed:

$$
\begin{aligned}
&\boldsymbol{e}_1 = \boldsymbol{e}_2 = \boldsymbol{e}_x \\
&\boldsymbol{k}_a = 2\boldsymbol{k}_1 - \boldsymbol{k}_2 = k_a \boldsymbol{e}_z \\
&P_{y,\mathrm{casc}} = 3A_1^2 A_2^* (4\pi/\varepsilon(\omega)) \chi_{1122}^{(2)\mathrm{Q}}(\Omega;\ \omega_1,\ -\omega_2)\, \chi^{(2)\mathrm{D}}(\omega_a;\ \omega_L,\ \Omega)\, (\boldsymbol{e}_1 \boldsymbol{e}_2^*)[\boldsymbol{e}_1 \boldsymbol{q}]_y.
\end{aligned}
\tag{23}
$$

'Cascade' processes can be amplified if the phase-matching conditions are satisfied at the first stage of the cascade, i.e. if the following equality is achieved:

$$
|\boldsymbol{q}| = |\boldsymbol{k}'| = (\omega'/c)\sqrt{\varepsilon(\omega')}.
$$

To evaluate the effectiveness of a 'direct' or a 'cascade' process, it is sufficient to estimate the magnitude of the corresponding dipolar, quadrupolar or hybrid non-linear susceptibility.

Experimental data show (see [8]) that in the concentrated solution of arabinose chiral molecules, $|\chi^{(2)\mathrm{D}}| \approx 10^{-10}$ esu; assuming in the OR scheme, by the order of magnitude, that $\mathrm{Im}|\chi^{(2)\mathrm{D}}| \approx 10^{-10}$ esu and using formula (11), we obtain $U_{\max} \approx$ 0.6 mV if $\mathcal{P} = 1$ MW, $\mathcal{E}(0) = 10$, $n(\omega) = 1.5$, $d = 2$ cm, $\lambda = 1\,\mu$m, $S = 1$ cm^2 and $\nu = \pm\pi/4$ (circularly polarized light wave).

In the case of the PEOE scheme with $U = 1$ kV, $\lambda = 0.5\,\mu$m, $n(\omega) = 1.5$, $|\chi^{(2)\mathrm{D}}| = 10^{-10}$ esu, we find $|E_y/A| \approx 0.3 \times 10^{-3}$.

A general consideration allows us to evaluate the magnitude of $|\chi^{(3)\mathrm{Q}}k|$, specifically $|\chi^{(3)\mathrm{D}}|\,a/\lambda$, where a/λ is a spatial dispersion parameter (a is a characteristic size of a molecule, and λ is the wavelength). In the visible range, $a/\lambda = 10^{-2}\ldots 10^{-4}$. Typical values of a 'background' dipolar susceptibility $|\chi^{(3)\mathrm{D}}| \approx 10^{-14}$ cm^3 erg^{-1}. Hence, $|\chi^{(3)\mathrm{Q}}k| \approx 10^{-16}\ldots 10^{-18}$ cm^3 erg^{-1}. (As far as we know, no direct measurements of the ratio $|\chi^{(3)\mathrm{Q}}k/\chi^{(3)\mathrm{D}}|$ have been carried out so far.) When analysing condensed substances, the sensitivity of up-to-date non-linear spectrometers allow us to detect signals due to non-linearities down to a minimum $\chi^{(3)\mathrm{D}} \approx 10^{-20}\ldots 10^{-21}$ cm^3 erg^{-1}. For this reason, the absolute magnitude of the NOA active spectroscopy signal is quite sufficient for a reliable detection (provided the concurrent CARS signal is fully suppressed).

The fourth-order dipolar susceptibility of a saturated solution of chiral molecules can be roughly estimated using the condition

$$
|\chi^{(4)\mathrm{D}}/\chi^{(3)\mathrm{D}}| \propto |\chi^{(3)\mathrm{D}}/\chi^{(2)\mathrm{D}}|.
$$

Therefore, if $|\chi^{(2)\mathrm{D}}| \approx 10^{-10}$ esu, $\chi^{(3)\mathrm{D}} \approx 10^{-14}$ esu, we obtain $|\chi^{(4)\mathrm{D}}| \approx 10^{-18}$ esu. Using then formula (15) and assuming $\mathcal{P}(\omega) = 1$ MW, $\lambda = 1\,\mu$m, $n = 1.5$, $w_0 = 30\,\mu$m, we obtain $\mathcal{P}(2\omega) \approx 10$ W; for $\tau = 10$ ns and a SH pulse, the energy is $W(2\omega) = 10^{-7}$ J (this implies $\sim 10^{12}$ photons per pulse, which is more than enough for normal detection).

5. Conclusions

It is thus realistic to expect that the proposed schemes of non-linear spectroscopy can be experimentally realized without excessive difficulties. Spectroscopic data which can be revealed through them on the magnitude and dispersion of dipolar even-order and quadrupolar odd-order hyperpolarizabilities of chiral molecules cannot be obtained using the ordinary linear optical methods. One can hope to select experimental arrangements in which the detected signals have been generated by chiral molecules and there is no undesirable contribution from centro-symmetrical solvent molecules.

Acknowledgments

The main ideas and results of the present paper were often discussed with the late Professor Akhmanov. His positive criticism is greatly acknowledged.

The author is also indebted to Yu P Svirko, V A Makarov, A P Shkurinov and V F Kamalov for helpful discussions of the results.

References

[1] Pasteur L 1907 *Uber die asymmetrie bei naturlich vorkommenden organischen verbindungen* (Leipzig: Teubner)

[2] Bentley B 1969 *Molecular Asymmetry in Biology* vol 1 (New York: Academic) ; 1970 vol 2

[3] 1979 *Origins of Optical Activity in Nature* ed D Walkered (Amsterdam: Elsevier)

[4] Koroteev N I 1985 *Spectroscopy of Isotopic Acentric Media Based on Measurements of Optical Rectification and Pockel's Electro-optic Effects* Preprint no 29, Physics Department, Moscow University (in Russian)

[5] Akhmanov S A, Kamalov V F and Koroteev N I 1987 Picosecond coherent Raman and fluorescent spectroscopy of biological objects *Laser Scattering Spectroscopy of Biological Objects* ed J Stepanek *et al* (Amsterdam: Elsevier) part 6, p 89

[6] Giordmaine J 1965 *Phys. Rev.* A **138** 1599

[7] Rentzepis P M, Giordmaine J and Wecht K W 1966 *Phys. Rev. Lett.* **16** 792

[8] Wolkenshtein M V 1975 *Molecular Biophysics* (Moscow: Nauka) (in Russian)

[9] Akhmanov S A and Koroteev N I 1981 *Methods of Nonlinear Optics in Light Scattering Spectroscopy* (Moscow: Nauka) (in Russian)

[10] Sirpotin Yu A and Shaskol'skaya E V 1978 *Principles of Crystallophysics* (Moscow: Nauka) (in Russian)

[11] Kielich S 1977 *Molecularna Optyca Nieliniowa* (Warsaw: PWN) (in Polish)

[12] Flytzanis C 1975 *Quantum Electronics, a Treatise* ed H Rabin and C Tang vol 1 Nonlinear Optics (New York: Academic) part A p 9

[13] Flytzanis C and Bloembergen N 1976 *Progress in Quantum Electronics* ed J H Sanders and S Stenholm vol 4 part III (Oxford: Pergamon) p 271

Coupled nonlinear cavities: new avenues to ultrashort pulses

F Mitschke, G Steinmeyer and H Welling

Institut für Quantenoptik, Universit"at Hannover, D-3000 Hannover 1, Germany

1. Prologue

Much of science consists of exploring the very big and the very small, be it in space or in time. This essay deals with very small times. In particular, we describe how a new method for the production of ultrashort pulses of light emerged recently. This technique utilizes optical nonlinearity in a resonator coupled to the laser resonator for enhancement of mode locking. Its beauty lies in its relative simplicity, which makes it very practical for many applications. But why would one want ultrashort light pulses?

A human life typically lasts 2–3 Gigaseconds. The human eye has a temporal resolution of the order of 0.1 s. Between these extremes lies the range of time spans which we can experience directly with our visual sense. Whatever is outside this range requires help through more or less indirect means. When Galileo investigated the rules of gravitational fall, he found that the process was too fast to be assessed correctly. He resorted to the trick of artificially slowing down the phenomenon: forgoing the free fall, he settled for sliding objects down an inclined plane. At another point in his life, he attempted to determine the speed of light. His well–known experiment involved two men with lanterns on two hills, sending signals back and forth. This technique was doomed to fail because light is just too fast, and in this case there was no way he could artificially slow down the phenomenon.

This sorry state of short time measurement technology persisted for a long time. Some 70 years later, Roemer succeeded in determining the speed of light only because he could use an enormously long distance. It took another two centuries until Fizeau and Foucault invented methods with improved temporal resolution which allowed a direct measurement.

The advent of photography brought answers to questions like whether galloping horses momentarily lift all four legs from the ground, an issue that was settled by Eadweard Muybridge. Standard photography covers the millisecond regime, due to limited shutter speed. With flashlamp illumination, there is microsecond capability, the limit being set by the duration of the light pulse.

The next stride forward came with the advent of electronics which made it possible to resolve a nanosecond at about the middle of this century. Then, optics saw tremendous progress after the invention of the laser. The first great successes in mode locking in 1965 (DiDomenico 1966) pushed the pulse duration to 80 ps and thus beyond what was accessible with electronic means. From then on, optics has always kept the lead: the ultimate limit of available temporal resolution today is set by the duration of the shortest laser pulses.

The present limit, achieved with skillfully compressed pulses from a colliding–pulse mode–locked laser, is pegged at 6 femtoseconds (Fork 1987). No non–optical technique comes even close. In that short span of time, light — which travels to the moon in about the time it takes to say "The Speed of Light" — will propagate about one tenth of the thickness of the sheet of paper this page is printed on.

In this contribution, we will concentrate on a relatively recent method to produce ultrashort pulses that relies on optical nonlinearity, typically provided by a fiber, incorporated into the laser feedback ("internal pulse shaping"). In comparison to external compression, this method has some appeal due to its straightforwardness and simplicity. It also has the advantage of yielding remarkably stable operation.

We outline the historical development of this method and present recent results, including some from our lab. At this point we have to apologize: we may have failed to give proper credit to some related work — the sheer volume of literature makes it impossible to assemble an exhaustive presentation. At the same time we apologize that certain variations on our main theme will be given only marginal consideration. This occurs for reasons of length limitation, and is not to be construed to imply that we attribute less importance to those studies.

2. The soliton laser

2.1. Some facts in a nutshell

An optical fiber differs from the ideal light pipe mainly in three respects: 1) There is linear loss which, although small (e.g. $0.2\,\text{dB/km}$ at $1.5\,\mu\text{m}$), leads to a gradual decay of signals sent through long spans of fiber. 2) There is dispersion of the group velocity which lets different Fourier components of a signal travel at different speeds and thus makes pulses spread out. 3) Silica fiber is a Kerr material, with an index of refraction given in very good approximation by $n = n_0 + n_2 I$, where n_0 is the usual (small signal) index, I the intensity, and $n_2 = 3.2 \cdot 10^{-20}\,\text{m}^2/\text{W}$ the coefficient of nonlinearity. Note that this assumes the nonlinearity to be instantaneous, which holds well down into the femtosecond regime.

In this paper, we deal with short lengths of fiber only and are therefore not concerned with linear loss. The other items, dispersion and index nonlinearity, deserve discussion.

An intense pulse propagating along a fiber creates a disturbance of the index that travels with it. At the center of the pulse, the index is raised by $n_2 \hat{I}$, where $\hat{I}$ is the peak intensity. This modulation of index serves to create a self phase modulation that varies over the pulse. This, in turn, implies a modulation of the instantaneous frequency: the leading edge is "red-shifted" (lowered in frequency), the trailing edge "blue-shifted" (raised in frequency). In the regime of anomalous dispersion, lower frequency components travel at slower speed, so that the trailing slope 'catches up' while the leading edge is retarded with respect to the pulse center. This can be used for pulse compression. One can also balance self phase modulation and dispersion, to keep the pulse width constant.

In a manner of speaking, the lens–like index modulation focuses the pulse onto itself (usually self-focusing refers to an action in the transverse coordinate; think here of the longitudinal coordinate only!). The result is a self–trapped pulse.

A formal treatment involving a nonlinear wave equation (the nonlinear Schrödinger equation) corroborates this intuitive argument: particular solutions exist that belong to

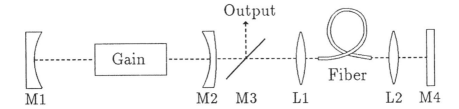

Figure 1. Schematic set–up for coupled cavity mode locking. M1, M4: high reflectors, M2, M3: partial reflectors, L1, L2: lenses. Gain cavity: M1–M2, control cavity: M2–M4.

stable pulses for which the action on the pulse shape, produced by self phase modulation and dispersion, are balanced out. These solitary solutions are called solitons.

If one launches pulses of only very roughly the right shape and power, a soliton emerges after some propagation distance; the balance of the initial pulse is spread off by way of dispersion. Fiber solitons are thus self–stabilizing light pulses and can travel long distances (Mollenauer 1991). They exist under conditions of anomalous dispersion, which in standard fibers means for wavelengths $\lambda > 1.3\mu$m. Due to their stability, solitons should thus be viewed upon as the natural bits of communication in fibers.

2.2. How it all began

In the early 80's, Mollenauer and others at Bell Laboratories in Holmdel performed a number of experiments on the compression of optical pulses in fibers in the soliton regime, see e.g. (Mollenauer 1982). There is tremendous potential for extreme pulse narrowing, particularly if one can start with already reasonably short pulses (Mollenauer 1983). This reasoning induced Mollenauer to construct a laser that made internal use of soliton compression as it had a fiber incorporated in the laser's feedback. The necessity of setting power in the laser cavity and in the fiber separately led to the arrangement of two coupled cavities, one for the gain and one for the soliton shaping (see figure 1).

The first experiments demonstrated the feasibility of this principle, and the device was called the soliton laser (Mollenauer 1984,1985). It was based on a KCL:Tl^0(1) color center laser. However, the requirement of stably keeping an interferometric match of the lengths of the two coupled cavities could not be met initially, and early attempts of active stabilization were less than convincing.

This changed after thorough analysis in Mollenauers lab revealed that it suffices to keep the average intensity in the control cavity at a constant value somewhere in between of the extremes corresponding to different relative phases. A suitable servo loop immediately improved the soliton laser's performance and usefulness dramatically (Mitschke 1986a).

Within a year, several experiments were performed that relied on the soliton laser as a stable source for femtosecond pulses, running the gamut from testing of ultrafast photo detectors (Downey 1986, Bowers 1986) through investigations of the propagation of ultrashort pulses through fibers. For example, the interaction force between light pulses, predicted a few years earlier (Gordon 1983), was experimentally demonstrated for the first time (Mitschke 1987a), and the self frequency shift of light pulses was discovered (Mitschke 1986b).

The shortest pulses obtained from the soliton laser were 60 fs wide, and with additional external soliton compression pulses of a mere 19 fs duration were achieved (Mitschke 1987b) — four cycle pulses that remain to this day among the shortest signals ever produced in the near infrared.

It was also discovered that the soliton laser, once the pulse formation had commenced, required no further synchronous pumping. In fact, when the pump energy was switched from mode locked to continuous, the soliton laser would just continue "business as usual". Only some strong external perturbation could interrupt the pulsing. Once the pulsing stopped, it would not resume by itself. In other words, the soliton laser was passively mode locked, but required a kick start (Mitschke 1986a). This observation was to become significant in the course of later development.

In time, there were several variations on the theme, of which we mention two: Islam *et al* created a fiber Raman amplification soliton laser ("FRASL") which was pumped by a color center laser and oscillated around 1.55 μm. Through a combination of Raman gain and soliton shaping, it produced pulses of 240 fs with some pedestal or, in an improved version, 280 fs with no pedestal. The all–fiber construction was meant to make the device particularly well compatible with optical communications purposes (Islam 1986).

A group at Cornell succeeded in operating a soliton laser with $F_2^+:O^{2-}$ color centers in NaCl (Pollock 1989), a material with wider tuning range and potentially higher output power. Indeed, the results from the soliton laser were improved upon in terms of output power and tuning range, if not pulse duration (Yakymyshyn 1989a).

3. Extension to non–soliton coupled cavity lasers

3.1. Historical background

Several attempts were undertaken by a number of authors to create a theoretical model of the soliton laser that would convincingly describe all experimental facts (Haus 1985, Blow 1986, If 1986, Berg 1987, Bélanger 1988). In a theoretical paper submitted in Aug. 1987, Blow and Wood of British Telecom raised the issue of whether the soliton laser really absolutely depended on solitons (Blow 1988a). They found through numerical work that a nonlinear cavity could lead to pulse width reduction even in total absence of dispersion. The nonlinearity was chosen as either a saturable absorber or a saturable amplifier. In either case, the combination of the pulse distorted by the nonlinearity and the one circulating in the main cavity could yield a shorter pulse — provided these pulses were superposed with a suitable phase shift. The principle is demonstrated in figures 2 and 3, with some exaggeration for clarity.

In an experiment to follow soon (Blow 1988b,c), it was proven that dispersion plays only a secondary role. The authors demonstrated that a soliton laser worked almost equally well if the fiber was replaced by a different one which had normal instead of anomalous dispersion. Surely, no solitons can exist when the dispersion has the wrong sign.

This result boosted the interest of others. Almost simultaneously, successful operation of such "non–soliton soliton lasers" was reported by three groups only months later: one of the University of St. Andrews, one at the MIT, and one at Cornell.

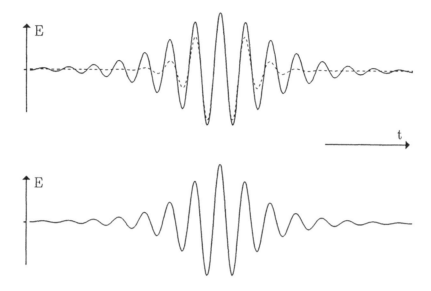

Figure 2.   Top: A pulse distorted by saturable absorbtion (dotted line) superimposed over the original pulse (solid line). Bottom: The interference yields a pulse shorter than the original.

The group in Scotland operated a KCl:Tl^0(1) and a LiF:F$_2^+$ color center laser, also with a fiber as the nonlinear element. Pulses of 260 fs and 1 ps, respectively, were obtained. If an InGaAsP diode amplifier was chosen as the nonlinear element, the pulse width became 250 fs. The pulses fed back into the main cavity were reported to be longer indeed than the output pulses (Kean 1988, 1989).

The MIT group also used a KCl:Tl^0(1) color center laser and fiber, but reached 127 fs at 50 mW average output power. In their paper (Mark 1989) they suggest a name for this technique: they call it "Additive Pulse Mode Locking".

At Cornell, again a KCl:Tl^0(1) laser and later a NaCl:O^{2-} laser were used. With the latter, 75 fs pulses were obtained. In the publication (Yakymyshyn 1989a) these authors adopt the name "Additive Pulse Mode Locking" with reference to the MIT work. Shortly thereafter, by the way, the Cornell group also demonstrated that these pulses can be efficiently frequency–doubled: with LiIO$_3$, 5 mW average power was obtained (Yakymyshyn 1989b).

3.2. Interference from coupled cavities

Clearly, a generalization of the concept of the soliton laser needed to be formulated. It had become obvious that the dispersion played a much lesser role than the fiber nonlinearity, and that both amplitude and phase nonlinearities were suitable for pulse shaping. In the case of a fiber as the nonlinear element, there is only phase nonlinearity, i.e. an intensity–dependent chirp. The emerging picture was that of an interference between an unchirped pulse from the gain cavity with a chirped one from the fiber cavity. The resulting pulse could be shorter than either because, suitable adjustment provided, interference is constructive at pulse center whereas the wings of the pulse would then be attenuated through destructive interference (see figure 4).

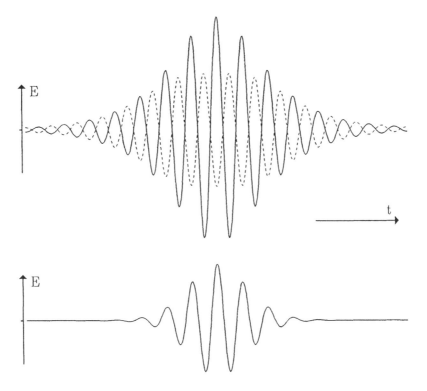

Figure 3. Top: A pulse distorted by saturable amplification (dotted line) superimposed over the original pulse (solid line). Bottom: The interference results in a reduction of pulse width.

Among the several names given to the phenomenon by various authors, like "Interferential Mode Locking" (Morin 1989) or "Coupled Cavity Mode Locking" (Barr 1989a,b, Zhu 1989), the neologism of "Additive Pulse Modelocking", or "APM", finally made it into accepted general use. At the Conference on Lasers and Electro–Optics (CLEO) in May 1989 there was a whole session by that name. At the same time a paper of this title by E.P. Ippen, H.A. Haus, and L.Y. Liu (1989) went to press that presented a detailed analysis of the process and established a theoretical model that became widely accepted.

4. An idea catches on

For quite some time, this new technique for the generation of ultrashort pulses had been restricted to color center lasers (with the exception of related work in Quebec on CO_2 lasers (Oullette 1986,1988)). This was to change at the CLEO 1989 in Baltimore.

Barr et al. reported work on a Nd:YAG laser coupled to an external cavity that contained a second–harmonic generating crystal as a nonlinear element. While this laser did not operate in a continuous mode, it was relevant for the further debate because here it was obvious that the pulsing did not require an extra push to start, but rather took off on its own (Barr 1989a,b).

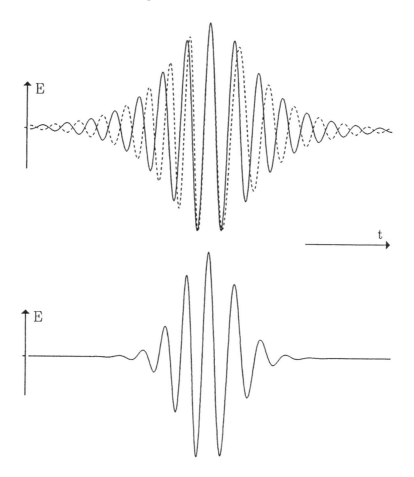

Figure 4.   The interference of a chirped (dotted line) and an unchirped pulse (solid line) creates a shorter pulse (bottom). This is the essential idea in additive pulse mode locking.

Goodberlet *et al* presented work on additive pulse mode locking of a Ti:sapphire laser. The addition of a fiber cavity (Goodberlet 1989a) allowed the generation of 1.6 ps pulses, a significant improvement over the 50 ps duration observed with conventional mode locking. At about the same time, a group at the Imperial College published very similar work (French 1989) in which almost transform–limited pulses of 770 fs duration were achieved.

Ti:sapphire is a very attractive material, and these results were even more appreciated when later that year Goodberlet *et al* followed up on their Ti:sapphire work with a report that in their laser the process was self–starting also (Goodberlet 1989b). No active mode locker was required to initiate the pulsing: the cw operation was unstable and gave way to a stream of fs pulses. With this attractive feature, the APM principle gained ever more appeal to many laboratories.

Now the events unfolded at a rapid pace. The sessions on APM at the CLEO 1990 in Anaheim drew large audiences. Goodberlet *et al* reported on a self–starting diode–pumped Nd:YAG laser (Goodberlet 1990a,b), and Huxley *et al* on lamp–pumped Nd:YAG lasers at both 1.06 and 1.32 μm with the same property (Huxley 1990).

Ti:sapphire turned out to be very special. French et al. reported (French 1990a,b) on such a laser with an external cavity which was empty and therefore certainly not nonlinear! (It had a moving mirror, though). In a postdeadline paper at the same conference, the St. Andrews group topped that by reporting that Ti:sapphire would spontaneously produce fs pulses with no external cavity, in fact no additional components at all (Spence 1990)! To some, this appeared as sheer magic.

This self mode locking (SML) involved the simultaneous excitation of the fundamental and some higher–order transverse resonator modes. It was conjectured (Spence 1991a) that it resulted from an interaction of (at least) two such resonator modes, in contrast to APM, where the respective fundamental modes of two resonators interact. The modes are modified by dispersion and the Kerr nonlinearity of the gain material (Spence 1991b). Later it was suggested (Keller 1991b) that the mode beats merely provided an aid for start–up, and that the mechanism really is what has been named Kerr Lens Mode Locking (KLM). KLM arises from a combination of Kerr–induced self focusing in the gain medium and an aperture somewhere in the resonator such that pulses, being more strongly focused, are preferred over cw light which experiences losses at the aperture.

At the conference on Ultrafast Phenomena VII that same year, a group from Vienna contributed a talk on a self–starting Nd:glass laser (Spielmann 1990, see also Krausz 1990a,b). Goodberlet *et al* reported on self–starting diode–pumped Nd:YAG and Nd:YLF lasers, and a Ti:sapphire laser (Goodberlet 1990c). Keller *et al* of Bell Labs set out to use a resonant nonlinearity, namely a multiple quantum well, as a nonlinear element in the external cavity, and called this technique "Resonant Passive Modelocking", or RPM (Keller 1990a,b, 1991a, Haus 1991a).

At much the same time, J.M. Liu and Chee studied a lamp-pumped Nd:YLF laser (1990a), and Malcolm *et al* took up diode–pumped Nd:YLF also (1990). The list of types of lasers capable of this kind of operation became longer when (after Bulushev *et al* had made such suggestions (1991)) Fermann *et al* (1991) and shortly thereafter Duling (1991) reported the operation of all–fiber lasers, relying on a nonlinear anti-resonant loop mirror as the nonlinear element. Fermann et al. were able to produce pulses of 125 fs with their device, Duling reports 2.1 ps.

Table 1 lists laser types that have been operated with coupled cavities, and indicates the shortest pulse widths that have been, to the best knowledge of these authors, obtained so far.

Several thorough and detailed studies were devoted to a clarification of the precise mechanisms involved in APM lasing, and related techniques like RPM and KLM. It is a practical impossibility to do justice to all these works. However, we will dwell a little on the mechanism for self–starting, since this is one of the most beautiful features of APM lasers.

5. Understanding self–starting

As mentioned, clean self–starting was first reported in (Goodberlet 1989b). Soon after that experiment a first theoretical explanation was put forward by Ippen, Liu, and Haus (1990). It can be understood as follows:

The fiber resonator, as viewed from the main laser cavity, acts like an effective mirror with intensity–dependent reflectivity. Suitable adjustment provided, an increase in intensity yields an increase in reflectivity, and thus the increase in intensity is enhanced. It follows that the cw emission is unstable to perturbations. On the other hand, there are mechanisms that damp the growth of perturbations. One is

Table 1. An overview of coupled cavity lasers. Color center lasers, various Nd lasers, and other lasers have been used for the generation of ultrashort pulses by this technique. In some cases the average output power is indicated.

Laser	Pulse width	Reference
$KCl:Tl^0(1)$	60 fs	(Mitschke 1987b)
with ext.compression:	19 fs	"
$NaCL:O^{2-}$	75 fs @ 300 mW	(Yakymyshyn 1989a)
$LiF:F_2^+$	1 ps	(Kean 1989)
Nd:YAG (lamp pumped)	6 ps @ 2.4 W (1064 nm)	(Liu 1990b)
	10 ps @ 0.7 W (1318 nm)	"
	6 ps (1318 nm)	(Gabetta 1991)
Nd:YLF (lamp pumped)	3.2 ps @ 10 W	(Chee 1990)
Nd:YAG (diode pumped)	1.7 ps @ 45 mW	(McCarthy 1991)
Nd:YLF (diode pumped)	1.4 ps @ 110 mW	"
Nd:Glass	88 fs	(Spielmann 1991a)
Nd:Fiber	38 fs	(Hofer 1992)
Ti:Sapphire	90 fs	(Spence 1991c)
Er:Fiber	3.3 ps	(Duling 1991)
CO_2	3 ns	(Oullette 1986)

dynamic gain saturation. If the gain is reduced due to saturation fast enough, i.e. before the perturbation is over, then the perturbation cannot grow, and the cw emission is stabilized. A condition expressing this thought was derived that involves the emission cross section of the gain medium. Lasers with low cross section media like the Nd laser family and Ti:sapphire lasers start easily, while color center lasers with their much larger cross sections do not. This explains why the soliton laser did not self–start. (A self–starting operation of a NaCl:O^{2-} color center laser could only be achieved when G. Sucha of Bell Labs in late 1990 used a ten times higher power in the fiber than required for operation after start–up. After this kick–start, he achieved pulse widths of 160 fs (Sucha 1991)).

Subsequent works have pointed out that this very simple picture does not describe all the facts. For example, it does not explain the observed intensity dependence of self–starting. The model has therefore been refined in (Krausz 1991a, Wang 1991, Haus 1991b).

It is worth pointing out that the start–up process of APM lasers differs from the start–up of an actively mode–locked laser in a crucial way. In an actively mode–locked laser, the pulse shortening per round trip is initially large because wide pulses "feel" the large losses created by the modulator away from the instant of maximum transmission. As the pulse width becomes shorter, the modulator is hit by the light only near its maximum transmission time where the temporal derivative of the modulator loss goes to zero. Therefore, the pulse shortening per round trip is reduced and "peters out" when the final pulse width is reached.

In contrast, the pulse shaping in the APM laser relies on the nonlinear effect of self–phase modulation. As explained above, an initial perturbation of the steady state is required. However, a small perturbation only creates a very mild self–phase modulation. Only after the pulse is already appreciably shortened does its peak become more intense. In other words, this process builds on itself and becomes stronger as the pulse width goes down. The final pulse width is, of course, given by a competition with dispersive pulse broadening.

The duration of the start–up phase has been studied experimentally. Three cases are to be distinguished: 1) to a synchronously pumped laser, a control cavity is added suddenly, 2) for an APM laser, a synchronous pump is turned on suddenly, and 3) for an APM laser, a cw pump is turned on suddenly. Case 1) leads to build-up times for ultrashort pulses of 200 ns (Mitschke 1986a) or between 200 ns and 4.5 μs (Zhu 1990, 1991). Case 2 is similar in build–up time to case 1 (Zhu 1990, 1991). In case 3), which concerns us here most, build–up times of 100 – 200 μs were observed in different APM lasers (Goodberlet 1990d,e, Spielmann 1991b). It makes sense that this is the same value as for a Ti:sapphire laser with passive mode locking due to a fast saturable absorber (Ishida 1991).

6. Current trends

After it had been understood that self phase modulation was the crucial factor for pulse shaping, the role of dispersion had taken somewhat of a back seat for a while. Nevertheless, for a thorough understanding of the process as well for technical optimization, it is required to study its influence. Several recent papers have therefore dealt with dispersion (Spence 1991c, Grant 1991, Brabec 1991, Spielmann 1991a).

There are other lines of thought how to further improve the practical realizations of APM lasers. The mechanism of pulse shaping boils down to the observation that the interference between two copies of one light pulse, chirped to a different degree, may result in a shorter pulse. Most realizations so far provided these two copies through two coupled but otherwise independent cavities, one creating strong chirp, the other practically none. That approach, on which this contribution concentrates, allows for a maximum of freedom in choosing parameters.

However, the requirement of an interferometric match appears as an inconvenience since length fluctuations due to environmental disturbances must be cancelled. The smarter way may be to have the perturbations create less path difference in the first place. The Vienna group (Spielmann 1991a, Krausz 1991b) reported on a Nd:glass APM laser where the effective output mirror was not a Fabry–Perot interferometer, but of the Michelson type. The fiber was placed in one arm, and dispersion–controlling prisms in the other. Stability was improved because the arm lengths could be kept short. Grant and Sibbett (Grant 1991) made a detailed comparison of lasers with Fabry-Perot and Michelson topology and confirm the improved stability of the latter. Haus (1991c,d) went further in suggesting the use of a single cavity formed from a birefringent fiber. The two polarization modes would experience differing degrees of chirp as long as they carry different powers. The stability should be perfect because all acoustical and thermal length fluctuations would cancel out more or less completely.

7. Working with the APM laser

7.1. The set-up

In our lab, we have built a Nd:YAG APM laser similar to the device described by Liu, Huxley, Ippen, and Haus in (1990b). We reproduced their experimental results very closely, and in the process learned a lot about what is critical in the design and what is not. It turns out that some aspects are much more critical than one would have guessed, while in other respect the system is remarkably forgiving.

Our experiments began with some frustration: we had set up everything we thought it ought to be, but neither with nor without an active mode locker could we get any pulse shortening. Rather, there was spontaneous Q–switching much of the time whenever we coupled the external to the main cavity.

The problem was eventually tracked down to ... the table the system was sitting on, of all components! This requires an explanation.

Our table was a simple optical bench of an old design with much less than optimum internal damping, in fact with a honeycomb made of plastic rather than steel. It rested on a rigid support and was thus coupled to building vibrations, which in our 3rd floor laboratory are severe enough to disturb an interferometric experiment. Of course, an APM laser is an interferometer.

And this explains our initial problems: on a table that does not sufficiently suppress vibrations, the relative cavity length can easily fluctuate enough to prevent the slow start–up whereas the same set–up may still allow active mode locking. An acoustical vibration of, say, 1 kHz with an amplitude of order λ would be sufficient to sweep the interferometric phase over a considerable portion of π during the critical first 200 μs. And lo and behold, after the set–up was transferred to a high–quality table, only a few minor corrections were required until it worked very well.

These corrections included an optimization of the suppression of spurious feedback. An APM laser is extremely sensitive to feedback of the wrong timing. Every surface in the system, and in fact outside, has to be considered. This refers especially to surfaces that follow the curvature of the wave fronts. Lenses are less critical; the fiber ends and the YAG rod are the most critical.

We first used a YAG rod which was perpendicularly cut. In spite of its antireflection coatings, the APM laser would only function well if the rod was tilted in the beam by as much as geometry would allow (1...2 degrees). A slightly wedged rod which we received later made the alignment much easier.

We tried first to suppress the reflections from the fiber ends with the use of index matching oil and butt coupling of the fiber to the end mirror. Just as in (Liu 1990a), this gave no satisfactory results. We eventually settled for cutting the fibers at an angle of a few degrees, thus steering the Fresnel reflections out.

On the other hand, it seems amazing that we could obtain good results with a fiber that was neither polarization–preserving nor even single mode at 1064 nm. In fact, we used ordinary telecommunications fiber that supported a higher order mode which was, in fact, excited somewhat during operation. Polarization in the fiber appeared to be nearly scrambled; at least did a quarter wave plate, inserted in front of the fiber, not do harm to the APM operation at any orientation.

Our set–up is shown in figure 5. Note that we use an adjustable output coupler, comprising of two polarizers and a half–wave plate so that we can precisely adjust how the power from the YAG is split between the output and the fiber cavity. (Note that in a Michelson geometry, this beam splitter would have to handle much higher powers, which is why, at this point, we prefer the Fabry–Perot geometry).

A stabilization circuit as described in (Mitschke 1986a) and widely used for APM lasers kept the operating point stable through suitable length adjustment of the fiber cavity by means of a piezoceramic transducer (PZT). Stability was even better if not the fundamental but rather the second harmonic was used for the servo loop. This

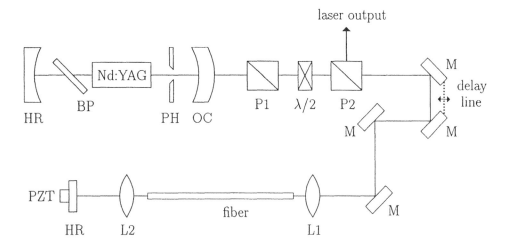

Figure 5. Experimental set–up of our APM Nd:YAG laser. HR: high reflector, BP: brewster plate, PH: pinhole, OC: output coupler, P1, P2: polarizers, $\lambda/2$: half wave retarder, L1, L2: lenses, PZT: piezoceramic transducer.

was pointed out for a color center laser by Sucha (1991). We find that it improves the operation of a Nd:YAG laser also. The reason is that the frequency–doubled output provides a cleaner error signal. We demonstrate this by an experiment which gives insight into the mechanism. With the servo turned off, the length of the fiber cavity is ramped slowly (≈ 1 free spectral range per $0.1\,\mathrm{s}$) by means of the PZT. Figure 6 shows the observed intensity in the laser output P_ω and the frequency–doubled output $P_{2\omega}$. Its interpretation is straightforward:

In the absence of the nonlinear pulse–shaping mechanism, P_ω would be sinusoidally modulated depending on the phase mismatch of the cavities. The point of maximum power $P_\omega^{(\mathrm{max})}$ obviously corresponds to constructive interference, or $\varphi = 0$, and $P_\omega^{(\mathrm{min})}$ occurs at $\varphi = \pm\pi$. Between $\varphi = \varphi_1$ and $\varphi = \varphi_3$, the measured signal deviates from the sinusoidal behavior (which is shown continued as a dotted line in order to guide the eye). This is the regime where significant nonlinear phase shift occurs. The second harmonic signal $P_{2\omega}$ shows that this is also the regime where the laser produces pulses. The precise value of φ_1, φ_2, and φ_3 depends on adjustment, but the pulsing regime always lies within the positive slope part of $P_\omega(\varphi)$. This is expected because the positive slope is a precondition for self-starting: only for positive slope does, say, a positive-going perturbation of the cw power create an increase in effective feedback from the fiber cavity so that the cw solution is destabilized in favor of the pulsing solution. It is also gratifying that the APM operation is in fact centered around $\varphi = -\pi/2$ as required by theory (Ippen 1989).

P_ω shows large spikes at φ_1 and φ_3, where pulsing starts and stops, respectively. This is a phenomenon caused by the ramping in this particular experiment. We will not be concerned with these spikes here any further. The shortest pulses must correspond to the maximum of $P_{2\omega}$ and occur at $\varphi = \varphi_1$. Here, the onset of pulses raises P_ω to $P_\omega^{(\mathrm{op})}$ which is higher because the interference is now more constructive.

For $\varphi_1 < \varphi < \varphi_2$, $P_{2\omega}$ decreases, the pulses become broader, and the nonlinear phase shift will be reduced. This trend of $P_{2\omega}$ should be compared to that of P_ω which

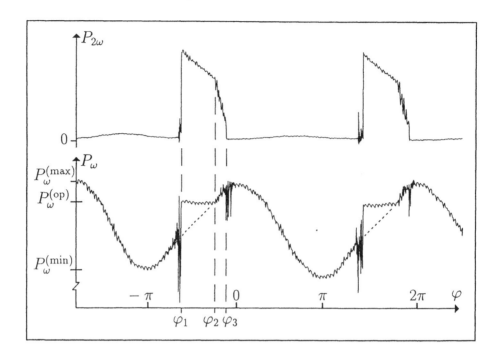

Figure 6.   Intensity of the fundamental and second harmonic versus relative phase φ of the fiber cavity.

is practically constant at $P_\omega = P_\omega^{(op)}$. This is why $P_{2\omega}$ provides a superior error signal for the servo loop. We will demonstrate below how the pulse width can be tuned by varying the set point for the servo loop. For $\varphi > \varphi_2$, the decrease in $P_{2\omega}$ is steep, and at φ_3, pulsing ceases.

From the height of the jump of P_ω at $\varphi = \varphi_1$, one can estimate an average nonlinear phase shift $\overline{\varphi}_{nl}$. In the wings of the interfering pulses the phase difference remains φ_1, while at pulse center the phase difference is modified by the peak nonlinear phase shift $\hat{\varphi}_{nl}$ to some φ_4 so that $\varphi_1 + \hat{\varphi}_{nl} = \varphi_4$. Since the resulting average power is $P_\omega^{(op)}$, at least part of the pulse must be shifted close to $\varphi = 0$, i.e. constructive interference. It follows that $\hat{\varphi}_{nl} \approx |\varphi_1|$, or 0.7π in the example shown.

7.2. Results

The pulse width depends on the fiber length. The dependence is not linear, though, as might be suspected. In fact, it seems to be closer to square root ($\tau \sim \sqrt{L}$), a dependence which is known for the soliton laser (Mollenauer 1984) but seems to be more general.

Shorter fibers require more intrafiber power which is why the shortest pulses are obtained with relatively low power available at the laser output. If one settles for slightly longer pulses, much more output power can be extracted. Our device gave, e.g., $\tau = 18\,\text{ps}$ at $P = 6.8\,\text{W}$ or $\tau = 6.8\,\text{ps}$ at $P = 3.6\,\text{W}$ (P: average output power at a repetition rate of 82 MHz). Figures 7 and 8 show a representative example of an

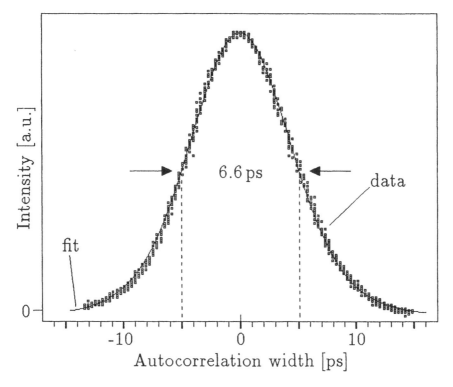

Figure 7. Autocorrelation of a Nd:YAG APM–laser pulse. Symbols: measured data, solid line: least squares fit assuming a hyperbolic secant pulse shape.

output pulse. The agreement with an assumed $sech^2$ pulse shape is excellent, both in autocorrelation and in the spectrum.

Figure 9 shows the observed pulse width as a function of the set point in the loop. As expected from figure 6, the shortest pulses occur at one end of the range. Since peak power times pulse width $\hat{P}\tau \approx$ const. throughout the range of stable operation (see $P_\omega^{(op)}$ in figure 6), and observing that the set point picks a $P_{2\omega}$ which is proportional to $\hat{P}^2\tau$, it follows immediately that $\tau \sim 1/P_{2\omega}$ — assuming that the pulse shape does not change too much. This relation is indicated by the solid lines. The agreement is reassuring.

We also measured the trend of the spectral width (see figure 10; here the solid line is intended to guide the eye only). It turns out that the *shortest* pulses have the *narrowest* spectrum. Moreover, we find that all but the shortest pulses are asymmetric (figure 11). The duration–bandwidth product finally reveals that the shortest pulses are Fourier limited within the bounds of experimental accuracy whereas longer pulses have a mild chirp (figure 12).

The peak phase shift finally can be estimated from the observation that about $P_f = 2.5\,\text{W}$ of average power had to circulate in the fiber for the shortest pulses (as measured at the fiber end). The peak nonlinear phase shift $\hat{\varphi}_{\text{nl}}$ is given by

$$\hat{\varphi}_{\text{nl}} = \frac{2\pi}{\lambda}\, 2L\, n_2 \hat{I}_f \tag{1.1}$$

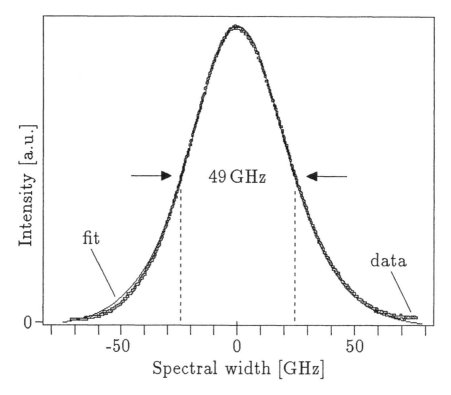

Figure 8.   Spectrum of the pulse in figure 7. Symbols: data measured with a scanning Fabry–Perot interferometer (FSR = 130 GHz, finesse > 100). Solid line: least squares fit, assuming a hyperbolic secant pulse shape.

with L = fiber length and $\hat{I}_f$ the peak intensity in the fiber which (for sech^2 pulses) is given by

$$\hat{I}_f = \ln\left(1 + \sqrt{2}\right) \frac{T}{\tau} \frac{P_f}{A_{\text{eff}}}. \tag{1.2}$$

Here, $T = 12.2$ ns is the inverse repetition rate, and $A_{\text{eff}} = 50\,\mu\text{m}^2$ is the fundamental mode field area. Putting in all the numbers, we obtain $\hat{\varphi}_{\text{nl}} = 2.9\pi$.

However, the intensity is overestimated since for the fiber used here, which was neither single mode nor polarization–preserving, only part of P_f contributed to the APM process. This reasoning is corroborated by the observation that wrapping the fiber around a 20 mm diameter mandrel (which damps higher order modes and introduces some birefringence which may help maintain polarization) cuts the intrafiber power requirement almost in half.

We believe that $\hat{\varphi}_{\text{nl}} \approx \pi$ is a realistic conclusion. Clearly, to remove the uncertainties from this calculation, further experimentation with a polarization–preserving, single–mode fiber is required — and is under way.

8. Epilogue

Coupled–cavity lasers come with the slight inconvenience of requiring active length

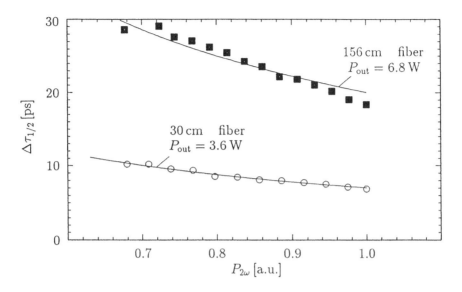

Figure 9. Dependency of the pulse width (FWHM) on the set point of the stabilization loop.

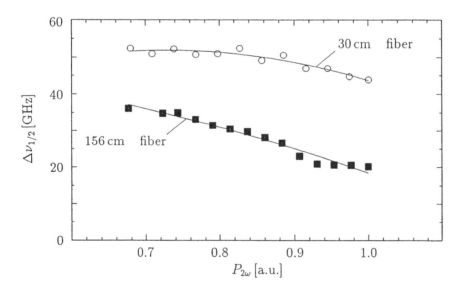

Figure 10. Dependency of the spectral width (FWHM) on the set point of the stabilization loop.

stabilization with an electronic servo loop which, however, is built easily and does not introduce any significant cost. While it might be suspected that this extra complication makes APM lasers inherently unstable, experience shows that indeed they are more stable than, e.g., synchronously pumped lasers. The requirement for a servo loop may even be overcome since inherently stable topologies are being discussed that would

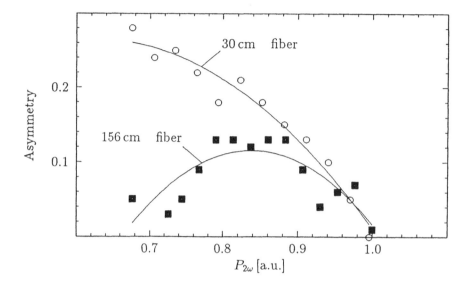

Figure 11. Dependency of the spectral asymmetry on the set point of the stabilization loop. This asymmetry is defined as $(\nu_+ - \nu_-)/(\nu_+ + \nu_-)$ where ν_+ and ν_- are the half widths at half maximum on the high and low frequency side, respectively.

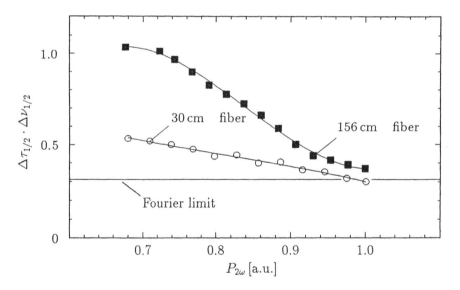

Figure 12. Dependency of the time–bandwidth–product on the set point of the stabilization loop. At the highest possible set point, the pulses are Fourier limited within experimental accuracy.

obviate its necessity.

Self–starting APM lasers have the particular advantage of not requiring a modulator with the pertinent RF electronics. This more than compensates for the expense of the servo loop. APM lasers are therefore an attractive source of stable ultrashort pulses.

Nonlinearity is a leitmotiv of laser physics, and it is not a new idea to shape pulses by nonlinear techniques. In APM lasers and similar designs, a well–defined nonlinearity is given responsibility for pulse shaping, and the gain medium and the nonlinear medium are formed into a tight–knit unit which allows to exploit the available gain bandwidth to its fullest.

Some twenty years after the advent of mode locking, its more sophisticated offspring called "Additive Pulse Mode Locking" basks in the limelight. Surely, it will find numerous and widespread applications. Testing of ultrafast electronics, e.g., is often done with the help of ultrashort laser pulses, and will benefit from this reliable source.

The limit to the generation of ever shorter pulses tends to be set by their *relative* bandwidth: it is hard to find materials with sufficiently small dispersion over ever wider frequency ranges. This is why optics, which deals with enormously high frequencies (typically 200 or 500 THz in the near infrared or the visible, respectively), has an edge over electronics when it comes to ultrashort signals. Optics is here to stay as the fastest technology around.

As a final thought, let us ponder the extremes of time scales, both the very big and the very small. A nanosecond, now considered not such a particularly short time, is as far away from our heartbeat as a Gigasecond — nine decimal orders. A post–doc is typically about 1 Gs old. As we proceed to consider picoseconds, we can contrast them with Teraseconds: one Ts ago, creatures of a new species called *homo sapiens* began to roam this planet. In time, they invented clothes, the fireplace, the wheel, and a couple more things. One of the latest gadgets is the few–femtosecond pulse laser. The production of a signal of one femtosecond duration is not far away from present achievement: its 15–orders-of–magnitude correspondent at the other end of the time scale is the petasecond. A petasecond ago most species of mammals just came into existence.

The attosecond, for now, stays out of reach for manmade technology. And one exasecond, or 10^{18} seconds ago, the universe was not born yet.

Acknowledgments

Our own experiments described here benefitted from the patience and skill of C. Fallnich. Financial support by the Deutsche Forschungsgemeinschaft is gratefully acknowledged.

References

Barr J R M and Hughes D W 1989a *Appl. Phys.* **B49** 323-5

Barr J R M, Hanna D C and Hughes D W 1989b *Conference on Lasers and Electro–Optics* FQ5, pp 442-3

Bélanger P A 1988 *J. Opt. Soc. Am.* **B5** 793-8

Berg P, If F, Christiansen P L and Skovgaard O 1987 *Phys. Rev.* **A35** 4167-74

Blow K J and Wood D 1986 *IEEE Journ. Quant. El.* **QE-22** 1109-16

Blow K J and Wood D 1988a *J. Opt. Soc. Am.* **B5** 629-32

Blow K J and Nelson B P 1988b *Ultrafast Phenomena VI* eds Yajima T, Yoshihara K, Harris C B and Shionoya S (New York: Springer) WC4, p 67-9

Blow K J and Nelson B P 1988c *Opt. Lett.* **13** 1026-8

Bowers J E, Burrus C A and Mitschke F M 1986 *Electron. Lett.* **22** 633-5

Brabec T, Krausz F, Spielmann Ch, Wintner E and Budil M 1991 *J. Opt. Soc. Am.* **B8** 1818-23

Bulushev A G, Dianov E M and Okhotnikov O G 1991 *Opt. Lett.* **16** 88-90

Chee J K, Kong M N, Cheung E C and Liu L Y 1990 *Ultrafast Phenomena VII* eds Harris C B, Ippen E P, Mourou G A and Zewail A H (New York: Springer) pp 8-10

DiDomenico Jr. M, Geusic J E, Marcos H M and Smith R G 1966 *Appl. Phys. Lett.* **8** 180-3

Downey P M, Bowers J E, Burrus C A, Mitschke F M and Mollenauer L F 1986 *Appl. Phys. Lett.* **49** 430-1

Duling III I N 1991 *Opt. Lett.* **16** 539-41

Fermann M E, Hofer M, Haberl F, Schmidt A J and Turi L 1991 *Opt. Lett.* **16** 244-6

Fork R L, Brito Cruz C H, Becker P C and Shank C V 1987 *Opt. Lett.* **12** 483-5

French P M W, Williams J A R and Taylor J R 1989 *Opt. Lett.* **14** 686-8

French P M W, Kelly S M J and Taylor J R 1990a *Opt. Lett.* **15** 378-80

French P M W, Kelly S M J, Wigley P G J and Taylor J R 1990b *Conference on Lasers and Electro–Optics* CFN2, pp 540-2

Gabetta G, Huang D, Jacobson J, Ramaswamy M, Haus H A, Ippen E P and Fujimoto J G 1991 *VII*th *International Symposium on Ultrafast Processes in Spectroscopy* TU 1-1

Goodberlet J, Wang J, Fujimoto J G, Schulz P A and Henion S 1989a *Conference on Lasers and Electro–Optics* FQ4, pp 442-3

Goodberlet J, Wang J, Fujimoto J G and Schulz P A 1989b *Opt. Lett.* **14** 1125-7

Goodberlet J, Jacobson J, Fujimoto J G, Schulz P A and Fan T Y 1990a *Conference on Lasers and Electro–Optics* CFN5, p 544

Goodberlet J, Jacobson J, Fujimoto J G, Schulz P A and Fan T Y 1990b *Opt. Lett.* **15** 504-6

Goodberlet J, Jacobson J, Wang J, Fujimoto J G, Fan T Y and Schulz P A 1990c *Ultrafast Phenomena VII* eds Harris C B, Ippen E P, Mourou G A and Zewail A H (New York: Springer) pp 11-3

Goodberlet J, Wang J, Fujimoto J G and Schulz P A 1990d *Conference on Lasers and Electro–Optics* CFN1, pp 540-1

Goodberlet J, Wang J, Fujimoto J G and Schulz P A 1990e *Opt. Lett.* **15** 1300-2

Gordon J P 1983 *Opt. Lett.* **8** 596-8

Grant R S and Sibbett W 1991 *Opt. Comm.* **86** 177-82

Haus H A and Islam M N 1985 *IEEE Journ. Quant. El.* **QE-21** 1172-88

Haus H A, Keller U and Knox W H 1991a *J. Opt. Soc. Am.* **B8** 1252-8

Haus H A and Ippen E P 1991b *Opt. Lett.* **16** 1331-3

Haus H A 1991c *Conference on Lasers and Electro–Optics* JMA1, p 2

Haus H A, Fujimoto J G and Ippen E P 1991d *J. Opt. Soc. Am.* **B8** 2068-76

Hofer M, Ober M H, Haberl F and Fermann M E 1992 "Characterization of ultrashort pulse formation in passively mode locked fiber lasers" *IEEE J. Quant. El.* (to appear)

Huxley J M, Liu L Y, Ippen E P and Haus H A 1990 *Conference on Lasers and Electro–Optics* CFN3, pp 542-3

If F, Christiansen P L, Elgin J N, Gibbon J D and Skovgaard O 1986 *Opt. Comm.* **57** 350-4

Ippen E P, Haus H A and Liu L Y 1989 *J. Opt. Soc. Am.* **B6** 1736-45

Ippen E P, Liu L Y and Haus H A 1990 *Opt. Lett.* **15** 183-5

Ishida Y, Sarukura N and Nakano H 1991 *Conference on Lasers and Electro–Optics* JMB2, pp 14-5

Islam M N, Mollenauer L F and Stolen R H 1986 *Ultrafast Phenomena V* eds Fleming G R and Siegman A E (New York: Springer) pp 46-50

Kean P N, Grant R S, Zhu X, Crust D W and Sibbett W 1988 *Conference on Lasers and Electro–Optics* PD7

Kean P N, Zhu X, Crust D W, Grant R S, Langford N and Sibbett W 1989 *Opt. Lett.* **14** 39-41

Keller U, Knox W H and Roskos H 1990a *Opt. Lett.* **15** 1377-9

Keller U, Knox W H and Roskos H 1990b *Ultrafast Phenomena VII* eds Harris C B, Ippen E P, Mourou G A and Zewail A H (New York: Springer) pp 69-71

Keller U, Woodward T K, Sivco D L and Cho A Y 1991a *Opt. Lett.* **16** 390-2

Keller U, 't Hooft G W, Knox W H, Cunningham J E 1991b *Opt. Lett.* **16** 1022-4

Krausz F, Spielmann Ch, Brabec T, Wintner E and Schmidt A J 1990a *Opt. Lett.* **15** 737-9

Krausz F, Spielmann Ch, Brabec T, Wintner E and Schmidt A J 1990b *Opt. Lett.* **15** 1082-4

Krausz F, Brabec T and Spielmann Ch 1991a *Opt. Lett.* **16** 235-7

Krausz F, Fermann M E, Hofer M, Spielmann Ch, Wintner E and Schmidt A J 1991b *Conference on Lasers and Electro–Optics* JMB4, pp 14-6

Liu J M and Chee J K 1990a *Opt. Lett.* **15** 685-7

Liu L Y, Huxley J M, Ippen E P and Haus H A 1990b *Opt. Lett.* **15** 553-5

Malcolm G P A, Curley P F and Ferguson A I 1990 *Opt. Lett.* **15** 1303-5

Mark J, Liu L Y, Hall K L, Haus H A and Ippen E P 1989 *Opt. Lett.* **14** 48-50

McCarthy M J, Maker G T and Hanna D C 1991 *European Quantum Electronics Conference* OIFr 8, p 355

Mitschke F M and Mollenauer L F 1986a *IEEE Journ. Quant. El.* **QE-22** 2242-50

Mitschke F M and Mollenauer L F 1986b *Opt. Lett.* **11** 659-61

Mitschke F M and Mollenauer L F 1987a *Opt. Lett.* **12** 355-7

Mitschke F M and Mollenauer L F 1987b *Opt. Lett.* **12** 407-9

Mollenauer L F and Stolen R H 1982 *LaserFocus* **4** 193-8

Mollenauer L F, Stolen R H, Gordon J P and Tomlinson W J 1983 *Opt. Lett.* **8** 289-91

Mollenauer L F and Stolen R H 1984 *Opt. Lett.* **9** 13-5

Mollenauer L F 1985 *Phil. Tr. R. Soc. Lond.* **A315** 437-50 +

Mollenauer L F, Nyman B M, Neubelt M J, Raybon G and Evangelides S G 1991 *Electron. Lett.* **27** 178-9

Morin M and Piché M 1989 *Opt. Lett.* **14** 1119-21

Oullette F and Piché M 1986 *Opt. Comm.* **60** 99-103

Oullette F and Piché M 1988 *J. Opt. Soc. Am.* **B5** 1228-36

Pollock C R, Pinto J F and Georgiou E 1989 *Appl. Phys.* **B48** 287-92

Spence D E, Kean P N and Sibbett W 1990 *Conference on Lasers and Electro–Optics* CPDP 10-1, pp 619-20

Spence D E, Kean P N and Sibbett W 1991a *Opt. Lett.* **16** 42-4

Spence D E, Evans J M, Sleat W E and Sibbett W 1991b *Opt. Lett.* **16** 1762-4

Spence D E and Sibbet W 1991c *J. Opt. Soc. Am.* **B8** 2053-60

Spielmann Ch, Krausz F, Wintner E and Schmidt A J 1990 *Ultrafast Phenomena VII* eds Harris C B, Ippen E P, Mourou G A and Zewail A H (New York: Springer) pp 72-4

Spielmann Ch, Krausz F, Brabec T, Wintner E and Schmidt A J 1991a

Appl. Phys. Lett. **58** 2470-2

Spielmann Ch, Krausz F, Brabec T, Wintner E and Schmidt A J 1991b *IEEE J. Quant. El.* **QE-27** 1207-13

Sucha G 1991 *Opt. Lett.* **16** 922-4

Wang J 1991 *Opt. Lett.* **16** 1104-6

Yakymyshyn C P, Pinto J F and Pollock C R 1989a *Opt. Lett.* **14** 621-3

Yakymyshyn C P and Pollock C R 1989b *Opt. Lett.* **14** 791-3

Zhu X, Kean P N and Sibbett W 1989 *Opt. Lett.* **14** 1192-4

Zhu X, Sleat W, Walker D and Sibbett W 1990 *Ultrafast Phenomena VII* eds Harris C B, Ippen E P, Mourou G A and Zewail A H (New York: Springer) pp 87-91

Zhu X, Sleat W, Walker D and Sibbett W 1991 *Opt. Comm.* **82** 406-14

Transverse rotating waves in the non-linear optical system with spatial and temporal delay

N G Iroshnikov†and M A Vorontsov‡

†Moscow State University, Physics Department, 119899 Moscow, Russia

‡Moscow State University, International Laser Center, 119899 Moscow, Russia

Abstract. A spatially distributed dynamic system with large-scale spatial and temporal interactions has been investigated. The model of the system is a non-linear optical ring resonator with field rotation around the optical axis and signal time delay in the feedback circuit. The parameters of large-scale spatio–temporal interactions determine the characteristics of the non-linear rotatory modes.

1. Introduction

The non-linear wave dynamics of spatially distributed optical systems is now presented with a set of different phenomena in light field self-organization: spatial bistability and multistability (Firth and Galbraith 1985, Ivanov *et al* 1990), diffractive solitons (Rosanov and Khodova 1990), various spatial structures and instabilities (Le Berre *et al* 1991, Vorontsov *et al* 1988, Lugiato and Oldano 1988, Giusfredi *et al* 1988, Arecchi 1991), spatio–temporal chaos (Arecchi *et al* 1988, Akhmanov *et al* 1991). This very incomplete list of phenomena that have been observed experimentally or analysed theoretically indicates the appearance of a new tendency in non-linear optics which is closely connected with synergetics (optical synergetics).

Unlike classical non-linear optics where general efforts are traditionally directed towards the creation and the investigation of new non-linear media, in optical synergetics the experimental implementation and investigation of different types of spatial and temporal interactions are the key problems. Taking as a basis one and the same type of optical non-linearity (for example, Kerr non-linearity) merely by changing the topology of spatial interactions, it is possible to obtain an absolutely different dynamic behaviour of the system (Vorontsov 1991).

In this paper we suggest physical systems which have spatial and temporal interactions with various scales. Taking as an example a one-dimensional (by spatial coordinate) model, the combined influence on the dynamic behaviour of the system of local and non-local (large scale) spatial interactions, and non-localities of temporal response (time delay) as well, have been considered. The article consists of the following sections.

In section 2, which is a brief review, for the example of a non-linear passive ring resonator it has been shown what changes in the dynamics of the system the successive introduction of spatial and temporal interactions of various scales result in.

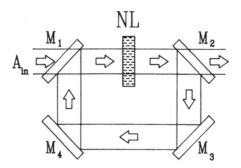

Figure 1. Non-linear passive ring resonator

The analysis of the system dynamics in the vicinity of the stability loss bifurcation of the spatially homogeneous solution and excitation of the rotatory modes has been carried out in section 3. The boundaries of the regions of the rotatory-mode excitation have been obtained on the basis of scale invariant solutions of the original non-linear equation (the technique of phase equations of Brand *et al* 1986).

In section 4 the expansion of the solution due into rotatory modes has been used. An infinite-dimensional system of differential equations describing the amplitudes of the interacting rotatory modes (the order parameters) has been obtained. On the basis of the slaving principle described in the work of Gang and Haken (1989) the reduction of the system of order-parameter equations to the finite-dimensional system of equations has been made. The possibilities of parametric excitation of the rotatory modes at the oscillatory bifurcation of the spatially homogeneous phase.

The competition of rotatory modes is analysed in section 5. It is shown that as a result of competition the suppression of all rotatory modes except one occurs. It corresponds to the WTA dynamics which is characteristic of artificial neural networks.

Section 6 presents the results of numerical modelling of the system dynamics. The comparison of the approximate solutions obtained at the point of bifurcation with the numerical analysis of the initial non-linear differential equation in partial derivatives with time delay and shifted spatial argument has been carried out.

The opportunities of experimental observation of the rotatory modes in the system with time delay have been discussed in section 7.

The analogy of the system dynamics of large-scale interactions with the models of developed neural networks is given in the conclusion.

2. Basic mathematical models

The prototype of the system is a non-linear ring resonator or interferometer. A number of papers (Firth and Galbraith 1985, Ikeda *et al* 1980, Otsuka and Ikeda 1989, Le Berre *et al* 1989) (figure 1) describe the study of different aspects of dynamics of these devices.

A coherent linearly polarized light wave with complex amplitude $A(\boldsymbol{r}, t)$, $\boldsymbol{r} = (x, y)$, having passed through a thin layer of non-linear medium NL, derives a supplementary phase shift u (non-linear phase modulation), proportional to the field intensity (the Kerr non-linearity). After passing through the feedback, formed by the beam splitters M_1 and M_2 and mirrors M_3 and M_4, the light beam interferes with the input field A_{in}. Taking into account one pass of the field through the resonator, and neglecting the

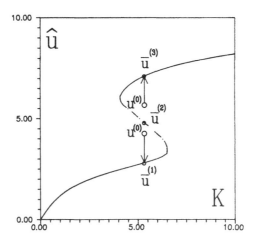

Figure 2. Stable (solid line) and unstable (dashed line) stationary solutions $\bar{u}$ of the system (1) as functions the parameter K; $u^{(0)}$ are the initial points, $\bar{u}^{(1)}$ and $\bar{u}^{(3)}$ are the stationary points

transverse interaction and time delay in the feedback as well, it is possible to obtain the equation for non-linear phase modulation at point $\boldsymbol{r}_i$ (Ikeda *et al* 1980):

$$\tau\frac{\mathrm{d}u(\boldsymbol{r}_i,t)}{\mathrm{d}t} + u(\boldsymbol{r}_i,t) = K[1 + \gamma\cos u(\boldsymbol{r}_i,t)] \tag{1}$$

where τ is the non-linearity relaxation time, K is the parameter, proportional to the input field intensity and γ is the visibility of the interference pattern. Note, that the dependence of the non-linear phase modulation on the spatial coordinate $\boldsymbol{r}$ has a parametric character (spatially transverse interactions are not taken into account).

Equation (1) describes the non-linear response of the spatially distributed (in the x–y plane) medium at every point $\boldsymbol{r}$. The dependence of the stationary phase modulation $\bar{u}$ from the parameter K (external influence) is multivalued (figure 2), which is the reason of optical bistability and hysteresis (Gibbs 1985).

The system dynamics governed by equation (1), is rather trivial in the region of bistability: relaxation to one for the two possible stable stationary states $\bar{u}^{(1)}$ or $\bar{u}^{(3)}$ (figure 2) occurs depending on the initial values $u(\boldsymbol{r},0) = u^{(0)}$. The stationary state $\bar{u}^{(2)}$ is unstable, and gives the boundary of the attraction basins of the initial values $u^{(0)}$ to the stable stationary points $u^{(1)}$ and $\bar{u}^{(3)}$.

The dynamical behaviour of the system changes considerably if the response to excitation is non-local in time (field delay in the feedback) (Ikeda *et al* 1980, Le Berre *et al* 1985):

$$\tau\frac{\mathrm{d}u(\boldsymbol{r}_i,t)}{\mathrm{d}t} + u(\boldsymbol{r}_i,t) = K[1 + \gamma\cos u(\boldsymbol{r}_i,t-T)] \tag{2}$$

where T is the time of field propagation in the feedback circuit. Note, that the system described by equation (2) can have an infinite number of degrees of freedom. With $K > K_{\mathrm{cr}}$ (K_{cr} is a certain critical value of non-linearity parameter) the increase in time delay T results in the appearance of a single-frequency mode of periodic phase oscillations, and the cascade of period doubling bifurcations becomes more and more complicated. As a result the transition to chaotic behaviour of the solution becomes possible.

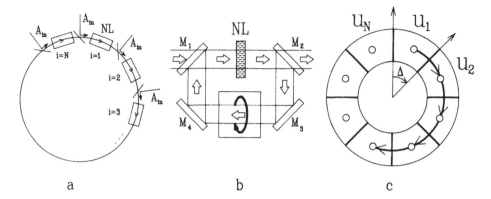

a   b   c

Figure 3. Conceptual model of a non-linear optical system with distributed elements: (*a*) chain of the coupled resonators investigated by Otsuka and Ikeda (1989); (*b*) two-dimensional non-linear passive ring resonator with field rotation for the angle $D = 2p/N$; (*c*) cross shift of the ray in one pass through the resonator.

Thus, temporal non-local interactions in the system with a hysteresis form of non-linearity lead to non-trivial dynamic behaviour.

Further complication of the model is associated with the introduction of a special type of non-local interactions with which N individual resonators prove to be coupled in succession. Figure 3(*a*,*b*) shows two schemes of possible implementation of such a coupling.

The chain of coupled non-linear resonators (figure 3(*a*)) has been studied in (Otsuka and Ikeda 1989) and is described by the following system of equations

$$\tau \frac{du_1}{dt} = K[1 + \gamma \cos u_N]$$

$$\dots$$

$$\tau \frac{du_i}{dt} = K[1 + \gamma \cos u_{i-1}] \tag{3}$$

$$\dots$$

$$\tau \frac{du_N}{dt} = K[1 + \gamma \cos u_{N-1}] \qquad I = 1, \dots, N.$$

Another scheme of the chain of coupled resonators is based on the use of a special unit implementing field rotation by the angle $\Delta = 2\pi/N$ (figure 3(*b*, *c*)), Vorontsov *et al* 1988, Ivanov *et al* 1991) in the two-dimensional feedback circuit. By denoting the non-linear phase modulation at the point $r_i = \{r_0, \theta_i\}$ of the annulus of radius r by $u_i(t)$ ($\theta_i = 2\pi i/N$, $i = 1, \dots, N$), figure 3(*c*), we obtain the scheme of the coupled resonators described by the same system of equation (3), implementation of which is significantly more simple. To rotate the field in the feedback it is possible to use, for example, a Dovet prism which makes it comparatively simple to change the number of connected resonators in the chain.

The system dynamics (3) without taking into account of time-delay has been considered in (Otsuka and Ikeda 1989). For large N, with the increase of the parameter K (the input field intensity) successive bifurcations have taken place, resulting in the complication of the spatial structure of the stationary solutions $\{\bar{u}_1, \dots \bar{u}_N\}$. At greater values of K there is a transition through the periodic oscillating solutions to spatio–temporal chaos. Thus, the availability of the simplest type of non-local spatial interactions coupling single resonators into a chain results in a considerable complication of the spatial structure of solutions.

The following step of the 'modernization' model under consideration is associated with the introduction of the supplementary local interactions. Actually this means the transition to the continuum model. In this case the vector $u(t) = \{u_1(t), \ldots u_N(t)\}$ is replaced by the spatially distributed function $u(r, t)$ (or in the example considered, by the function describing the non-linear modulation of the light beam phase). In the resonator shown in figure 3(b), local transverse interactions are stipulated by the excitation diffusion in the non-linear medium or by the diffraction light field in the feedback. Taking into account the transverse interactions of the diffuse type, the equation for the non-linear phase modulation takes the form (Vorontsov *et al* 1988):

$$\tau \frac{\partial u}{\partial t} + u = D\Delta_\perp + K[1 + \gamma \cos u(\rho, \theta + \Delta, t)]. \tag{4}$$

Here D is the diffusion coefficient characterizing the strength of local interactions. In this case non-local interactions are described by the transformation of the transverse coordinates (ρ, θ) on the right-hand side the equation, Δ is the rotation angle of the field around the optical axis (figure 3(c)). In the spatially distributed system the angle of rotation of the field D can be arbitrary, i.e., not obligatory to be $2\pi/N$-fold as in the case of the chain of coupled resonators. A model with one spatial dimension is the simplest kind, and it can be comparatively easily implemented in the optical scheme shown in figure 3(b), by placing into the feedback an amplitude mask $T(r)$ absorbing the radiation everywhere, except a narrow ring layer of radius ρ_0. The system is described by the equation (Ivanov *et al* 1991)

$$\tau \frac{\partial u(\theta, t)}{\partial t} + u = D\frac{\partial^2 u}{\partial \theta^2} + K[1 + \gamma \cos u(\rho + \Delta, t)] \qquad D = d/r_0^2 \tag{5}$$

with periodic boundary conditions.

The system dynamics with local and large-scale transverse interactions for relatively small values of the parameter K is theoretically and experimentally analysed in (Akhmanov *et al* 1991, Ivanov *et al* 1991). It has been shown that the solution has the form of moving in periodic space rotatory modes: $u(\theta, t) = f(\omega_n t + n\theta)$, the frequency of rotation ω_n and the spatial period of the rotatory wave n (the number of petals) depend on the angle of field rotation in the feedback and the diffusion coefficient D:

$$\tau \omega_n = (1 + Dn^2) \tan(n\Delta). \tag{6}$$

The estimation of dimension of the system attractor depending on the parameters K and Δ is given in (Vorontsov and Razgulin 1991).

By taking as a basis the system model (5) with long-range and local interactions we introduce the supplementary time-delay. The system (5) is transformed into the form

$$\tau \frac{\partial u(\theta, t)}{\partial t} + u = D\frac{\partial^2 u}{\partial \theta^2} + K[1 + \gamma \cos u(\theta + \Delta, t - T)]. \tag{7}$$

Equation(7) has the following characteristic scales of spatial and temporal interactions: diffusion-type interactions with the length $l_d \cong (D\tau)^{1/2}$, long-range interactions fulfilling the direct coupling of fields at the points spaced at a distance Δ, interactions with characteristic temporal scale τ and a non-local coupling of fields in time with delay T. Note, that at $\Delta = 0$, $D = 0$ equation (7) transfers into (2), describing Ikeda chaos (Ikeda *et al* 1980), and at $D = 0$, $T = 0$, $\Delta = 2\pi/N$ it coincides with (3) (the chain of coupled resonators) (Otsuka and Ikeda 1989).

3. The analysis of the system dynamics at the bifurcation point

Like the analysis carried out by Ivanov *et al* (1991) for the case $T = 0$ we shall look for the solution to equation (7) in the form of a harmonic wave (rotatory mode):

$$u_n(\theta, t) = A_n(t) \cos(\omega_n t + n\theta) + \bar{u}(t). \tag{8}$$

Substituting (8) into (7) and assuming the amplitudes of the rotatory waves to be small, we obtain the equation for the functions $A_n(t)$, $u_n(t)$ and the frequency ω_n (phase equations). The same technique for the analysis of the system in the vicinity of the bifurcation point has been described in the work of Brand *et al* (1986) (the technique of phase equation):

$$\tau \frac{\mathrm{d}\bar{u}}{\mathrm{d}\bar{t}} + \bar{u} = K[1 + \gamma J_0(A_n(t - T)] \cos \bar{u}(t - T)] \tag{9a}$$

$$\tau \frac{\mathrm{d}A_n}{\mathrm{d}t} + A_n(1 + Dn^2) = -2K\gamma \sin \bar{u}(t - T) \cos(\omega_n T - n\Delta)(A_n(t - T)) \tag{9b}$$

$$\tau \omega_n = (1 + Dn^2)\tan(\omega_n T - n\Delta) \tag{9c}$$

where $J_0(.)$, $J_1(.)$ are Bessel functions of zero and first order. Note, that equation (9a) describing the spatial homogeneous component of the phase modulation, is similar to equation (2) for resonators with delay, considered by Ikeda *et al* (1980). At small amplitudes, i.e., at the bifurcation of initialing the rotatory mode, equations (9a) and (9b) are uncoupled. This means that at excitation the spatially inhomogeneous mode does not influence the dynamics of the spatially homogeneous phase which is described by the equation

$$\tau \frac{\bar{u}}{t} + \bar{u} = K[1 + \gamma \cos \bar{u}(t - T)].$$

Let us assume that the parameter K is less than the critical value K_{cr}, at which the transition from the stationary spatially homogeneous solution $\bar{u}(t - T) = \bar{u}_0$ to the oscillating regime appears. In this case it is not difficult to obtain from (9b) the condition of excitation of the rotatory mode A_n

$$\Lambda \leq \Lambda_n = \frac{1 + Dn^2}{\cos(n\Delta - \omega_n T} < 0 \tag{10}$$

where $\Lambda - K\gamma \sin \bar{u}_0$ is the controlling parameter. Note that for $T = 0$ condition (10) is transformed into the corresponding condition of excitation of the rotatory modes obtained in (Vorontsov *et al* 1988). In figure 4 the boundaries of the excitation regions of the rotatory modes with different wave parameter n depending on the scale of non-local interactions Δ (the angle of field rotation). From (10) it follows that for

$$K < \frac{1 + D}{\gamma \cos \bar{u}_0}$$

there is only spatially homogeneous solution. The rotation mode excitation u is possible only if $K \geq K_n = (1 + Dn^2)/(\gamma \sin \bar{u})$, with the critical value of the excitation threshold depending on the value of the delay parameter T. Note, that for $T = 0$ the rotatory mode $u_n(\theta, t)$ can exist only for a limited range of D-values (this range is determined by the condition $\cos(n\Delta) < 0$).

The introduction of the time delay radically changes the picture of excitation of rotatory modes. In figure 5 the dependence of boundaries on the excitation of the

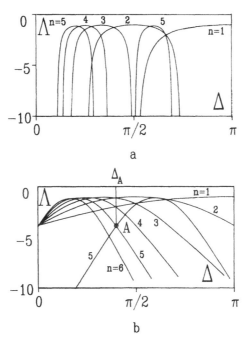

a

b

Figure 4. The boundaries of rotation instability for structures with different wave numbers: (*a*) $T = 0$, (*b*) $T = 0.5\tau$

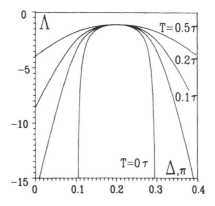

Figure 5. The influence of the time-delay T on the geometry of the boundary of the mode excitation $u_5(q,t)$.

rotatory mode $u_5(\theta, t)$ at the change of T parameter is shown. The time-delay results in the possibility of exciting at any value of the field rotation angle Δ the rotatory mode $u_n(\theta, t)$ with an arbitrary value n and at a corresponding value of the coefficient K; with the increase of T the boundary value Λ_n (the threshold of the mode $u(q,t)$ excitation) decreases (figure 6). It is interesting to note that for $D = 0$ and small values of the parameter D the threshold of excitation is generally determined by the time-delay parameter (figures 4, 6).

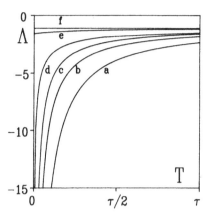

Figure 6.   The dependence of the excitation threshold on the angle of field rotation
Δ: (a) $\Delta = 0$, (b) 0.05π, (c) 0.1π, (d) 0.15π, (e) 0.175π, (f) 0.2π).

Another feature of the dynamics of the rotatory waves in the system with time-delay is the ambiguity of the dependence $\omega_n(\Delta)$ (figure 7), i.e., for the same rotation angle Δ conditions of excitation are fulfilled for the modes with the same value of spatial wave number n, but different frequencies ω. Actually, the mode having the lowest excitation threshold will be really excited. Note, that for the rotation angle $\tilde{\Delta} = 2\pi/n$ the excitation thresholds of the corresponding modes are equal (point A in figure 4(b)). In this case the solution is a superposition of two modes having the same spatial period but rotating with the same frequency in the opposite directions.

As was shown by numerical simulations of the equation (7) their interaction results in the appearance of a quasi-stable dissipative structure that fails at an insignificant variation of the parameter Δ.

4. Finite-dimensional dynamics models. A set of equations for order parameters

The analysis of conditions of the stability loss of the spatially homogeneous solution (10) showed that at a point of bifurcation the excitation conditions may be fulfilled simultaneously for several rotatory modes and their number rises with the increase of time delay T and the decrease of the diffusion coefficient D. Here the question about the interaction of rotatory modes rises, and it cannot be solved on the basis of the analysis of the phase equations (9).

Let us represent the solution $u(\theta, t)$ of the initial problem (7) in the vicinity of a critical point in the form of an expansion into the rotatory modes:

$$u(\theta, t) = \bar{u}(t) + \sum_{n=1}^{\infty} A_n(t) \cos[\omega_n t + n\theta]. \tag{11}$$

We shall consider the amplitudes of the rotatory modes to be small, and superpose the condition (9c) on the frequencies ω_n. Substituting (11) into equation (7) and restricting ourselves to terms of first order of magnitude relative to A_n, we obtain the infinite chain of coupled equations determining the dynamics of the rotatory modes

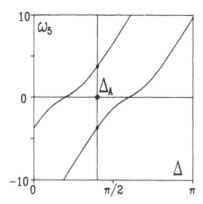

Figure 7. Angular velocity ω_5 versus rotation angle Δ. The corresponding boundaries of the mode excitation $u_5(\theta, t)$ are shown in figure 4 by dashed lines.

near a point of stability loss of a spatially homogeneous solution:

$$\tau + \frac{d\bar{u}}{dt}u = K\{1 + \gamma \prod_{n=1}^{\infty} J_0[A_n(t - T)]\cos[\bar{u}(t - T)]\} \tag{12a}$$

$$\tau\frac{dA_n}{dt} + A_n(1 + Dn^2) = -2K\gamma \sin\bar{u}(t - T)\cos(\omega_n T - n\Delta)$$

$$\times \prod_{j \neq n}^{\infty} J_0[A_j(t - T)] J_1[A_n(t - T)]. \tag{12b}$$

It is easy to see that the infinite products are convergent. Note that the reduction procedure used for the reduction of the system of equations (12) is close to the formalism of the amplitude equations (Newell and Whitehead 1969, Bestehorn and Haken 1990).

The amplitudes of the rotatory modes $A_n(t)$ and the spatially homogeneous phase $\bar{u}(t)$ are actually the order parameters describing the system dynamics in the vicinity of the critical points. And the system of equations (12) can be related to the systems of order-parameter equations, well known in synergetics (Haken 1983, Haken 1987).

Let us investigate the problem of reduction of an infinite dimensional system of equations (12) to some subsystem having a finite dimension N. This reduction can be accomplished on the basis of the slaving principle, well known in synergetics (Gang and Haken, 1989, Haken 1983).

Let us consider at first the case $T = 0$. Let us find the eigenvalues λ_n of the system of equation (12) linearized in the vicinity of the stationary point

$$\bar{u} \equiv \bar{u}(t \to \infty) \qquad A = 0 \qquad n = 1, \dots. \tag{13}$$

The eigenvalues of λ_n are determined by the following expressions

$$\lambda_0 = -I$$
$$\lambda = -(1 + Dn^2) - K\gamma \sin(\bar{u}_0)\cos(n\Delta). \tag{14}$$

Taking into account (13) and (14) the system (12) can be rewritten in the following form suitable for the application of the slaving principle:

$$\tau \frac{\mathrm{d}\bar{u}}{\mathrm{d}\bar{t}} = \lambda_0 \bar{u} + K\{1 + \gamma \cos(\bar{u}(t))\} \tag{15a}$$

$$\tau \frac{\mathrm{d}A_n}{\mathrm{d}t} = \lambda_n A_n + f_n(\bar{u}A_1, A_2, \dots, A_{j\neq n}, \dots, A_N, \dots) \tag{15b}$$

where the function f_n has the terms of the magnitude order $O(A_n)$. A stationary spatially homogeneous solution is stable if all the eigenvalues λ_n are negative.

Variation of the parameters K, d, γ and Δ results in a change of λ_n. And one or more of the values of λ_n can become positive, which leads to the loss of stability for the spatially homogeneous solution. Note that for arbitrary values of K, γ, Δ it is always possible to find such a value of the rotatory mode number N, such that for $n > N_{\max}$ all λ_n are negative. From (14) it is easy to obtain an estimate of the number $N_{\max}$:

$$N_{\max} < \mathrm{int}\{[(K\gamma - 1)/D]\}^{1/2} + 1$$

where by int{ } the integer part of the number is designated. Using the slaving principle in the vicinity of a critical point it is possible to reduce the initially infinite-dimensional system of equations (12) to a finite-dimensional system assuming in (12) $T = 0$ and $n = 0, 1 \dots N_{\max}$:

$$\tau \frac{\mathrm{d}A_n}{\mathrm{d}t} + A_n(1 + Dn^2) = -2K\gamma \sin(\bar{u}_0) \cos(n\Delta) \prod_{j\neq n}^{N_{\max}} J_0[A_j(t)] J_1[A_n(t)]. \tag{16}$$

This reduction means that if the initial conditions for the functions $\bar{u}(t)$ and $A_n(t)$ are chosen in a small neighbourhood of the stationary point,

$$|\bar{u}(0) - \bar{u}| \ll 1 \qquad |A(0)| \dots, |A(0)| \dots \ll 1 \tag{17}$$

for $t \gg \tau_1$, where $\tau_1 = \tau / \min(|\lambda_n|)$, the solution of the infinite-dimensional system (12) asymptotically approaches the solution of the finite-dimensional system of equations (16).

Among the remaining $N_{\max}$ eigenvalues λ_n there are both positive and negative ones. Let us denote by λ_m^u the positive eigenvalues to which the unstable solutions correspond and by λ_m^s the negative ones, $m = M + 1, \dots, N_{\max}$. In this case we have enumerated all the eigenvalues, arranging them in decreasing order:

$$\lambda_1^u > \lambda_2^u >, \dots, > \lambda_M^u > 0 \qquad |\lambda_{M+1}^u| > |\lambda_{M+2}^u| >, \dots, > |\lambda_{N_{\max}}^s| > 0.$$

The dynamics of a finite-dimensional system of coupled differential equations of the type in (16) has been analysed in (13) on the basis of the slaving principle. Let us use some results of this work.

Suppose that the initial conditions $A_n(0)$, $n = 0, 1, \dots, N_{\max}$ are in the vicinity of the bifurcation point (condition (17)). In this case the system attractor is an M-dimensional unstable manifold. This allows us to make further reductions of the system of equations (15, 16), decreasing its dimension. The reduced system includes only M equations corresponding to positive eigenvalues λ_n^u, $n = 1, \dots, M$:

$$\tau = \lambda_n A + f(u, A \dots, A, A \dots, A) \tag{18}$$

where the functions $A = \Phi(\bar{u}_0, A_1 \dots, A_M)$, $(j = M+1, \dots, N)$ determine the unstable manifold (Gang and Haken 1989). In the case under consideration this means that it is sufficient to take M equations (16) corresponding to the positive values of λ_n

$$\tau 1 + Dn^2 \bar{u}_0 \prod_{j\neq n}^{M} J_0 j J_{1n}. \tag{19}$$

Now consider the general case $T \neq 0$. Let $K < K_{cr}$ (in this case the existence of a time delay T does not result in the spatially homogeneous phase oscillations) then from (12a) with $A_n = 0$ it is possible to determine the stationary value $\bar{u}_0 = \bar{u}(\infty)$. It is not difficult to show that near the bifurcation point the eigenvalues λ_n of the systems of equations with delay (12) are determined by the relations:

$$
\begin{aligned}
\lambda_n &= -I \\
\lambda_n &= -(1 + Dn^2) - K\gamma \sin(u) \cos(\omega_n T - n\Delta) \exp(-\lambda_n T)
\end{aligned}
\tag{20}
$$

($n = 1, \dots$) and can take on complex values. Nevertheless, even in this case it is possible to use the slaving principle and in analogy to (19) to obtain the reduced system of equations:

$$
\tau \frac{dA_n}{dt} + A_n(1 + Dn^2) = -2K\gamma \sin \bar{u}(t - T) \cos(\omega_n T - n\Delta)
$$

$$
\times \prod_{\substack{j=1 \\ j \neq n}}^{M} J_0[A_j(t - T)] J_1[A_n(t - T)] \qquad n = 1, \dots, M.
\tag{21}
$$

Thus, setting the initial conditions in the vicinity of bifurcation (13) and fulfilling the conditions $K < K_{cr}$ and $La \leq La_n$ (condition (10)) the following hierarchy of instabilities arises. The spatially homogeneous solution loses stability—rotatory modes $A_n(t)$, $n = 1, \dots M$ for which the inequality (10) is true. Here, in accordance with the condition $K < K_{cr}$ equation (12a) has a stationary solution $\bar{u}_0$, and the dynamics of the system is determined as a whole by the reduced system of equations (21) in which instead of $\bar{u}(t - T)$ we must take $\bar{u}_0$.

With the increase of K the dimension of the reduced system grows (more and more rotatory modes lose stability). At $K = K_{cr}$ the oscillatory bifurcation of the spatially homogeneous phase $\bar{u}(t)$ occurs. As a result the stationary point $\bar{u}_0$ loses stability. At the expense of the multiplier $\sin[\bar{u}(t - T)]$ in equations (21) the oscillations of the function $\bar{u}(t - T)$ result in an essential complication of the system dynamics. In this case there can occur a parametric excitation of the spatial rotatory modes that were not included in the reduced system at first (the slaving principle stops functioning).

On this example we can follow how the introduction of the time parameter, which the time delay T is, can lead through the successive oscillatory bifurcations of the spatially homogeneous phase to the parametric excitation of the rotatory modes (the enrichment of the spatial spectrum of the solution).

Note that equations (21) have a restricted field of application (the vicinity of the point of stability loss of the spatially homogeneous solution), nevertheless they demonstrate the possible mechanism of the appearance of the spatio-temporal chaos; the cascade of period-doubling bifurcations results in the stochastic parametric excitation of the spatial modes.

5. Further reduction of the system. Competition of rotatory modes. WTA dynamics

Consider the problem of interaction of the rotatory modes. The reduced system of equations (21) can be associated with systems with competition that are actively investigated in the theory of artificial neural networks (Grossberg 1988). Rewrite (21) in the form typical for the most of models of neural networks:

$$\tau \frac{\mathrm{d}A_n}{\mathrm{d}t} = (1 + Dn^2)A_n + f_N(\bar{u}_0, A_1, \dots, A_M) \tag{22a}$$

$$f_n(\bar{u}_0, A_1, \dots, A_M) = -2K\gamma \sin(\bar{u}_0)\cos(\omega_n T - n\Delta) \prod_{j \neq n}^{M} J_0(A_j)J_1(A_n) \tag{22b}$$

(it is supposed that $K < K_{\mathrm{cr}}$).

In case the conditions

$$\mathrm{d}f_n/\mathrm{d}A_k \leq 0 \tag{23}$$

are fulfilled for all $k \neq n$ in the system with competition, the WTA condition (Winner Takes All) is implemented. The increase of amplitude of one of the rotatory modes (for example A_k) in fulfilling the condition (23) results in the decrease of the derivatives dA_n/dt $(n \neq k)$, i.e., in the slowing of growth or the decrease of all the other A_n amplitudes. As a result of the competition in the WTA system only one of the variables becomes the winner.

Let us turn to the condition of competition (23). From (22b) it follows immediately

$$\mathrm{d}f_n/\mathrm{d}A_k = 2K\gamma \sin(\bar{u}_0)\cos(\omega_n T - m\Delta)B_{n,k} \tag{24a}$$

where

$$B_{n,k} = \prod_{\substack{j \neq n \\ j \neq k}}^{M} J_0[A_j(t-T)]J_1[A_k(t-T)]J_1[A_n(t-T)] \qquad n = 1, \dots, M. \tag{24b}$$

Only the equations for the rotatory modes having positive eigenvalues λ_n constitute the reduced system

$$\begin{aligned} \lambda_n &= -(1 + Dn^2) - K\gamma \sin(\bar{u}_0)\cos(\omega_n T - n\Delta) > 0 \\ K\gamma \sin(\bar{u}_0)&\cos(\omega_n T - nD) < -(1 + Dn^2) < 0. \end{aligned} \tag{25}$$

At small amplitudes of the rotatory modes, which is exactly the case under consideration, the expressions for B in (24b) are positive, and taking account of (25) the partial derivates $df_n/dA_k \leq 0$. Thus, for the reduced system (22a) the conditions of competition with suppression are fulfilled. It means that the system possesses WTA dynamics and as a result of the competition only one rotatory mode 'survives'.

As the system starts from a small neighbourhood of the stationary point (condition (17)), mode A wins the competition as a result of the evolution process. This mode has maximum value of the expansion coefficient into the circular harmonics of the function $u(\theta, t)$ taken at the initial moment of time $t = 0$:

$$u(\theta, t) = \sum_{j=1}^{\infty} a_j(t)\cos j\theta. \tag{26a}$$

The amplitudes of the circular harmonics $a_j(t)$ are determined from the relation (Arsenault 1989):

$$a_n(t) = \int_0^{2\pi} u(\theta, t)\cos(n\theta)\,\mathrm{d}\theta \qquad n = 1, \dots, N. \tag{26b}$$

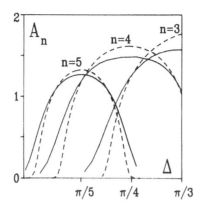

Figure 8. Stationary amplitudes of the rotatory modes versus Δ: solid line — the solution of (7), dashed lines — (9).

The impact of fluctuations may change the situation. As is shown in Gang and Haken (1989) starting from a neighbourhood of a stationary point in the presence of fluctuations, the most unstable mode wins the competitions, and further reduction of the system of equations for the rotatory modes is possible, as a result of which there are left only equations (19) for the one most unstable mode. If we suppose that fluctuations are absent, then such a reduction proves to be ineligible, because with selection of initial conditions due to the competition any of M unstable modes can become a winner.

6. Numerical investigation of the dynamics of the system with spatio–temporal delay

The results given above are valid only near the point of stability loss of the spatially homogeneous solution. Consider the characteristic features of the system evolution, based on the numerical investigation of the initial non-linear equation in partial derivatives (7).

Assume the initial condition for the function $u(\theta,t)$ in the form of expansion (26). First of all, assume that the function $u(\theta,0)$ has only one n-harmonic component (in (26a) $a_j = a_0\delta_{jn}$ where δ_{jn} is the Kronecker symbol). In this case the dynamics of the rotatory mode $u_n(\theta,t)$ near the critical point is described by the system of equations (9).

Compare the solution $u_n(\theta,t) = A_n(t)\cos(\omega_n t + n\theta) + \bar{u}(t)$ obtained from the approximate equations (9) with the solution of the non-linear equation (7) (see figure 8). The numerical investigations carried out showed that for relatively small increase of the stability threshold $\delta_n = |\Lambda - \Lambda_n|/|\Lambda_n|$ when the phase u does not exhibit oscillations ($K < K_{\mathrm{cr}}$), the system of equations (9) describes well the dynamics of development of a certain rotatory mode. With the increase of δ_n the rotatory mode shape is distorted first of all due to the appearance of second harmonics $2\omega_n$ (figure 9).

Close agreement of values for rotation speeds ω_n obtained on the basis of numerical solution of the equation (7) (see figure 10) and from the relation (9c) should be noted as well. Let us investigate the problem of non-linear interaction of the rotatory modes. Let a few circular harmonic components a_i for which the excitation conditions (10)

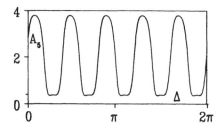

Figure 9.   The shape of the rotating wave $u_5(\theta, t)$ distorted by the second harmonic $n = 10$.

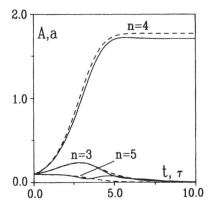

Figure 10.   Competition of the rotatory modes with $n = 3$, 4 and 5. The solid line plots the solution of (7), the dashed line that of (21).

are fulfilled be present in the initial conditions. For example, in (26) the coefficients a_1, a_2, a_3 and a_4 differ from zero. As the numerical studies of the equation (7) have shown only one rotatory mode 'survives' as a result of the evolution process as a rule. The winner is a rotatory mode with a maximum initial value of $a_n(0)$. Figure 10 illustrates the competition of modes. It shows time evolution of the amplitudes of circular harmonic components $a_n(t)$ of the function $u(\theta, t)$ obtained by the numerical simulation of equation (7) (solid line) and the corresponding time-dependences of the amplitudes $A_n(t)$ obtained from the reduced system (21). The amplitudes of circular harmonics $a_n(t)$ were calculated according to (26b).

Figure 11 shows how due to selecting different initial conditions the WTA dynamics of rotatory mode has been implemented. The essential distinction of competition in the system considered, compared with the WTA-type competition in the artificial neural networks, is that the WTA-dynamics of the artificial neural networks is realized on the basis of the dynamic systems with the simplest attractor structure containing only a stable focus (Grossberg 1989). At the same time in the example considered the system attractor is more highly complicated and the non-linear WTA-type competition is realized between the dynamic non-linear modes.

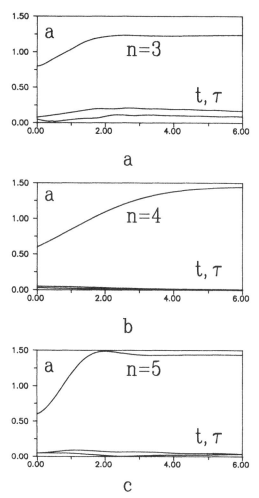

Figure 11. WTA dynamics in the competition of the rotating waves $n = 3$, 4 and 5 [Ivanov *et al* (1991)].

7. The opportunities of the experimental implementation of the optical system with spatial and time delay

The rotatory modes in the system without time-delay have been found experimentally in the work of Vorontsov *et al* (1988). Detailed investigation of their dynamics has been given in Akhmanov *et al* (1991), Vorontsov (1991).

In these experiments they used a liquid-crystal light valve modulator (LCLV) as a non-linear medium; that modulator possesses a significant Kerr type non-linearity (the equivalent non-linearity parameter $n_2 = 1.2 \times 10^{-2}$ m^2 W^{-1}). A typical example of the rotatory mode arising in such a system is shown in figure 12.

From the view point of time-delay implementation in the system of the schemes on the basis of the LCLV are not suitable because of the relatively slow speed of liquid-crystal response ($\tau \simeq 0.1$–0.01 s).

Figure 12.   An example of rotating structure investigated in the work of Ivanov *et al* (1991).

Consider another opportunity. Suppose the length of the optical path of the light beam in the system $L = 3$ m, and the delay time T is $\simeq 10^{-8}$ s. This means that the characteristic time for relaxation of non-linearity τ must not be greater than 10^{-7}–10^{-8} s, which is the admissible value for a number of non-linear semiconductor materials: InSb, GaAs (bulk), GaAs (MQW) and others (Walker 1991). The typical parameter values are taken from, for example, for GaAs (MQW): $n_2 = 2 \times 10^{-8}$ m^2 W^{-1}, $\tau = 2 \times 10^{-2}$ s and $\lambda = 0.8$ mm. The radiation intensity required for the development of rotatory instability is determined from $K = k \ln_2 I \simeq 1$ (l is the length of the non-linear medium). Let us take for estimation $\lambda_n = 2$ mm, then the required level of radiation intensity will be 1 W cm^{-2}. The estimations show the possibility of practical realization of the system considered.

8. Conclusion

Let us pay attention to one aspect of investigations of non-linear dynamics of spatially distributed systems with different scales of spatial and temporal interactions. Consider a certain analogy between the structure of such systems and the basic characteristics confined within the architecture of the developed neuron ensembles.

From the classical physical viewpoint, developed neural networks present themselves as very unusual objects. And probably the main feature distinguishing them from traditional physical models lies in the successive implementation of the principle of long-range or spatial non-locality of interactions. Each of the neurons has numerous 'long' connections with other neurons that allow for a comparatively short time to transfer excitation to the distances compared with the dimensions of the system itself.

In the developed neural networks there are also traditional extension mechanisms of the diffusive type of excitation when the excitation propagates locally from point to point. A very important feature is also the temporal delay of signals extending along axons coupling neurons. The delay of signals actually means the temporal non-locality of interactions of certain neural networks elements. Moreover, every system element (neuron) is strongly non-linear and the number of elements can be large ($\simeq 10$ neurons).

How do we approach the analysis of such a complex formation ? One possible way is the construction of relatively simple basic models of physical systems possessing on the functional level the general features inherent in the developed neural networks: non-linearity of every element, great number of elements, the availability of local and non-local spatial and temporal interactions.

From this point of view the system considered above can be regarded as one of the simplest models of neuromorphic systems of a special type. Every point of the transverse section of the laser beam r may be compared with a certain formal neuron and the non-linear phase modulation $u(r,t)$ can be compared with the excitation function of a neural network. In this case equation (7) simulates some general physical features inherent in the developed neuron structures: a great number of non-linear elements with a threshold non-linearity (figure 2), a local mechanism of excitation transfer between the elements (a diffusion term in (7)), large-scale 'long' connections (spatial shift Δ, temporal delay T).

General features of behaviour indicate a certain analogy between the dynamics of artificial neural networks and spatially distributed system with large-scale interactions: a great variety of possible solutions (non-linear mode), strong competition due to which the selection of solutions occurs, an important influence of the initial conditions on the system dynamics. WTA-dynamics of non-linear modes (page 5 of this paper) has a direct analogy with the behaviour of the artificial neural networks of WTA-type.

Non-linear optical systems with long-range interactions of different type are an extremely interesting object of investigation for synergetics as a whole. As has been shown the dynamics of even the simplest system of this type displays a surprising variety of behaviours.

An important feature of such systems is non-factorization of space and time. The investigations carried out showed that the variation of time delay (temporal parameter) can lead to the transformation of the spatial structure of solution and at the same time the alternating of spatial scale of interactions Δ essentially changes such temporal characteristics of the dynamic process as the structure movement speed. As a result the process of transfer to the spatio-temporal chaos occurs in such systems in quite a different way than it takes place in the classical dynamical systems described by the equation in ordinary derivatives; temporal stochastization of the solution inevitably causes spatial instability.

References

Akhmanov S A, Vorontsov M A, Ivanov V Yu, Larichev A V and Zheleznych N I 1991 *J. Opt. Soc. Am.* B to be published

Arecchi F T 1991 Physica D, 51, 450–464

Arecchi F T, Gadomski W, Lapucci A, Mancini H, Meucci R and Roversi J A 1988 *J. Opt. Soc. Am.* B **5** 1153-1159

Arsenault H H 1989 *Optical Processing and Computing* (New York: Academic) 315-342

Bestehorn M and Haken H 1990 *Phys. Rev.* A **42** 7195-7203

Brand H R, Lombdahl P S and Newel A C 1986 *Physica* D **23** 345–361

Firth W J and Galbraith I 1985 *IEEE, J. Quant. Electron.* **21** 1399–1403

Gang H and Haken H 1989 *Z. Phys.* B **76** 537–545

Gibbs H M 1985 *Optical Bistability: Controlling Light with Light* (New York: Academic)

Giusfredi G, Valley J F, Pon R, Khitrova G and Gibbs H M 1988 *J Opt.Soc.Am.* B **5** 1181–1191

Grossberg S 1988 Nonlinear Neural Networks: Principles, Mechanisms, and Architectures *Neural Networks* **1** 17–61

Haken H 1983 *Synergetics, an Introduction* 3rd edn (New York: Springer)

—— 1987 *Advanced Synergetics* (New York: Springer)

Ikeda K, Daido H and Akimoto O 1980 *Phys. Rev. Lett.* **45** 709–712

Ivanov V Yu, Larichev A V and Vorontsov M A 1991 *Proc. SPIE* no 1402 145–153

Ivanov V Yu, Zheleznykh N I and Vorontsov M A 1990 *OQE* **22** 505–515

Le Berre M, Ressayre E and Tallet A 1989 *Opt. Commun.* **72** no 1, 2

—— 1991 *Phys. Rev. A* **43** no 9

Lugiato L A and Oldano C 1988 *Phys. Rev. A* **37** 3896–3908

Newell A C and Whitehead J A 1969 *J. Fluid. Mech.* **38** 279

Otsuka K and Ikeda K 1989 *Phys. Rev. A* **39** 5209–5228

Rosanov N N and Khodova G V 1990 *JOSAB* **7** *1057–1065*

Vorontsov M A 1991 *Proc.SPIE* no 1402, 116–144

Vorontsov M A, Ivanov V Yu and Shmalhauzen V I 1988 *Laser Optics of Condensed Matter* (New-York: Plenum) pp 507–517

Vorontsov M A and Razgulin A V 1991 *Photonics and Optoelectronics* no 1 1991 to be published

Walker A C 1991 *Optical Computing and Processing* **1** no 1 91–106

The effect of intraspectral harmonic correlation of broad-band fields on the excitation of quantum resonance systems

A M Bonch-Bruevich and S B Przhibelskii

Undoubtedly S A Akhmanov was one of the first investigators of the role of field statistics in non-linear processes of the interaction of matter with radiation. One of his and R V Khokhlov's achievements in that area was the discovery of strong suppression of phase fluctuations of random fields in non-linear systems—the parametron. It turned out that classical fields formed in parametrons have the features of squeezed quantum ones [1]. Those classical fields are periodic non-stationary stochastic processes with harmonic correlation. The role of that correlation in non-linear optical processes has not been clarified yet and investigations connected with this field attract much attention nowadays. Akhmanov's work on classical squeezed fields may be considered the pioneering works in that area.

For a long time our interests in the non-linear interaction of non-monochromatic fields with matter were close to Akhmanov's. Due to scientific contacts we know Akhmanov to be a bright and highly industrious person. We dedicate this article to the memory of S A Akhmanov.

Among non-linear optical phenomena those caused by fluctuations play a significant role. In fact fluctuations are the result of both material processes and irregularities in radiation fields. The interest in the effects caused by the latter is for a number of reasons, two of them being the most important. First, strong radiation often turns out to be pulsed and multimode. Second, in precise measurements the limit of accuracy is often determined by the quantum fluctuations of the field.

Nowadays a new aspect to the interest in the effects of field fluctuations is coming about. The interest is aroused by the effects in which one can find the optimal required multi-photon action on matter or a device. This aspect is connected also with the problems of generation and formation of the fields with required properties, for example, the creation of a squeezed field and the formation of pulses exciting a concrete system selectively in multiphoton processes.

In the present article the effects connected with the statistics of fields, regarded as stationary noise are going to be discussed. Mainly the attention will be paid to the influence of the continuous-spectrum field harmonics correlation on excitation of a two-level system.

A stationary stochastic process that models a fluctuating field can't have arbitrary correlation of spectrum harmonics. The limitation of allowed correlations is defined by the following consideration. Optical signals are usually presented as

$$F(t) = E(t)e^{-i\omega t} + E^*(t)e^{i\omega t} \tag{1}$$

where ω is an optical frequency, $E(t)$ and $E^*(t)$ are complex conjugate amplitudes changing little the period $2\pi/\omega$. The spectral representation

$$E(t) = \sum_\nu \epsilon_\nu e^{i\nu t} \tag{2}$$

connects the stochastic function $E(t)$ with random numbers ϵ_ν which are the amplitudes of spectrum harmonics. A stationary stochastic process is characterized by two correlators: the normal $\langle E(t_1)E^*(t_2)\rangle = K(t_1 - t_2)$ and the anomalous $\langle E(t_1)E(t_2)\rangle = Q(t_1 - t_2)$. K and Q are mutually independent although they are linearly functionally connected. The stationary conditions determine the allowed harmonics correlations $\langle \epsilon_\nu \epsilon_\mu^* \rangle = s_\nu \delta_{\nu\mu}$ and $\langle \epsilon_\nu \epsilon_\mu \rangle = \sigma_\nu \delta_{\nu-\mu}$ where s_ν is the spectrum intensity at frequency ν and $\delta_{\nu\mu}$ is the Kronecker symbol. σ determines the degree of correlation between the conjugate harmonics and frequencies ν and $-\nu$. If the process is real, that is, $E(t) = E^*(t)$, then correlation is maximal and $Q(t) = K(t)$. When correlation is absent then $Q(t) = 0$.

The field statistics are commonly determined by correlators of all ranks. However, we believe the process is Gaussian when K and Q fully define the whole stochastic process.

The influence of stationary noise field fluctuations on the TLS excitation character has been recognised in many works. The role of fluctuations in amplitude and phase, their spectral characteristics, the difference of effects of fields with different statistics have been investigated. The appearance of the peculiarities of field fluctuations was found in absorption spectra and TLS secondary radiation.

The results of this work lead, roughly speaking to the following conclusions: the effect of the field with an asymptotically wide spectrum depends slightly on field statistics. The action of broadband fields is defined by its spectral density at the resonance frequency of TLS.

The field with narrow spectrum, the Raby frequency being much more than the spectrum width, acts according to its statistics. For example, the influence of intraband correlations is revealed in the absorption spectrum form of the weak probe field in the three-level registration scheme (figure 1) where the strong field is regarded as a Gaussian process. In the case of a real process when correlation is maximal the absorption line is non-split (figure 2). The splitting occurs when there is no harmonic correlation, that is, when the driving field is a complex Gaussian process with no addition of the real part. For the given scheme and narrowband spectra of a driving field the form of absorption lines is determined by the intensity distribution of stochastic fields. In the first case, for the real process this distribution is Gaussian and the line form is $x(\varepsilon) \sim \exp\{-\varepsilon^2/K(0)\}$; in the second case it is Rayleighan and $x(\varepsilon) \sim |\varepsilon/K(0)| \exp\{-\varepsilon^2/K(0)\}$.

The results of these conclusions mean that harmonic correlation is revealed in qualitative effects that are characterized by the higher-rank field correlations.

For the first time this qualitative effect has been observed in two-photon TLS excitation by broadband radiation. In excitation scheme (figure 3) the harmonics with frequency $\bar{\omega} + \nu$ and $\bar{\omega} - \nu$ were correlated in pairs which means that random phases were coupled by the condition $\phi_{\bar{\omega}+\nu} + \phi_{\bar{\omega}-\nu} = $ constant. The influence of correlation on the TLS excitation is found in lower ranks of non-linearity when the upper-level population of TLS is small. This case is described by the equation

$$\dot{\rho} = -\gamma_1 \rho + \left\{ V^2(t) \int_{-\infty}^{t} V^{*2}(t')e^{-(t-t')(\gamma_2 - 2i\bar{\omega} + i\omega_0)}dt' + \text{cc} \right\} \qquad (3)$$

where $\gamma_1; \gamma_2$ are population relaxation rates and TLS coherence relaxation rates respectively. $V^2(t)$ is the instant complex Raby frequency which is proportional to $E^2(t)$, ω_0 is the frequency of atomic transition.

The average population value $\langle \rho \rangle$ is determined by fourth-rank correlator $\langle V^2(t_1)V^{*2}(t_2)\rangle$ summed up of two terms one is proportional to $K(t_1 - t_2)$ and the

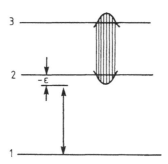

Figure 1. The three-level scheme of registration of the strong non-monochromatic field effect. Levels 2 and 3 are empty, only the first level is populated. The probe field is near the resonant 1–2 transition.

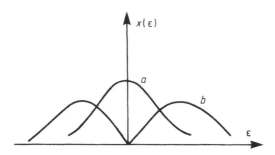

Figure 2. The absorption spectrum in the case of a narrow driving field spectrum. (a) without and (b) with maximal correlation harmonics.

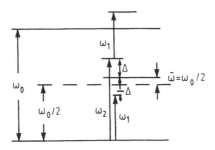

Figure 3. The two-photon excitation scheme by two harmonics with frequency ω_1 and ω_2. $\omega_{0/2}$ is the central frequency for all conjugated pairs of correlated harmonics.

other is to $|Q(0)|^2$. The first gives a non-resonant and small contribution the same as the field without harmonics correlation. The correlation provides resonance excitation of TLS by the broadband field. This correlation contribution is defined by expression

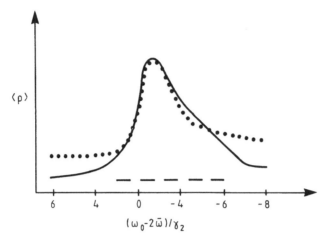

Figure 4. The two-photon excitation spectrum by a maximally correlated field with its spectrum width 11 δ_2. The dots are the theoretical results. The dotted line shows the level of excitation without harmonics correlation.

$$\langle\rho\rangle = 2|Q(0)|^2\gamma_2/\gamma_1\Big[(\omega_0 - 2\bar{\omega})^2 + \gamma_2^2\Big] \tag{4}$$

The resonance effect depends on the degree of harmonic correlation and power excitation that determines shift and broadening of the resonance line. It is interesting to note that the wider the spectrum of correlated field, the sharper the resonance action.

The resonance effect of a wide-band correlation field was observed for the first time in NMR experiments on Zeeman transitions in the ground state of Cd^{113} with resonance frequency $\omega_0 = 5$ kHz. In this experiment deviation from the predictions of approximate theory was found. This discrepancy shows that the experiment gives ample information about the dynamics of TLS in irregular fields.

In optics the effect of the correlation harmonics was observed for the first time [5] in the process of two-photon excitation of the transition $^6S_{1/2}-\ ^6D_{1/2;3/2}$ in atomic Cs vapour. In these experiments, the independence of the efficiency of the TLS excitation on the width of driving field spectrum was registered. And at the same time the low efficiency of the same but uncorrelated field was detected.

The above examples show that the correlation effect is connected with the abnormal correlator Q which determines resonant action. The same action can be in other multiphoton processes of even orders. The odd-order processes can also be made resonant by 'switching' the monochromatic component in consequence of the elementary acts in the excitation scheme.

Besides these real excitation effects which are connected with the population changing, there are other possible ones, which are connected with the changing of the energy spectrum of the quantum systems. In particular the correlation effects can be manifested in the dynamical Stark effect. Let us consider the effect of correlation on so-called radiation-induced phase relaxation in a quantum resonance system. It is known that in the above scheme (figure 1) the effect of driving field fluctuations is to cause the broadening of the absorption line in the probe field canal. In the case of the broad-band spectra of the driving field a Lorentzian line shape is formed. The width of this line is proportional to the spectral density of intensity field at the resonance frequency of the upper channel. The harmonic correlation of the exciting field is found

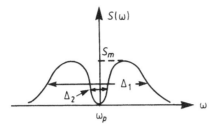

Figure 5. The two-hump-like driving field spectrum. ω_p is the resonance frequency of TLS.

when the above-indicated intensity is zero, in particular when field spectrum is double-hump-like (figure 5). When the width of spectrum Δ_1 and the width of dip $Delta_2$ are rather large i.e. much larger than the effective Raby frequency proportional to the intensity S_m, an absorption Lorentz emerges that stands on a broad pedestal. The width of this line is determined by

$$\Gamma \sim (1 - \eta^2) \int_0^\infty \mathrm{d}\omega\, S(\omega)/\omega^2 \sim (1 - \eta^2)\Delta_2^2/S_m \tag{5}$$

In this expression, the degree of correlation $0 \leq \eta \leq 1$ is conditionally defined by the value $Q(0)\,K(0)$. When correlation is maximal, that is $\eta = 1$, the non-broad component is formed in the absorption spectrum.

A more complicated combined effect of harmonic correlation is caused in the single-photon process of TLS excitation by the same double-hump-like spectrum radiation. In this case the growth of the degree of correlation reduces the efficiency of TLS excitation. The situation is possible when the field with harmonic correlation is weakly exciting and the same field without correlation is strongly exciting. This effect is explained by the specificity of the field-induced phase relaxation and non-resonant TLS excitation.

For realization of the harmonic correlation effects the ability to create the correlation and control its destruction is required. One of the methods of harmonic-correlation generation is based on *heterodyning*. Formally, in this method the noise signal is modulated by the monochromatic one whose frequency is twice that of the carrier of the first one. In the modulated signal

$$F(t)G\cos(2\omega t + \phi)$$

the component with frequencies near ω is extracted. In the resulting signal

$$F(t) + g[E(t) + E^*(t)]\cos \omega t$$

harmonic correlation is determined by the correlation degree g. In another method the harmonic correlation is achieved by the formation of the standing waves. We pay attention to the analogy between formation of the harmonic-correlation and the reverse in time of the stochastic process that resembles the well-known phase conjugation.

In the creation of the correlation in optical fields, one encounters a lot of practical difficulties. One of them is connected with frequency dispersion in media where the correlation is created, where the radiation propagates also in detectors. While the light travels the distance l through a medium with dispersion, where $n(\omega)$ is the index of refraction near the carrier frequency $\bar{\omega}$, there occurs a relative phase shift of harmonics which is equal to

$$\delta_\nu = l\nu^2\bar{\omega}/c\{(2/\omega)\partial n/\partial\bar{\omega} + \partial^2 n/\partial\bar{\omega}^2\}$$

This expression shows that decorrelation is inhomogeneous: the nearer the frequency of harmonics is to $\bar{\omega}$ the weaker is the decorrelation. This decorrelation has a systematic behaviour and can be compensated by a device with inverse dispersion to that of $n(\omega)$. Compensation of decorrelation can cause an effect that is similar to photon echo: the inversion of phase shifts converts the weak efficiency noise signal into one that strongly affects the resonance of the two-photon detector.

The correlation destruction in processes of light scattering from inhomogeneities in media is more complicated than that caused by destruction dispersion. It is natural to suppose that correlation disappears absolutely when the scattering is strong. By experiment [6] has shown the surprising stability of correlation in the case of light scattering on impurities and rough glass surfaces. In the case of diffusive scattering the two-photon efficiency of TLS excitation was shown to be independent of the spectrum width of scattering in the initially harmonic-correlated light. The phenomenon was explained by strong deformation of the wavefronts of all harmonics of stochastic field, the deformation being similar for all wavefronts. The equal deformation of all harmonic wavefronts is possible under two conditions.

(i) the absence of frequency dispersion and

(ii) no multiple scattering.

The experimental data [7] now do not allow one to find out the interesting features of that phenomenon such as the evolution of the correlation spectrum ∂_ν in scattering process.

The effects considered are those of classical fields. Quantum fields with correlated harmonics manifest their own peculiar features in resonance systems.

Formally, there is an analogy between the correlated noise signals considered above and the so-called non-classical states of quantum fields in particular: the squeezed vacuum. This analogy is based on the representation of the field correlator of the fourth rank by the lower ones. But unlike the above representation expressed by $K(t)$ and $Q(t)$ the quantum correlator has no unified representation and is determined in quite a different way by different field quantum statistics. So the indicated fourth-rank correlator may not be proportional to the square of the field intensity. The latter assertion explains the predicted effects [8, 9] of the linear dependence of the TLS excitation on field intensity in two-photon resonance processes. This paradoxical effect can be described by a simple two-mode model of the field whose state is close to vacuum and is defined by a function $a_0|0, 0\rangle + a_1|1_{\omega_1}, 0\rangle + a_2|0, 1_{\omega_2}\rangle + b|1_{\omega_1}, 1_{\omega_2}\rangle$. The energy of the field in this state is equal to $W = a_1^2\hbar\omega_1 + a_2^2\hbar\omega_2 + b^2\hbar(\omega_1 + \omega_2)$ and the amplitude A_2 of the two-photon TLS excitation is proportional to b. From this it is obvious that if $b \gg a_1$ and a_2 then $A_2^2 \sim b^2 \sim W$. The dominance of the term whose amplitude is b means that the field is in the state of correlated photons of different modes.

Another interesting effect of quantum harmonic correlation is revealed in the case of the interaction of the TLS and the squeezed vacuum [10, 11]. The effect of the latter induces the relaxation of TLS coherence. But unlike the above-mentioned induced relaxation, in this case the rates of relaxation of different so-called quadrature components of polarization are different. In other words the relaxation rates of different Bloch-vector components are different and the rate of one of them can become arbitrarily small in inverse proportion to the intensity of the squeezed vacuum.

The creation and destruction of quantum correlation is a more complicated process than in classical case. Thus, the squeezed vacuum is formed in optical parametric amplifier but unlike the classical correlation this one is created only in the below-threshold regime. The destruction of the quantum correlation occurs when light passes through the boundary of transparent media without frequency dispersion.

In conclusion we shall admit that the effects of intraband correlation may become the basis of a new trend in spectroscopy. Its peculiarity is defined by the problems of revealing the processes which influence the creation and destruction of intraspectral correlation. It is possible that correlational effects can be used for solving the problems of information transmission in the presence of noise.

References

[1] *Novie fizicheskie principi opticheskoi obrabotki informacii* 1990 ed S A Akhmanov and M A Vorontzov (Moscow: Nauka)
[2] Przhibelskii S G and Khodovoi V A 1972 *Opt. Spectrosk.* **32** 237
[3] Przhibelskii S G 1973 *Opt. Spectrosk.* **35** 71
[4] Bonch-Bruevich A M, Przhibelskii S G and Chigir N A 1981 *Sov. Phys.-JETP* **80** 565
[5] Bonch-Bruevich A M, Przhibelskii S G and Chigir N A 1987 *Sov. Phys.-JETP* **92** 781
[6] Przhibelskii S G 1990 *Sov. Phys.-JETP* **98** 105
[7] Bonch-Bruevich A M, Przhibelskii S G, Mak An A and Chigir N A 1988 **94** 60
[8] Gea-Banachloche J 1989 *Phys. Rev. Lett.* **62** 1603
[9] Javanainen J and Gould P L 1990 *Phys. Rev.* A **41** 5088
[10] Gariner C W 1986 *Phys. Rev. Lett.* **56** 1917
[11] Palma G M and Knight P L 1989 *Opt. Commun.* **73** 131

Injection-locked femtosecond parametric oscillators on LBO crystal; towards 10^{17} W cm^{-2}

V M Gordienko, S A Magnitskii and A P Tarasevitch

Nonlinear Optics Laboratory, Moscow State University, 119899 Moscow, Russia

Abstract. The parametric down conversion of femtosecond laser pulses at high intensity is discussed. An injection-locked parametric oscillator, producing pulses up to 10^{14} W/cm^2, based on a new non-linear crystal LBO is presented. The n_2 and surface breakdown threshold (400 fs pulses) of LBO were measured to be 1.3×10^{-16} cm^2/W, and 3.8×10^{13} W/cm^2 respectively.

1. Introduction

In this paper we present our latest results on the development of femtosecond parametric oscillators, proposed by S A Akhmanov, which produce superintense light field (superintense OPOs) and are based on a new non-linear crystal LBO. All these results were obtained at the Nonlinear Optics Laboratory of Moscow State University.

There is a growing interest now concerning the behaviour of matter in a superintense field. Ultrahigh-brightness laser systems based on the amplification of femtosecond pulses in broadband amplifiers have been developed in the UV—$\lambda = 0.308$, 0.248 μm (excimer), at $\lambda \simeq 0.6$ μm (dye), $\lambda \simeq 0.8 - 0.9$ μm (Ti:Al$_2$O$_3$), and at $\lambda \simeq 1$ μm (Nd–glass). At the same time it is highly desirable to develop frequency-tunable sources of superintense field in other spectral ranges.

From our point of view optical parametric oscillators are one of the most promising tools for solving this problem, and the superintense OPO should be injection-locked. Unlike commonly used OPOs (Danielius et al 1983), the injection-locked oscillators will make it possible to achieve high spatial and temporal quality of light pulses (Boichenko et al 1984, Magnitskii et al 1986, Bayanov et al 1990a), which are of great importance for obtaining high intensity on the target, placed in the focus of a lens. We report the development of a superintense OPO, based on a LBO crystal, pumped by high-power femtosecond pulses from a dye and XeCl systems.

The choice of LBO is determined by its unique characteristics. Measurements, carried out by Chen et al (1989), Lin et al (1990), Kato (1990), Dyakov et al (1991), have shown the extremely broad spectral range of transparency of this crystal (0.16–2.6 μm), relatively strong non-linearity ($d_{32} \approx 3d_{36}$ KDP), and very high optical breakdown threshold (3 times higher than in KDP).

Ebrahimsadeh et al (1990) demonstrated the first, to our knowledge, OPO on LBO crystal, pumped by UV nanosecond pulses from an excimer laser, tunable from the UV to the IR.

At the same time femtosecond pulse generation was demonstrated by Laenen *et al* (1990a, 1990b), Gagel *et al* (1990), Fickenscher *et al* (1990), Danielius *et al* (1989). Pulses shorter then 100 fs were obtained in synchronously pumped OPOS.

In this paper we discuss femtosecond pulse generation at very high pump intensities (up to breakdown threshold). Our first results with LBO pumping at $\lambda = 308$ nm and $\lambda = 570$–630 nm were reported at CLEO 91 (Magnitskii *et al* 1991), and UPS 91 (Akhmanov *et al* 1991a).

2. Basic parameters

Let us consider a single-crystal travelling-wave parametric amplifier. It is the group-velocity dispersion that limits the efficiency of parametric conversion of short laser pulses (Akhmanov *et al* 1991b).

For parametric conversion of high-power pulses at least four basic parameters should be taken into account: an avalanche breakdown threshold I_{br}, non-linear length L_{nl}, group velocity mismatch length L_g, and B-integral. Here

$$L_{nl} = (k\chi^{(2)}E_p)^{-1} \qquad L_g = \frac{\tau}{(U_p)^{-1} - (U_{s,i})^{-1}} \qquad B = \frac{2\pi}{\lambda} \int_0^L n_2 I_p \, dz$$

where k is a wave vector, $\chi^{(2)}$ is a non-linear susceptibility, E_p, I_p are pump field and intensity, τ is a pulse duration, U_p, U_s, U_i are the group velocities of the pump, signal, and idler pulses respectively, and L is a crystal length.

One can see that L_{nl} diminishes with growing pump intensity. The lower limit for L_{nl} is determined by I_{br}. As soon as for an avalanche breakdown $I_{br} \sim (\tau)^{-1}$,

$$L_{nl}^{min} \sim (I_{br})^{1/2} \sim (\tau)^{1/2}.$$

Thus one can obtain an extremely short L_{nl} for high-power femtosecond pulses. This seems to be the way to make L_{nl} so small (smaller than L_g) that one need not take into account the group velocity mismatch. But as L_g diminishes faster then L_{nl} with shortening pulse ($L_g \sim \tau$), for pulses short enough we shall have $L_g < L_{nl}$, and the efficiency of parametric conversion will be small.

If we take into account self-focusing and fix a certain value of B, we also get a limit for pump intensity:

$$I_{max} \simeq \frac{\lambda}{2\pi} \frac{B}{n_2} (L)^{-1}$$

and for $L \simeq L_g$:

$$I_{max} \sim (\tau)^{-1}.$$

So the problem is to determine the pulse duration range of effective parametric conversion. And to do this one has to know the optical breakdown and self-focusing thresholds, group-velocity dispersion, and effective non-linearity. Our experimental results and calculations are presented below.

3. Experimental setup

Our experimental setup (figure 1) produces ~ 400 fs, ~ 3 mJ pulses, for pumping an OPO in the range 0.57–0.63 μm, or ~ 300 fs, 10 mJ pulses for pumping at $\lambda = 0.308$ μm. It

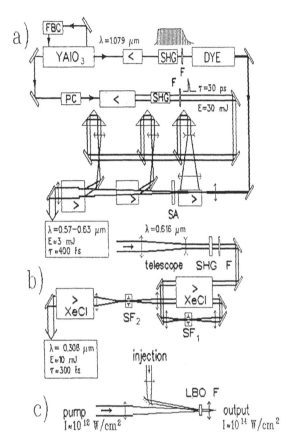

Figure 1. Experimental setup: (a) dye system, (b) excimer amplifiers, (c) OPO on LBO crystal, YalO₃ - picosecond master oscillator with feedback control FBC, DYE - synchronously pumped dye laser, SHG - second harmonic generator, PC - Pockels cell, SA - saturable absorber, F - filter, SF - spatial filter, XeCl - excimer amplifier.

is based on a hybrid mode-locked dye laser pumped by the second harmonic of a pulsed active-passively mode-locked Nd:YAP picosecond laser with feedback control (described in detail by Bayanov *et al* 1990b, and Akhmanov *et al* 1991c). The latter generates 5 μs trains of 40 ps pulses ($\sim$ 500 pulses in each train) at 2 Hz repetition rate. The dye laser output pulse train with energy $\sim$ 1 nJ in each pulse is directed into a three-stage dye amplifier, pumped by a single picosecond pulse selected from the pulse train of the Nd:YAP laser (amplified and frequency-doubled). The total pump energy is 30 mJ.

The dye laser was tuned from 0.57 to 0.63 μm with the help of a Liot plate. The dye solutions in the amplifier stages were changed to cover all this frequency range.

The 3 mJ femtosecond pulses from the dye amplifier could be used directly to pump OPO on the LBO crystal. To obtain the UV pumping of the OPO we tuned our dye laser to $\lambda = 0.616$ μm, doubled the frequency of the femtosecond pulse and amplified it in a XeCl chain up to 10 mJ ($\lambda = 0.308$ μm, figure 1). In both cases the pumping and injected waves were combined non-collinearly in the LBO crystal.

4. Results and discussion

We carried out computer calculations of tuning curves, group-velocity dispersion and conversion efficiency of LBO crystal. These results were obtained starting from the

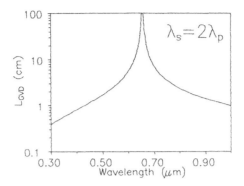

Figure 2. LBO group-velocity mismatch length.

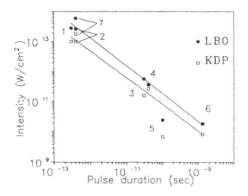

Figure 3. Surface damage threshold of LBO and KDP crystals. 1: $\lambda = 308$ nm, $\tau_p = 300$ fs; 2: $\lambda = 616$ nm, $\tau_p = 400$ fs; 3: $\lambda = 539$ nm, $\tau_p = 30$ ps; 4: $\lambda = 1.079$ μm, $\tau_p = 40$ ps; 5: $\lambda = 1.064$ μm, $\tau_p = 0.1$ ns; 6: $\lambda = 1.064$ μm, $\tau_p = 1.3$ ns; 7: $\lambda = 1.43$ μm, $\tau_p = 400$ fs.

Sellmeier equations and non-linear coefficients from the papers of Chen et al(1989), Lin *et al* (1990), Kato (1990), Dyakov *et al* (1991). The group-velocity mismatch curve is shown in figure 2. It is important to note, that this curve indicates the possibility of group synchronism at $\lambda_p \sim 0.6$–0.7 μm.

The damage threshold was determined simultaneously for LBO and KDP crystals. The measurements were held at different wavelengths and pulse durations. The breakdown onset was determined in a long series of shots. The results are presented in figure 3. Points 1, 2, 3, 4 are obtained with the dye system second harmonic and fundamental, Nd:YAP second harmonic and fundamental respectively. Points 5 and 6 are taken from Chen *et al* (1989), Lin *et al* (1990). All points except 5 lie very well on the $1/\tau$ line, revealing I_{br}^{LBO} to be about three times higher than I_{br}^{KDP}. Note that the points 5 also give $I_{br}^{LBO}/I_{br}^{KDP} \simeq 3$. Points 7 are obtained with femtosecond OPO on LBO crystal.

We determined n_2 for LBO by measuring the spectrum broadening due to self-phase modulation of a femtosecond pulse ($\lambda = 0.616$ μm, 400 fs), passing through the crystal. From these measurements we get $n_2 = 1.3 \times 10^{16}$ cm²/W.

Starting with the damage and self-focusing thresholds, L_g, and LBO effective non-linearity we can define the time duration range of efficient parametric conversion.

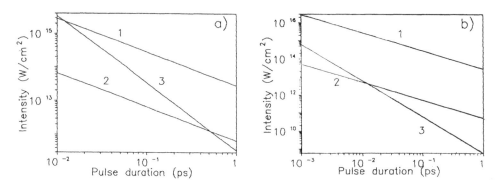

Figure 4. Pulse duration range of efficient parametric conversion in LBO crystal: (a) $\lambda_p = 308$ nm, (b) $\lambda_p = 616$ nm.

Figure $4(a)$ and $4(b)$ are plotted for $\lambda_p = 308$ nm and 616 nm respectively ($\lambda_s = 2\lambda_p$, $L = L_g$). Here 1 is surface damage threshold intensity, 2 is self-focusing intensity, 3 is the pump level corresponding to approximately 20% intensity conversion to the signal wave (constant pump-field approximation, 1 kW/cm^2 injection).

From these figures we can define the range of efficient parametric amplification—4. One can see, that in spite of the fact that 308 nm pumping allows one to cover a much wider frequency range, for $\lambda_p = 616$ nm we obtain a range of efficient conversion stretching up to much shorter pulses.

Our first results on parametric conversion in an LBO crystal were obtained with 308 nm pumping (injection wavelength 539 nm) (Magnitskii *et al* 1991). At low intensity of the injected signal this OPO operates near the self-focusing threshold. Efficient conversion was possible only at a high injection level ($\sim 10^9$ W/cm^2). With $E_p \approx 10$ mJ, ($\tau_p \approx 300$ fs) the femtosecond pulses at $\lambda = 717$ nm with energy 0.5 mJ were generated.

The use of a dye system as a tunable source allows one to get a tunable OPO operation with a fixed frequency of an injected wave. The detailed results of our experiments on parametric conversion are presented elsewhere (Akhmanov *et al* 1991a). We used a 0.9 cm long LBO crystal with $\theta = 85°$, $\phi \simeq 9°$.

We also injected an energetic wave at $\lambda = 1.08$ μm (~ 40 ps) in the crystal (signal wave) to achieve a strongly non-linear regime of parametric conversion ($> 10\%$). The idler wave was tuned throughout the range 1.2–1.5 μm by changing the pump frequency from 0.57 to 0.63 μm.

For ~ 1 TW/cm^2 pumping we achieve 10% efficiency of conversion into the idler pulse. It corresponds to $\sim 25\%$ total quantum efficiency. The autocorrelation functions measured for the signal pulse show a pulse duration of 400 fs.

At 2 mJ pumping ($\lambda = 0.616$ μm) the energy of the idler pulse ($\lambda = 1.4$ μm) was as high as 0.2 mJ. With F/20 focusing the beam waist diameter turned out to be 25 μm (figure 5). Thus in the focal plane of the lens we obtain the intensity of 10^{14} W/cm^2.

5. Conclusion

We have developed an efficient parametric converter on LBO crystal pumped in the UV ($\lambda = 0.308$ μm), and in the visible ($\lambda = 0.57$–0.63 μm) producing a light field with

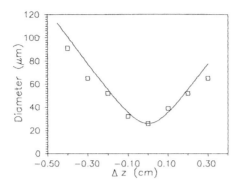

Figure 5. The beam waist profile (idler wave, $\lambda = 1.43$ μm, F/20 focusing), squares represent experimental results, the solid curve represents a Gaussian beam profile with waist diameter 26 μm.

an intensity up to 10^{14} W/cm^2. To obtain high-level parametric conversion and high beam quality we use injection locking.

We have measured the n_2, and surface damage threshold of this crystal at different pulse durations.

We have calculated the group-velocity dispersion curve and efficiency of parametric conversion in LBO crystal.

It should be stressed that this crystal may be very useful for parametric conversion of high-power femtosecond pulses. The use of dye amplifiers, of course, limits the pump energy to a level of several millijoules. But the relatively large value of L_g at $\lambda \simeq 0.8$ μm mm indicates that this is a proper crystal for the conversion of much more energetic Ti:Al$_2$O$_3$ radiation. For LBO crystal with a cross-section of 1 cm^2, according to our results one can expect a parametric generation of femtosecond pulses with a power up to 0.1 TW, and the intensities up to 10^{17} W/cm^2.

Acknowledgments

This investigation was initiated by the late Professor S Akhmanov. We would like to acknowledge his valuable discussions and support of this work.

References

Akhmanov S A, Bayanov I M, Gordienko V M *et al* 1991a *Tech. digest of the International Symposium on Ultrafast Processes in Spectroscopy* (Bayreuth, Germany, 1991) Paper Mo1-3; *UPS'91 Proceedings* 1992 (Bristol: IOP Publishing) to be published

Akhmanov S A, Vysloukh V A and Chirkin A S 1991b *Optics of Femtosecond Pulses* (New York: AIP)

Akhmanov S A, Bayanov I M, Gordienko V M *et al* 1991c *Sov. J. Quantum Electron.* **21** 248

Bayanov I M, Biglov Z A, Gaivoronskii V Ya, and Gordienko V M 1990a *Sov. J. Quantum Electron.* **19** 1049

Bayanov I M, Biglov Z A, Gordienko V M *et al* 1990b *Izvestia Acad. Nauk SSSR Ser Fiz* **54** 161

Boichenko V L, Zasavitskii I I, Kosichkin Yu V *et al* 1984 Sov. J. Quantum Electron. **14** 141

Chen C, Wu Y, Jiang A *et al* 1989 *J. Opt. Soc. Am.* B **6** 616

Danielius R, Grigonis R, Piskarskas *et al* 1989 *Proc. 6th UPS Int. Symp.* (Berlin: Springer) p 40

Danielius R, Piskarskas A, Sirutkaitis V *et al* 1983 *Optical Parametric Oscillators and Picosecond Spectroscopy* (Vilnius: Mokslas)

Dyakov V A, Dzhafarov M Kh, Lukashev *et al* 1991 *Sov. J. Quantum Electron.* **21** 339

Ebrahimsadeh M, Robertson G and Dunn M 1990 *CLEO Tech. Digest* (Washington: OSA) post-deadline paper CPDP26

Fickenscher M, Purucker H and Laubereau A 1990 *Appl. Phys.* B **51** 207

Gagel R, Angel G and Laubereau A 1990 *Opt. Commun.* **76** 239

Kato K 1990 *IEEE J. Quantum Electron.* **QE-26** 1173

Laenen R, Graener H and Laubereau A 1990a *Opt. Lett.* **15** 971

Laenen R, Graener H and Laubereau A 1990b *Opt. Commun.* **77** 226

Lin S, Sun Z, Wu B *et al* 1990 *J. Appl. Phys.* **67** 634

Magnitskii S A, Akhmanov S A, Bayanov I M *et al* 1991 *CLEO Tech. Digest* (Washington: OSA) paper CTuB4

Magnitskii S A, Malachova V I, Tarasevitch A P *et al* 1986 *Opt. Lett.* **11** 18

KEYWORD INDEX

Printed and bound by CPI Group (UK) Ltd, Croydon, CR0 4YY

17/10/2024

01775696-0012